Bacterial Survival in the Hostile Environment

Developments in Applied Microbiology and Biotechnology

Bacterial Survival in the Hostile Environment

Edited by

ASHUTOSH KUMAR, Ph.D

Department of Microbiology, Tripura University (A Central University), Agartala, Tripura, India

SHIVENDRA TENGURIA, Ph.D

Department of Pathology and Laboratory Medicine, Cedars-Sinai Medical Center (UCLA), Los Angeles, CA, United States

ELSEVIER

ACADEMIC PRESS

An imprint of Elsevier

Academic Press is an imprint of Elsevier
125 London Wall, London EC2Y 5AS, United Kingdom
525 B Street, Suite 1650, San Diego, CA 92101, United States
50 Hampshire Street, 5th Floor, Cambridge, MA 02139, United States
The Boulevard, Langford Lane, Kidlington, Oxford OX5 1GB, United Kingdom

Notices
Knowledge and best practice in this field are constantly changing. As new research and experience broaden our understanding, changes in research methods, professional practices, or medical treatment may become necessary.

Practitioners and researchers must always rely on their own experience and knowledge in evaluating and using any information, methods, compounds, or experiments described herein. In using such information or methods they should be mindful of their own safety and the safety of others, including parties for whom they have a professional responsibility.

To the fullest extent of the law, neither the Publisher nor the authors, contributors, or editors, assume any liability for any injury and/or damage to persons or property as a matter of products liability, negligence or otherwise, or from any use or operation of any methods, products, instructions, or ideas contained in the material herein.

ISBN: 978-0-323-91806-0

For Information on all Academic Press publications
visit our website at https://www.elsevier.com/books-and-journals

Publisher: Stacy Masucci
Acquisitions Editor: Linda Versteeg-Buschman
Editorial Project Manager: Tim Eslava
Production Project Manager: Sajana Devasi PK
Cover Designer: Christian J. Bilbow

Typeset by MPS Limited, Chennai, India

Working together
to grow libraries in
developing countries

www.elsevier.com • www.bookaid.org

Dedication

We dedicate this book to Dr. B. R. Ambedkar

"To develop scientific temper, humanism and the spirit of inquiry and reform" is one of the fundamental duties of the people of India.

Article 51A of the Constitution of India

Contents

11. Biofilm: a coordinated response of bacteria against stresses **149**

Roopshali Rakshit, Aayush Bahl, Ashutosh Kumar, Deeksha Tripathi and
Saurabh Pandey

12. The bacterial communication system and its interference as an antivirulence strategy **163**

Suruchi Aggarwal, Pallavi Mahajan, Payal Gupta, Alka Yadav, Gagan Dhawan,
Uma Dhawan and Amit Kumar Yadav

13. Microbial adaptations in extreme environmental conditions   193

Jayshree Sarma, Aveepsa Sengupta, Mani Kankana Laskar, Shatabdi Sengupta,
Shivendra Tenguria and Ashutosh Kumar

14. Adaptation strategies of piezophilic microbes   207

Somok Banerjee, Swatilekha Pati, Aveepsa Sengupta, Shakila Shaheen,
Jayshree Sarma, Palla Mary Sulakshana, Shivendra Tenguria and Ashutosh Kumar

List of contributors

Suruchi Aggarwal
Translational Health Science and Technology Institute, NCR Biotech Science Cluster, Faridabad, Haryana, India

Faizan Ahmed
Department of Medical Sciences, Cedars-Sinai Medical Center (UCLA), Los Angeles, CA, United States

Rajani Chowdary Akkina
Department of Microbiology and Food Science and Technology, GITAM Institute of Science, GITAM (Deemed to be University), Visakhapatnam, Andhra Pradesh, India

Gunjan Arora
Section of Infectious Diseases, Department of Internal Medicine, Yale University School of Medicine, New Haven, CT, United States

Md. Asadulghani
Biosafety and BSL3 Laboratory, Biosafety Office, International Centre for Diarrhoeal Disease Research (icddr,b), Dhaka, Bangladesh

Aayush Bahl
Microbial Pathogenesis and Microbiome Lab, Department of Microbiology, School of Life Sciences, Central University of Rajasthan, Ajmer, Rajasthan, India

Aniruddha Banerjee
Department of Microbiology, Tripura University (A Central University), Agartala, Tripura, India

Chaitali Banerjee
Vidyasagar College for Women (Affiliated to University of Calcutta), Kolkata, West Bengal, India

Shukla Banerjee
Dolphin PG Institute of Biomedical and Natural Sciences, Dehradun, Uttarakhand, India

Somok Banerjee
Department of Microbiology, Tripura University (A Central University), Agartala, Tripura, India

Kolluru Viswanatha Chaitanya
Department of Microbiology and Food Science and Technology, GITAM Institute of Science, GITAM (Deemed to be University), Visakhapatnam, Andhra Pradesh, India

Taane G. Clark
London School of Hygiene & Tropical Medicine, London, United Kingdom

Gagan Dhawan
Department of Biomedical Science, Acharya Narendra Dev College, University of Delhi, New Delhi, Delhi, India

Uma Dhawan
Department of Biomedical Science, Bhaskaracharya College of Applied Sciences, University of Delhi, New Delhi, Delhi, India

Soumyadip Ghosh
Department of Microbiology, Tripura University (A Central University), Agartala, Tripura, India

Anil Kumar Gupta
IRCH, All India Institute of Medical Sciences, New Delhi, Delhi, India

Kuldeepkumar Ramnaresh Gupta
Department of Microbial Pathogenesis, Yale University School of Medicine, New Haven, CT, United States

Payal Gupta
Translational Health Science and Technology Institute, NCR Biotech Science Cluster, Faridabad, Haryana, India

Arif Hussain
Laboratory Sciences and Services Division, International Centre for Diarrhoeal Disease Research (icddr,b), Dhaka, Bangladesh

Sana Ismaeel
Department of Pathology and Laboratory Medicine, Cedars-Sinai Medical Center (UCLA), Los Angeles, CA, United States

Arti Singh Katiyar
Department of Chemistry, MVGU, Jaipur, Rajasthan, India

Ashutosh Kumar
Department of Microbiology, Tripura University (A Central University), Agartala, Tripura, India

Mani Kankana Laskar
Department of Microbiology, Tripura University (A Central University), Agartala, Tripura, India

Pallavi Mahajan
Translational Health Science and Technology Institute, NCR Biotech Science Cluster, Faridabad, Haryana, India

Anitha Mamillapalli
Department of Biotechnology, GITAM Institute of Science, GITAM (Deemed to be University), Visakhapatnam, Andhra Pradesh, India

Razib Mazumder
Laboratory Sciences and Services Division, International Centre for Diarrhoeal Disease Research (icddr,b), Dhaka, Bangladesh

Dinesh Mondal
Laboratory Sciences and Services Division, International Centre for Diarrhoeal Disease Research (icddr,b), Dhaka, Bangladesh

Nishant Nandanwar
Division of Infectious Diseases, Department of Pediatrics, Children's Hospital Los Angeles, Los Angeles, CA, United States

Monika Pandey
Cytogenetics Laboratory, Department of Zoology, Banaras Hindu University, Varanasi, Uttar Pradesh, India

Saurabh Pandey
Department of Biochemistry, School of Chemical and Life Sciences, Jamia Hamdard, New Delhi, Delhi, India

Swatilekha Pati
Department of Microbiology, Tripura University (A Central University), Agartala, Tripura, India

Vidyullatha Peddireddy
Department of Nutrition Biology, School of Interdisciplinary and Applied Sciences, Central University of Haryana, Mahendragarh, Haryana, India

Palkar Omkar Prakash
Department of Microbiology and Food Science and Technology, GITAM Institute of Science, GITAM (Deemed to be University), Visakhapatnam, Andhra Pradesh, India

Roopshali Rakshit
Microbial Pathogenesis and Microbiome Lab, Department of Microbiology, School of Life Sciences, Central University of Rajasthan, Ajmer, Rajasthan, India

Keerthi Rayasam
Department of Microbiology and Food Science and Technology, GITAM Institute of Science, GITAM (Deemed to be University), Visakhapatnam, Andhra Pradesh, India

Andaleeb Sajid
Section of Infectious Diseases, Department of Internal Medicine, Yale University School of Medicine, New Haven, CT, United States

Jayshree Sarma
Department of Microbiology, Tripura University (A Central University), Agartala, Tripura, India

Aveepsa Sengupta
Department of Microbiology, Tripura University (A Central University), Agartala, Tripura, India

Shatabdi Sengupta
Department of Microbiology, Tripura University (A Central University), Agartala, Tripura, India

Shakila Shaheen
Department of Microbiology, Tripura University (A Central University), Agartala, Tripura, India

Bipin Kumar Sharma
Department of Microbiology, Tripura University (A Central University), Agartala, Tripura, India

Divakar Sharma
Department of Microbiology, Maulana Azad Medical College, New Delhi, Delhi, India

Amit Singh
Centralized Core Research Facility (CCRF), All India Institute of Medical Sciences, New Delhi, Delhi, India; Department of Gastroenterology and HNU, All India Institute of Medical Sciences, New Delhi, Delhi, India

Palla Mary Sulakshana
Department of Pharmacy, Raghu College of Pharmacy, Visakhapatnam, Andhra Pradesh, India

Shivendra Tenguria
Department of Pathology and Laboratory Medicine, Cedars-Sinai Medical Center (UCLA), Los Angeles, CA, United States

Deeksha Tripathi
Microbial Pathogenesis and Microbiome Lab, Department of Microbiology, School of Life Sciences, Central University of Rajasthan, Ajmer, Rajasthan, India

Alka Yadav
Department of Biomedical Science, Bhaskaracharya College of Applied Sciences, University of Delhi, New Delhi, Delhi, India

Amit Kumar Yadav
Translational Health Science and Technology Institute, NCR Biotech Science Cluster, Faridabad, Haryana, India

CHAPTER 1

Mycobacterium tuberculosis adaptation to host environment

Aniruddha Banerjee[1], Shatabdi Sengupta[1], Nishant Nandanwar[2], Monika Pandey[3], Deeksha Tripathi[4], Saurabh Pandey[5], Ashutosh Kumar[1] and Vidyullatha Peddireddy[6]

[1]Department of Microbiology, Tripura University (A Central University), Agartala, Tripura, India
[2]Division of Infectious Diseases, Department of Pediatrics, Children's Hospital Los Angeles, Los Angeles, CA, United States
[3]Cytogenetics Laboratory, Department of Zoology, Banaras Hindu University, Varanasi, Uttar Pradesh, India
[4]Microbial Pathogenesis and Microbiome Lab, Department of Microbiology, School of Life Sciences, Central University of Rajasthan, Ajmer, Rajasthan, India
[5]Department of Biochemistry, School of Chemical and Life Sciences, Jamia Hamdard, New Delhi, Delhi, India
[6]Department of Nutrition Biology, School of Interdisciplinary and Applied Sciences, Central University of Haryana, Mahendragarh, Haryana, India

Introduction

Mycobacterium tuberculosis, a highly pathogenic bacterium and the causative agent of tuberculosis worldwide. It majorly infects the host lungs and can survive in hostile microenvironment of the lungs resulting even in latent infection. Upon colonization in the host, *M. tuberculosis* modulates and adapts itself to withstand and counteract various unfavorable environmental conditions and stress factors such as acidic conditions, hypoxic stress, heat shock, immune system stress, and oxidative stress. There is a high ratio of lipid in the cell wall of mycobacteria, it consists of lesser pores, DNA contains a high GC content and there is a single ribosomal RNA operon, all these factors participate in the slow growth rate of mycobacteria (Kumar et al., 2017). Almost every environmental bacteria consist a variety of Toxin-antitoxin (TA) systems which helps them to withstand different stress conditions in the environment. Usually the number and presence of TA-system is much higher in the environmental bacteria as compared to the intracellular pathogens. Similarly, *M. tb* also has to face different types of stress conditions like acidic environment, hypoxia, oxidative stress, immune stress etc. in its pathogen cycle. For the survival and adaptation in these unfavorable conditions, it also consist various types of TA-systems (Kumar et al., 2019a). For the survival in different conditions inside the host dormancy is one of the well-developed approach of *M. tuberculosis*. *M. tuberculosis* DosR regulates a wide array of proteins. Among all the proteins belong to DosR regulon, DATIN (Dormancy Associated Translation Inhibitor) can activates PBMC (peripheral blood mononuclear cells), macrophages etc. that have a major role in various immune responses like formation and maintenance of granulomas (Kumar et al., 2013). It was observed that DATIN plays a major role in the regulation

Bacterial Survival in the Hostile Environment
DOI: https://doi.org/10.1016/B978-0-323-91806-0.00005-9

of latent phase in *M. tuberculosis* and it can inhibit bacterial translation by binding with ribosome (Kumar et al., 2012). This resuscitated bacteria usually become tolerant to different antibiotics by means of biofilm formation. *M. tuberculosis* has wide number of proteins which includes cyclophilins. One such cyclophilin is the PpiB (peptidyl-prolyl isomerase) which plays a role in the formation of biofilms and drug tolerance (Kumar et al., 2019b). *M. tuberculosis* Cyclophilins or Peptidylprolyl cis-trans isomerases are involved in the induction of various types of cytokine secretion, folding of proteins, peptide bond isomerization for overcoming different stress conditions (Pandey et al., 2016, 2017). *M. tuberculosis* initially interconnects with the host immune cells such as the lung macrophages. These are the phagocytic cells that engulf any invading pathogen by the mechanism known as phagocytosis and degrade them by forming a phagolysosome. Conversely, *M. tuberculosis* relies on various mechanisms involving proteins and gene functions that prevent fusion of phagosome and lysosome subsequently inhibiting phagolysosome formation. One such mechanism is the expression of mannose receptors and fusion of the mannose-capped lipoarabinomannan, which is among the unique molecular motifs located within the cell wall of infective *M. tuberculosis* (Brennan and Nikaido, 1995). This prevents the fusion of phagosome and lysosome resulting in inhibition of phagolysosome formation. Another mechanism that facilitates phagosome-lysosome arrest, is controlled by mycobacterial-specific proteins such as TB9.8 (encoded by esxG) and TB10.4 (encoded by *esxH*) that are released by Esx-3 type VII secretion system of *M. tuberculosis* (Mehra et al., 2013). *M. tuberculosis* can also withstand acidic pH inside the phagolysosome. It contains a periplasmic protease known as MarP that promotes survival in the acidified phagosomes. It cleaves and activates the peptidoglycan hydrolases RipA (Botella et al., 2017). *M. tuberculosis* also survives an acidic environment by maintaining the cytoplasmic pH homeostasis. *M. tuberculosis* membrane-bound serine protease (Rv3671c) and the pore-forming protein OmpA also facilitate maintenance of pH homeostasis and resistance to acid (Vandal et al., 2008; Raynaud et al., 2002).

M. tuberculosis also encounters hypoxic conditions because of the profound change in the oxygen concentrations from the high level of environmental oxygen echelons while transmitting from aerosol to the microaerobic or hypoxic surroundings inside necrotizing granulomas (Tsai et al., 2006). Mycobacterial proteins such as the Lsr2 protein facilitate survival during hypoxic conditions. Lsr2 is a nucleotide-associated protein (NAP) and serves as a histone-like nucleoid-structuring protein (H-NS) functional homolog. Lsr2 regulates DNA transcription during hypoxic conditions. Another strategy to overcome hypoxic stress is by regulating the rTCA cycle utilizing the enzyme citrate lyase (Watanabe et al., 2011; Fang et al., 2012). Citrate lyase is the chief regulator of rTCA cycle and helps in switching from TCA cycle to rTCA cycle. *M. tuberculosis* responds to the increased temperatures and heat shocks by utilizing specialized proteins recognized as heat shock proteins (Hsp) such as Hsp70, Hsp22.5, HrcA, and HspR. The gene *acr2*, an active member of the α-crystalline family, shows greater activation during a heat shock (Stewart et al., 2002). After infection,

M. tuberculosis exploits and manipulates the metal ion trafficking inside the macrophages. *M. tuberculosis* requires iron, which gets depleted due to the host proteins such as ferritin and transferrin or via a divalent metal cation transporter efflux from the phagosome which is known as natural resistance-associated macrophage protein 1 (Nramp1) (Cellier et al., 2012a, b; Li et al., 2011). This iron depletion for *M. tuberculosis* is accomplished by its siderophores, MBT (mycobactin) and cMBT (carboxymycobactin) (Ratledge, 2004). During infection, response of the host immune system is carried out by overloading the phagosomes with excess of zinc and copper which are toxic to *M. tuberculosis* when present in excess concentration. *M. tuberculosis* tackles this zinc overload by utilizing unique proteins such as metallothioneins with the help of a special mechanism known as "zinc proteome." Zinc overload is also regulated by expressing the zinc detoxification genes like zinc efflux transporter-encoding genes (*znt1/slc30a1*) & the metallothionein-encoding genes *mt1* and *mt2* (Lichten and Cousins, 2009). Similarly, regulation of copper overload is carried out by an essential *M. tuberculosis* enzyme, CtpV which is a copper-transporting ATPase, and is expressed during excess copper levels (Ward et al., 2008).

M. tuberculosis adaptation to host immune system and oxidative stress

M. tuberculosis resides intracellularly in macrophages which are their primary host cells. Macrophages behave as the initial defense mechanism opposing the invading disease-causing organisms. To survive these immune challenges, *M. tuberculosis* has developed various survival strategies such as the modulation of the macrophage activation and effectors functions. In addition, during infection, the development of antigen-specific T cells response also takes place. *M. tuberculosis* exploits a variety of machineries to protect itself from reactive nitrogen species (RNS) and reactive oxygen species (ROS) comprising unswerving foraging of the reactive species and the restoration and safeguarding of its DNA and proteins (Ehrt and Schnappinger, 2009). The battle of *M. tuberculosis* against ROS could be partially attributed to the presence of profuse cell wall of *M. tuberculosis* comprising phenolic glycolipid I (PGL-1), lipoarabinomannan (LAM), and cyclo-propanated mycolic acids which are effective foragers of oxygen radicals (Flynn and Chan, 2001a,b). Furthermore, it synthesizes the ROS foraging enzymes such as superoxide dismutases, catalase, peroxidase and peroxy-nitrite reductase complex of Lpd, SucB (DlaT), AhpC, and AhpD (Bryk et al., 2002). Resistance to peroxides in *M. tuberculosis* can be attributed to the presence of the thioredoxin/ thioredoxin reductase systems (Zhang et al., 1999). An antioxidant mycothiol with low-molecular-weight plays the role of glutathione and is significant in preserving an abridged milieu and protecting many bacteria against oxidative stress (Buchmeier et al., 2003). The sulfur accumulative pathway was demonstrated to play a vital role in nitrosative and oxidative stress since the cysH mutant lacking methionine and cysteine synthesis was found to have enhanced sensitivity toward these stresses (Senaratne et al., 2006). *M. tuberculosis* DNA is unswervingly shielded from ROS by Lsr2, the DNA-binding protein (Colangeli et al., 2009). Other reactions which protect against RNS comprise the catalytic detoxification of nitric oxide (NO) accomplished by the abridged hemoglobin (trHbN) in an oxygen-dependent mechanism (Ouellet et al., 2002; Pathania et al., 2002). The aerobic respiration in *Mycobacterium smegmatis* is protected by trHbN from reticence by NO (Pathania et al., 2002). *M. tuberculosis* gene cluster Rv0014c-Rv0019c [PknA (encoded by Rv0014c) and FtsZ-interacting protein A (FipA) (encoded byRv0019c)] and FtsZ and FtsQ, form the cell wall division cluster which facilitates *M. tuberculosis* survival during oxidative stress. During the interaction of FipA with FtsZ and FtsQ, establishment of PknA-dependent phosphorylation of FipA at T77 and FtsZ at T343, is required for the division of cells under oxidative tension (Sureka et al., 2010).

Strategies to counter microbicidal effect of myeloid cells

During illness, the dendritic cells and resident macrophages are usually the first cells interacting with *M. tuberculosis*. Dendritic cells are a type of antigen-presenting cells that present foreign pathogens to other immune cells such as T cells. Macrophages, on the other hand, engulf invading foreign pathogens and degrade them with the help of some unique organelles called phagosomes and lysosomes, which fuse to form a phagolysosome. After phagocytosis, phagosomes engulf the pathogen and carry the phagocytized pathogen toward the lysosomes, which then fuses to form the phagolysosome, eventually degrading the pathogen with the help of enzymes in acidic conditions. *M. tuberculosis* on a contrary, has a unique trait of inhibiting the formation of phagolysosome. *M. tuberculosis*, also inhibits autophagy, a mechanism in which the body removes the unwanted and dysfunctional components. It is regulated by autophagy-related genes (Atg) and depends on autophagosome. This autophagosome works by engulfing cytoplasmic components and delivering them to the lysosomes for degradation (Mizushima et al., 2011). In the instance of primary infection, the phagocytosis of *M. tuberculosis* mainly occurs by nonosmotic mechanisms (Schafer et al., 2009). This happens primarily because in the alveolar space, serum and complements are very much deficient when the initial contact occurs between the pathogen (*M. tuberculosis)* and macrophages (Schafer et al., 2009; Schluger, 2001). Phagocytosis of *M. tuberculosis* using mannose receptor and fusion of the mannose-capped lipoarabinomannan are among the best notable molecular models present on the pathogenic *M. tuberculosis* cell wall (Brennan and Nikaido, 1995), which results in the arrest the fusion of phagosome-lysosome (Kang et al., 2005). Some other critical mycobacterial proteins such as TB9.8 (encoded by *esxG*) and TB10.4 (encoded by *esxH*) are also responsible for the phagosome-lysosome arrest and their secretion is carried out by the Esx-3 type VII secretion system of *M. tuberculosis* (Mehra et al., 2013). The TB9.8 and TB10.4 are heterodimers that can show interaction with the hepatocytic growth factor-regulated tyrosine kinase substrate (Hrs/Hgs). It is a part of the endosomal sorting complex required for transport (ESCRT) (Mehra et al., 2013). This ESCRT transportation system comprises four protein complexes which are conscripted to the endosomal membrane sequentially & promote phagocytized mycobacteria delivery into the lysosomes and subsequent restriction of intracellular microbial expansion (Mehra et al., 2013). This Esx-1 type VII exudation system has great involvement in the extremely antigenic T-cell proteins secretion which then inhibits the intracellular vesicular trafficking like the early secretory antigen target 6 kDa (ESAT-6, encoded by *esxA*) and the culture filtrate antigen 10 kDa (CFP-10, encoded by *esxB*) (Tan et al., 2006). In *M. tuberculosis*, these proteins are accountable for the properties of permeabilization of phagosome, which result in the rupturing and translocation of *M. tuberculosis* from phagosome to cytosol (van der Wel et al., 2007), initiation of autophagy (Watson et al., 2012) and also inflammasome (Mishra et al., 2013).

Microenvironment modulation by *M. tuberculosis*

Host immune cells, such as macrophages, can activate apoptotic cell death programs to avoid further intracellular replication of microbes and increase the disclosure of infective antigens. Apoptosis, a highly controlled cellular activity of cell destruction minimizes the effect of inflammation and pathology by the formation of apoptotic bodies. These contain the dismembered dead cell contents within the membrane-bound vesicles (Taylor et al., 2008). The apoptotic bodies are responsible for the emission of signals, signaling engulfment, and digestion in the phagocytes by a mechanism known as efferocytosis (Martin et al., 2014). Necrosis, alternatively is defined as the forfeiture of membrane integrity and dispersal of intracellular contents into the extracellular space and involves increased inflammation in comparison to apoptosis (Moraco and Kornfeld, 2014).

One of the effective mechanisms in virulent *M. tuberculosis* to prevent apoptosis and promote necrosis of infected macrophages is the regulation of eicosanoid fabrication (Chen et al., 2008). It has been observed that an avirulent strain of *M. tuberculosis* H37Ra stimulates the formation of prostaglandin E2 (PGE2), required for the protection of the plasma membrane by the regulation of the calcium sensor called synaptotagmin 7, which is engaged in the repair mechanism facilitated by lysosome (Divangahi et al., 2009). In contrast, the virulent strain of *M. tuberculosis* H37Rv stimulates the formation of lipoxins such as lipoxin A4 (LXA4), resulting in the downregulation of the cyclooxygenase (COX)2 mRNA, eventually decreasing the production of PGE2 production and prevents its membrane shielding effects, which results in necrosis of the contaminated cell (Chen et al., 2008; Divangahi et al., 2009). Similarly, rupture of the vacuole of *M. tuberculosis* is induced by Esx-1 secretion system (van der Wel et al., 2007), and by pore-forming activity of ESAT-6 (Smith et al., 2008) which are vital in this process. A further contender involved in the host necrotic cell death is the CpnT, a NAD^+ glycohydrolase involved in depleting cellular NAD^+ pools (Danilchanka et al., 2014; Sun et al., 2015).

M. tuberculosis adaptations to acidic environment of phagolysosomes

The acidic pH can cause a widespread change to the *M. tuberculosis* functioning, including the induction of numerous stress genes and the PhoPR regulons. At an acidic pH, *M. tuberculosis* remodels its metabolic activities when its growth is dependent on solitary carbon foundations at acidic pH, it needs carbon bases such as pyruvate, acetate, and oxaloacetate to fuel the anaplerotic node (Baker et al., 2014). Acidic pH might lead to an alteration of the redox status of the cytoplasm, which results in the formation of a plummeting atmosphere inside *M. tuberculosis* cytoplasm. The macrophage compartment residing *M. tuberculosis* have a pH ranging from pH 6.2—4.5 which depends on the macrophage activation state (MacMicking et al., 2003; Schaible et al., 1998; Via et al., 1998). *M. tuberculosis* consists of a cell wall that is rich in lipids and contains a classical bilayered plasma membrane. It has a layer of covalently linked peptidoglycan-arabinogalactan with mycolic acids. This intricate cell envelope serves as an intimidating permeability barricade for antibacterial effectors comprising protons which have a significant role in resistance to acids (Botella et al., 2017). *M. tuberculosis* can persist in a latent state for years within human hosts. Several studies have revealed that the periplasmic-protease MarP plays a key role in the endurance of *M. tuberculosis* in the acidic environment of the phagosomes. MarP cleaves and activates the peptidoglycan hydrolase RipA during acid stress (Botella et al., 2017). It was observed that *M. tuberculosis* lacking MarP (DmarP) is highly sensitive to acidic environment (Vandal et al., 2008). The mutants lacking MarP or RipA share similar phenotypes; displaying an increase in cell length and chain formation in acidic condition. It is hypothesized that strain with mutated RipA reiterate the phenotype of MarP-deficient cells suffused with the acidic medium (Botella et al., 2017).

In resting macrophages, the maturation of phagosome and prevention of the fusing phagolysosome is inhibited by *M. tuberculosis*. The moderately acidic pH (6.1—6.4) of the phagosome is due to the elimination of the vacuolar proton-ATPase in the *M. tuberculosis* encompassing vacuole. The *Mycobacterium*-mediated blockade is removed by activating macrophages by IFN-γ, leading to acidification of the phagosome (pH 4.5—5.4) (MacMicking et al., 2003; Schaible et al., 1998; Via et al., 1998). Studies showed that induction of pH-responsive genes takes place in macrophages. Membrane-bound serine protease (Rv3671c) has role in the survival at acidic pH (Biswas et al., 2010). The pore-forming protein OmpA also contributes in acid resistance. At a pH of 5.5, OmpA expression is induced, and when these genes are mutated, the growth within macrophages and mice is impaired (Raynaud et al., 2002). The presumed magnesium transporter (MgtC) is vital for *M. tuberculosis* growth at a moderately acidic pH of 6.25, which requires Mg^{2+} concentrations. Disruption of MtgC can lead to an attenuated growth in macrophages. Hence, Mg^{2+} procurement is imperative when *M. tuberculosis* encounters decreased pH in the phagosomal partition.

Mycobacterial adaptations to hypoxic environment

M. tuberculosis encounters reactive nitrogen and oxygen stress within the host and needs to withstand ROS and RNS. *M. tuberculosis* regulates different metabolic pathways for survival in hypoxic conditions (Fig. 1.1). *M. tuberculosis* Lsr2 protein plays a vital role in reactive oxygen defense by DNA protection, providing the capability to adapt to high and deficient oxygen levels (Bartek et al., 2014). Lsr2 is very much involved in organizing bacterial chromosomes and global transcription regulation. Chromatin immunoprecipitation-sequencing (ChIP-seq) data have revealed that Lsr2 primarily behaves as a repressor that controls gene expression either unswervingly by binding in promoter regions or meanderingly by forming loop and coating of DNA. One of the oppressed genes of Lsr2 encrypts lipooligosaccharide, synthesizing polyketide synthase (Bartek et al., 2014).

   M. tuberculosis can change its metabolic pathway to a low-energy-conserving state for adapting to hypoxic conditions, leading to lower ATP levels in hypoxic cells (Shi et al., 2005). One of the vital parts of this strategy is the reductive edge of the tricarboxylic acid (rTCA) cycle, which allows carbon fixation in an aerobic milieu (Watanabe et al., 2011; Fang et al., 2012). Citrate lyase is a type of cytoplasmic enzyme that can catalyze the transformation of CoA and citrate into acetyl-CoA and oxaloacetate which serves as a key

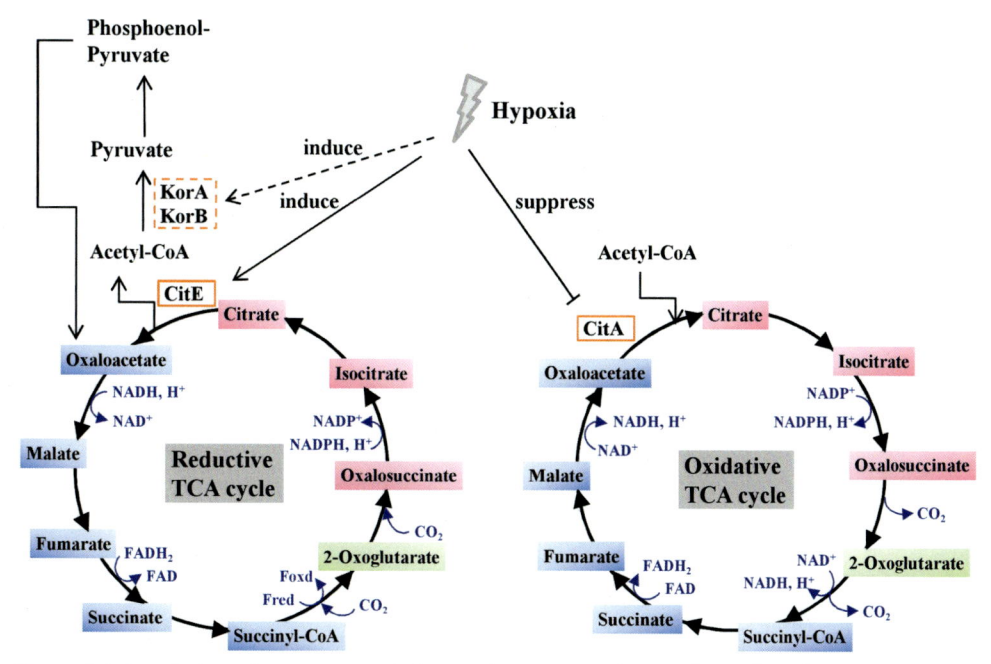

Figure 1.1 Regulation of energy metabolism in hypoxic condition of *M. tuberculosis. Adapted from Hu, J., Jin, K., He, Z.G., Zhang, H., 2020. Citrate lyase CitE in Mycobacterium tuberculosis contributes to mycobacterial survival under hypoxic conditions. PLoS ONE 15 (4), e0230786. https://doi.org/10.1371/ journal.pone.0230786.*

molecule in cellular metabolism for the biogenesis of a diversified group of particles (e.g., fatty acids and cholesterols), production of energy and acetylation of protein (Chypre et al., 2012). In optimal condition, *M. tuberculosis* initiates the TCA cycle to produce additional energy for its development, whereas the rTCA cycle is triggered when the abundance of ATP molecules inactivates citE catalyze the breakdown of citrate to acetyl-CoA and acetate (Hu et al., 2020). Further, when *M. tuberculosis* is prone to a hypoxic milieu, a change in oxygen tension occurs inside the host (Tsai et al., 2006), the bacterium lowers the rate of energy uptake and produces a lower amount of CitE, which helps in the mycobacterial survival in hypoxic conditions. The expression of CitE during hypoxic conditions increases in *M. tuberculosis* which is essential in the case of rTCA. Therefore, CitE has a major responsibility for the survivability of *M. tuberculosis* in hostile conditions (Hu et al., 2020).

M. tuberculosis adaptations to subsist heat shock

Heat shock proteins serve as molecular chaperones to maintain different cellular functions both in normal and stressed conditions. In general, under stressed state, upregulation of several heat shock proteins takes place in *M. tuberculosis* (Monahan et al., 2001; Sherman et al., 2001; Stewart et al., 2002; Voskuil et al., 2004). When *M. tuberculosis* is exposed to a heat shock it increases the hsp70 regulon expression, groES, Acr protein, and groEL protein (Stewart et al., 2002). Another heat shock protein Hsp22.5, belonging to heat shock regulons, gets triggered during hypoxic conditions, with the endurance of *M. tuberculosis* in the macrophages and murine lungs through the latent phase of the infection (Abomoelak et al., 2010). In *M. tuberculosis*, heat shock also triggers the expression of Acr2, which belongs to novel α-crystalline group of molecular chaperons (Wilkinson et al., 2005). Even though there is no notable temperature change when *M. tuberculosis* enters the host cell, the heat shock proteins get induced chiefly because of the adaptation of the bacteria to the hypoxic atmosphere in the phagosome (during intracellular growth and survival) (Stewart et al., 2002). During infection, it is observed that (i) *M. tuberculosis* is protected from macrophages by these heat shock proteins (Lee and Horwitz, 1995; Stewart et al., 2001) and (ii) it strongly influences the virulence mechanisms of *M. tuberculosis* (Stewart et al., 2002). DNA microarray experiments demonstrated that the gene acr2 (Rv0251c), an active member of the α-crystalline group of molecular chaperone genes, exhibits greater activation during a heat shock (Stewart et al., 2002). The deletion of heat shock protein HspR has minimal repercussions on the groE-hsp60 proteins (Stewart et al., 2002). Experiments showed that transcription regulator of a high quantity of heat shock-receptive genes is probably controlled by heat shock protein HrcA locus, which serves as the principal controller. A huge family of the heat shock-inducible α-crystalline family is greatly dependent on the phoP locus (Singh et al., 2014). Collectively, it can be concluded that the most crucial regulatory circuit demonstrates communications of the phoP with the heat shock-repressors (HrcA and hspR) and in what manner they possess a coordinated mechanism of controlling heat shock-responsive genes transcriptionally (Sevalkar et al., 2019).

M. tuberculosis adaptation to metal stress

Exploitation and manipulation of metal cation trafficking inside infected macrophages is a significant characteristic of *M. tuberculosis* that attributes to its intracellular survival. To protect against metal toxicity, a multifaceted resistance mechanism has been developed by *M. tuberculosis*, including controlling the uptake, confinement inside the cell, efflux, and oxidation. The host adopts these metal poisoning strategies through nutritional immunity mechanisms that cause deprivation of metals such as manganese and iron that prevents replication of bacteria. During phagosome maturation, both resistant mechanisms depend on the movement of metal transporter proteins carried to the phagosome membrane. Essential micronutrients such as manganese and iron are shielded from intracellular *M. tuberculosis* through the concealment of host proteins such as ferritin and transferrin or and through the outflow of the phagosome via divalent metal positive ion carrier known as Nramp1 (Cellier et al., 2012a,b; Li et al., 2011). To prevail over iron distress, *M. tuberculosis* employs well-organized iron capture systems based on siderophores MBT and cMBT (Snow, 1970; Luo et al., 2005) and capability to use heme (Jones and Niederweis, 2011; Tullius et al., 2011). Metal cations such as zinc and copper ions may get accumulated inside the vacuole of *M. tuberculosis* to a toxic level (Botella et al., 2011; Wolschendorf et al., 2011). *M. tuberculosis* exhibits resistance to metal intoxication by using metal efflux and purification systems, such as oxidases, P-type ATPases, and sequestration (Gold et al., 2008; Rowland and Niederweis, 2012, 2013).

For the growth of *M. tuberculosis*, MBT biosynthesis is very crucial once the inside iron supplies are exhausted. cMBT are proficient in eliminating iron from the ferritin and transferrin (Gobin and Horwitz, 1996). The encoding of cytoplasmic synthases synthesizes siderophores by two mbt operons (Rodriguez, 2006; Chavadi et al., 2011) and their transport are conjoined (Wells et al., 2013). They are dependent on the membrane proteins such as MmpS4 and MmpS5, which are linked with the transporters MmpL4 and MmpL5 of the resistance-nodulation-cell division superfamily. The cMBT that are secreted binds to iron, and are recaptured and transported across the outer membrane. IrtA or IrtB protein complexes aid in the transport of ferric-cMBT across the inner membrane (Rodriguez, 2006). This imported cMBT released iron through a reductive mechanism in place of enzymatic degradation (Ryndak et al., 2010). Hence the iron is made available, and the siderophores are left intact to be recycled by the MmpL4/MmpS4 and

MmpL5/MmpS5 export systems (Jones et al., 2014). To acquire iron, *M. tuberculosis* requires the type VII protein secretion system Esx-3 (Siegrist et al., 2009, 2014; Serafini et al., 2013).

Zinc is a kind of metal that shows toxicity inside a cell. With the availability of surplus free zinc, the eukaryotic cells react by translocating the zinc-sensing metal transcription factor MTF-1 to the nucleus, which subsequently results in the induction of expression of genes responsible for detoxification of zinc like the zinc efflux transporter-encoding gene *znt1/slc30a1* and the metallothionein-encoding genes *mt1* and *mt2* (Lichten and Cousins, 2009). Confocal microscopy revealed that small intracellular compartments are present in the *M. tuberculosis*-infected macrophages known as zincosomes (Ballestin et al., 2011). These zincosomes help in the storage and buffering of zinc to avoid excess zinc in the cytosol (Beyersmann and Haase, 2001).

Copper, an important micronutrient for *M. tuberculosis*, serves as the cofactor of two enzymes—the Cytochrome C oxidase and the CU/Zn-superoxide dismutase SodC, vital for its growth. The role of CtpV is very vital in the case of *M. tuberculosis* copper stress response. CtpV is an ATPase for copper transportation, the expression of which is triggered by the presence of excess copper (Ward et al., 2008). In *M. tuberculosis* the most recognized copper-binding regulator is the CsoR (Talaat et al., 2004) which in the absence of copper, serves as a repressor of the *cso* operon (Liu et al., 2007). When bound to copper CsoR loses its capability of repressing *cso* expression. This results in the expression of *cso* gene expression induced by copper, which occurs at a level that is proportional to the intracellular copper levels. This CtpV gene is included in this *cso* operon and it encodes for a *M. tuberculosis* copper response associated putative metal transporter (Ward et al., 2008). Therefore, in *M. tuberculosis*, CtpV is crucial for the detoxification of copper and has a major contribution to *M. tuberculosis* virulence.

Conclusion

M. tuberculosis is an acid-fast bacteria that primarily infect humans in their respiratory tract, usually lungs. It enters through our respiratory tract, and the immune system tries to clear it by recruiting lungs macrophages for phagocytosis and degrading it through lysosomes or phagolysosomes. For its survival, *M. tuberculosis* has developed various means to escape the host immune system and endure unfavorable conditions such as hypoxia, acidic environment, metal stress, and immune challenges. Due to the various specialized mechanisms in its arsenal of virulence attributes, *M. tuberculosis* withstands different hostile environments and is regarded as a highly seasoned pathogenic bacterium causing tuberculosis and latent infections in humans.

References

Baker, J.J., Johnson, B.K., Abramovitch, R.B., 2014. Slow growth of *Mycobacterium tuberculosis* at acidic pH is regulated by phoPR and host-associatedcarbon sources. Mol. Microbiol. 94, 56−69.

Ballestin, R., Molowny, A., Marín, M.P., Esteban-Pretel, G., Romero, A.M., Lopez-Garcia, C., et al., 2011. Ethanol reduces zincosome formation in cultured astrocytes. Alcohol Alcohol. 46 (1), 17−25.

Bartek, I.L., Woolhiser, L.K., Baughn, A.D., Basaraba, R.J., Jacobs Jr, W.R., Lenaerts, A.J., et al., 2014. *Mycobacterium tuberculosis* Lsr2 is a global transcriptional regulator required for adaptation to changing oxygen levels and virulence. mBio 5 (3), e01106−e01114. Available from: https://doi.org/10.1128/mBio.01106-14.

Beyersmann, D., Haase, H., 2001. Functions of zinc in signaling, proliferation and differentiation of mammalian cells. Biometals 14 (3), 331−341.

Biswas, T., Small, J., Vandal, O., Odaira, T., Deng, H., Ehrt, S., et al., 2010. Structural insight into serine protease Rv3671c that protects *M. tuberculosis* from oxidative and acidic stress. Structure (London, England: 1993) 18 (10), 1353−1363. Available from: https://doi.org/10.1016/j.str.2010.06.017.

Botella, H., Peyron, P., Levillain, F., Poincloux, R., Poquet, Y., Brandli, I., et al., 2011. Mycobacterial P1-type ATPases mediate resistance to zinc poisoning in human macrophages. Cell Host Microbe 10 (3), 248−259.

Botella, H., Vaubourgeix, J., Lee, M.H., Song, N., Xu, W., Makinoshima, H., et al., 2017. *Mycobacterium tuberculosis* protease MarP activates a peptidoglycan hydrolase during acid stress. EMBO J. 36 (4), 536−548.

Brennan, P.J., Nikaido, H., 1995. The envelope of mycobacteria. Annu. Rev. Biochem. 64, 29−63. Available from: https://doi.org/10.1146/annurev.bi.64.070195.000333.

Bryk, R., Lima, C.D., Erdjument-Bromage, H., Tempst, P., Nathan, C., 2002. Metabolic enzymes of mycobacteria linked to antioxidant defense by a thioredoxin-like protein. Science 295, 1073−1077.

Cellier, M.F., 2012a. Nramp: from sequence to structure and mechanism of divalent metal import. CurrTop Membr. 69, 249−293.

Cellier, M.F., 2012b. Nutritional immunity: homology modeling of Nramp metal import. Adv. Exp. MedBiol 946, 335−351.

Chavadi, S.S., Stirrett, K.L., Edupuganti, U.R., Vergnolle, O., Sadhanandan, G., Marchiano, E., et al., 2011. Mutational and phylogenetic analyses of the mycobacterial mbt gene cluster. J. Bacteriol. 193 (21), 5905−5913.

Chen, M., Divangahi, M., Gan, H., Shin, D.S., Hong, S., Lee, D.M., et al., 2008. Lipid mediators in innate immunity against tuberculosis: opposing roles of PGE2 and LXA4 in the induction of macrophage death. J. Exp. Med. 205, 2791−2801. Available from: https://doi.org/10.1084/jem.20080767.

Chypre, M., Zaidi, N., Smans, K., 2012. ATP-citrate lyase: a mini-review. BiochemBiophys Res. Commun. 422, 1−4. Available from: https://doi.org/10.1016/j.bbrc.2012.04.144. PMID: 22575446.

Colangeli, R., Haq, A., Arcus, V.L., Summers, E., Magliozzo, R.S., McBride, A., et al., 2009. The multi-functional histone-like protein Lsr2 protects mycobacteria against reactive oxygen intermediates. Proc. Natl. Acad. Sci. U. S. A. 106, 4414−4418.

Danilchanka, O., Sun, J., Pavlenok, M., Maueröder, C., Speer, A., Siroy, A., et al., 2014. An outer membrane channel protein of *Mycobacterium tuberculosis* with exotoxin activity. Proc. Natl. Acad. Sci. U. S. A. 111, 6750−6755. Available from: https://doi.org/10.1073/pnas.1400136111.

Divangahi, M., Chen, M., Gan, H., Desjardins, D., Hickman, T.T., Lee, D.M., et al., 2009. *Mycobacterium tuberculosis* evades macrophage defenses by inhibiting plasmamembrane repair. Nat. Immunol. 10, 899−906. Available from: https://doi.org/10.1038/ni.1758.

Ehrt, S., Schnappinger, D., 2009. Mycobacterial survival strategies in the phagosome: defence against host stresses. Cell. Microbiol. 11, 1170−1178.

Fang, X., Wallqvist, A., Reifman, J., 2012. Modeling phenotypic metabolic adaptations of *Mycobacterium tuberculosis* H37Rv under hypoxia. PLoSComput Biol. 8, e1002688. Available from: https://doi.org/10.1371/journal.pcbi.1002688. PMID: 23028286.

Flynn, J.L., Chan, J., 2001a. Tuberculosis: latency and reactivation. Infect. Immun. 69, 4195–4201. Available from: https://doi.org/10.1128/IAI.69.7.4195-4201.2001. PMID: 11401954.

Flynn, J.L., Chan, J., 2001b. Immunology of tuberculosis. Annu. Rev. Immunol. 19, 93–129.

Gobin, J., Horwitz, M.A., 1996. Exochelins of *Mycobacterium tuberculosis* removes iron from human iron-binding proteins and donates iron to mycobactins in the *M. tuberculosis* cell wall. J. Exp. Med. 183, 1527–1532.

Gold, B., Deng, H., Bryk, R., Vargas, D., Eliezer, D., Roberts, J., et al., 2008. Identification of a copper-binding metallothionein in pathogenic mycobacteria. Nat. Chem. Biol. 4 (10), 609–616.

Hu, J., Jin, K., He, Z.-G., Zhang, H., 2020. Citrate lyase CitE in *Mycobacterium tuberculosis* contributes to mycobacterial survival under hypoxic conditions. PLoS ONE 15 (4), e0230786. Available from: https://doi.org/10.1371/journal.pone.0230786.

Jones, C.M., Niederweis, M., 2011. *Mycobacterium tuberculosis* can utilize heme as an iron source. J. Bacteriol. 193, 1767–1770.

Jones, C.M., Wells, R.M., Madduri, A.V., Renfrow, M.B., Ratledge, C., Moody, D.B., Niederweis, M., 2014. Self-poisoning of *Mycobacterium tuberculosis* by interrupting siderophore recycling. Proc. Natl. Acad. Sci. U. S. A. 111 (5), 1945–1950.

Kang, P.B., Azad, A.K., Torrelles, J.B., Kaufman, T.M., Beharka, A., Tibesar, E., et al., 2005. The human macrophage mannose receptor directs *Mycobacterium tuberculosis* lipoarabinomannan-mediated phagosome biogenesis. J. Exp. Med. 202, 987–999. Available from: https://doi.org/10.1084/jem.20051239.

Kumar, A., Alam, A., Bharadwaj, P., Tapadar, S., Rani, M., Hasnain, S.E., 2019a. Toxin-antitoxin (TA) systems in stress survival and pathogenesis. *Mycobacterium tuberculosis*: molecular infection biology, pathogenesis, diagnostics and new interventions. Springer, Singapore, pp. 257–274.

Kumar, A., Alam, A., Grover, S., Pandey, S., Tripathi, D., 2019b. Kumari M.,*et al.* Peptidyl-prolyl isomerase-B is involved in *Mycobacterium tuberculosis* biofilm formation and a generic target for drug repurposing-based intervention. npj Biofilms Microbiomes 53. Available from: https://doi.org/10.1038/s41522-018-0075-0.

Kumar, A., Lewin, A., Rani, P.S., Qureshi, I.A., Devi, S., Majid, M., et al., 2013. Dormancy associated translation inhibitor (DATIN/Rv0079) of Mycobacterium tuberculosis interacts with TLR2 and induces proinflammatory cytokine expression. Cytokine 64 (1), 258–264.

Kumar, A., Majid, M., Kunisch, R., Rani, P.S., Qureshi, I.A., Lewin, A., et al., 2012. Mycobacterium tuberculosis DosR regulon gene Rv0079 encodes a putative, 'dormancy associated translation inhibitor (DATIN)'. PLoS One 7 (6), e38709.

Kumar, A., Rani, M., Ehtesham, N.Z., Hasnain, S.E., 2017. Commentary: modification of Host responses by mycobacteria. Front. Immunol. 8, 466.

Lee, B.Y., Horwitz, M.A., 1995. Identification of macrophage and stressinducedproteins of *Mycobacterium tuberculosis*. J. Clin. Invest. 96, 245–249. Available from: https://doi.org/10.1172/JCI118028.

Li, X., Yang, Y., Zhou, F., Zhang, Y., Lu, H., Jin, Q., Gao, L., 2011. SLC11A1 (NRAMP1) polymorphisms and tuberculosis susceptibility: updated systematic review and meta-analysis. PLoS One 6 (1), e15831.

Lichten, L.A., Cousins, R.J., 2009. Mammalian zinc transporters: nutritional and physiologicregulation. Annu. Rev. Nutr. 29, 153–176.

Luo, M., Fadeev, E.A., Groves, J.T., 2005. Mycobactinmediated iron acquisition within macrophages. Nat. Chem. Biol. 1, 149–153.

MacMicking, J.D., Taylor, G.A., McKinney, J.D., 2003. Immune control of tuberculosis by IFN-gamma-inducible LRG-47. Science 302, 654–659.

Martin, C.J., Peters, K.N., Behar, S.M., 2014. Macrophages clean up: efferocytosis and microbialcontrol. Curr. Opin. Microbiol. 17, 17–23. Available from: https://doi.org/10.1016/j.mib.2013.10.007.

Mehra, A., Zahra, A., Thompson, V., Sirisaengtaksin, N., Wells, A., Porto, M., et al., 2013. *Mycobacterium tuberculosis* type VII secreted effector EsxH targets host ESCRT to impair trafficking. PLOS Pathog. 9, e1003734. Available from: https://doi.org/10.1371/journal.ppat.1003734.

Mizushima, N., Yoshimori, T., Ohsumi, Y., 2011. The role of Atg proteins in autophagosome formation. Annu. Rev. Cell Dev. Biol. 27, 107–132. Available from: https://doi.org/10.1146/annurev-cellbio-092910-154005.

Monahan, I.M., Betts, J., Banerjee, D.K., Butcher, P.D., 2001. Differential expression of mycobacterial proteins following phagocytosis by macrophages. Microbiology 147 (2), 459–471.

Moraco, A.H., Kornfeld, H., 2014. Cell death and autophagy in tuberculosis. Semin. Immunol. 26, 497–511. Available from: https://doi.org/10.1016/j.smim.2014.10.001.

Ouellet, H., Ouellet, Y., Richard, C., Labarre, M., Wittenberg, B., Wittenberg, J., et al., 2002. Truncated hemoglobin HbN protects *Mycobacterium bovis* from nitric oxide. Proc. Natl. Acad. Sci. U. S. A. 99, 5902–5907.

Pandey, S., Sharma, A., Tripathi, D., Kumar, A., Khubaib, M., Bhuwan, M., Chaudhuri, T.K., et al., 2016. Mycobacterium tuberculosis peptidyl-prolyl isomerases also exhibit chaperone like activity invitro and in-vivo. PLoS One 11 (3), e0150288.

Pandey, S., Tripathi, D., Khubaib, M., Kumar, A., Sheikh, J.A., Sumanlatha, G., et al., 2017. Mycobacterium tuberculosis peptidyl-prolyl isomerases are immunogenic, alter cytokine profile and aid in intracellular survival. Front. Cell. Infect. Microbiol. 7, 38.

Pathania, R., Navani, N.K., Gardner, A.M., Gardner, P.R., Dikshit, K.L., 2002. Nitric oxide scavenging and detoxification by the *Mycobacterium tuberculosis* haemoglobin, HbN in Escherichia coli. Mol. Microbiol. 45, 1303–1314.

Ratledge, C., 2004. Iron, mycobacteria and tuberculosis. Tuberculosis 84, 110–130.

Raynaud, C., Papavinasasundaram, K.G., Speight, R.A., Springer, B., Sander, P., Bottger, E.C., et al., 2002. The functions of OmpATb, a pore-forming protein of *Mycobacterium tuberculosis*. Mol. Microbiol. 46, 191–201.

Rodriguez, G.M., 2006. Control of iron metabolism in *Mycobacterium tuberculosis*. Trends Microbiol. 14, 320–327.

Rowland, J.L., Niederweis, M., 2012. Resistance mechanisms of *Mycobacterium tuberculosis* against phagosomal copper overload. Tuberculosis 92, 202–210.

Rowland, J.L., Niederweis, M., 2013. A multicopper oxidase is required for copper resistance in *Mycobacterium tuberculosis*. J. Bacteriol .

Ryndak, M.B., Wang, S., Smith, I., Rodriguez, G.M., 2010. The *Mycobacterium tuberculosis* high-affinity iron importer, IrtA, contains an FAD-binding domain. J. Bacteriol. 192, 861–869.

Schafer, G., Jacobs, M., Wilkinson, R.J., Brown, G.D., 2009. Non-opsonic recognition of *Mycobacterium tuberculosis* by phagocytes. J. Innate Immun. 1, 231–243. Available from: https://doi.org/10.1159/000173703.

Schaible, U.E., Sturgill-Koszycki, S., Schlesinger, P.H., Russell, D.G., 1998. Cytokine activation leads to acidification and increases maturation of *Mycobacterium avium*-containing phagosomes in murine macrophages. J. Immunol. 160, 1290–1296.

Schluger, N.W., 2001. Recent advances in our understanding of human host responses to tuberculosis. Respir. Res. 2, 157–163.

Senaratne, R.H., De Silva, A.D., Williams, S.J., Mougous, J.D., Reader, J.R., Zhang, T., et al., 2006. 5′-Adenosinephosphosulphate reductase (CysH) protects *Mycobacterium tuberculosis* against free radicals during chronic infection phase in mice. Mol. Microbiol. 59, 1744–1753.

Serafini, A., Pisu, D., Palu, G., Rodriguez, G.M., Manganelli, R., 2013. The ESX-3 secretion system is necessary for iron and zinc homeostasis in *Mycobacterium tuberculosis*. PLoS ONE 8, e78351.

Sevalkar, R.R., Arora, D., Singh, P.R., Singh, R., Nandicoori, V.K., Karthikeyan, S., et al., 2019. Functioning of mycobacterial heat shockrepressors requires the master virulenceregulator PhoP. J. Bacteriol. 201, e00013–e00019. Available from: https://doi.org/10.1128/JB.00013-19.

Sherman, D.R., Voskuil, M., Schnappinger, D., Liao, R., Harrell, M.I., Schoolnik, G.K., 2001. Regulation of the Mycobacterium tuberculosis hypoxic response gene encoding α-crystallin. Proc. Nat. Acad. Sci. U. S. A. 98 (13), 7534–7539.

Shi, L., Sohaskey, C.D., Kana, B.D., Dawes, S., North, R.J., Mizrahi, V., et al., 2005. Changes in energy metabolism of *Mycobacterium tuberculosis* in mouse lung and under in vitro conditions affecting aerobic respiration. Proc. Natl. Acad. Sci. U. S. A. 102, 15629–15634. Available from: https://doi.org/10.1073/pnas.0507850102. PMID: 16227431.

Siegrist, M.S., Unnikrishnan, M., McConnell, M.J., Borowsky, M., Cheng, T.Y., Siddiqi, N., et al., 2009. Mycobacterial Esx-3 is required for mycobactin-mediated ironacquisition. Proc. Natl. Acad. Sci. U. S. A. 106 (44), 18792–18797.

Siegrist, M.S., Steigedal, M., Ahmad, R., Mehra, A., Dragset, M.S., Schuster, B.M., et al., 2014. Mycobacterial Esx-3 requires multiple components for iron acquisition. mBio 5 (3), e01073–01014.

Singh, R., Anil Kumar, V., Das, A.K., Bansal, R., Sarkar, D., 2014. A transcriptionalco-repressor regulatory circuit controlling the heat-shock response of *Mycobacterium tuberculosis*. Mol. Microbiol. 94, 450–465. Available from: https://doi.org/10.1111/mmi.12778.

Snow, G.A., 1970. Mycobactins: iron-chelating growth factors from mycobacteria. Bacteriol. Rev. 34, 99–125.

Stewart, G.R., Snewin, V.A., Walzl, G., Hussell, T., Tormay, P., O'Gaora, P., et al., 2001. Overexpression of heat-shock proteinsreduces survival of *Mycobacterium tuberculosis* in the chronic phaseof infection. Nat. Med. 7, 732–737. Available from: https://doi.org/10.1038/89113.

Stewart, G.R., Wernisch, L., Stabler, R., Mangan, J.A., Hinds, J., Laing, K.G., et al., 2002. Dissection of the heat-shock response in *Mycobacterium tuberculosis* using mutants and microarrays. Microbiology 148, 3129–3138. Available from: https://doi.org/10.1099/00221287-148-10-3129.

Sun, J., Siroy, A., Lokareddy, R.K., Speer, A., Doornbos, K.S., Cingolani, G., et al., 2015. The tuberculosis necrotizing toxin kills macrophages by hydrolyzing NAD. Nat. Struct. Mol. Biol. 22, 672–678. Available from: https://doi.org/10.1038/nsmb.3064.

Sureka, K., Hossain, T., Mukherjee, P., Chatterjee, P., Datta, P., Kundu, M., et al., 2010. Novel role of phosphorylation-dependent interaction between FtsZ and FipA in mycobacterial cell division. PLoS ONE 5, e8590.

Taylor, R.C., Cullen, S.P., Martin, S.J., 2008. Apoptosis: controlled demolition at the cellular level. Nat. Rev. Mol. Cell Biol. 9, 231–241.

Tsai, M.C., Chakravarty, S., Zhu, G., Xu, J., Tanaka, K., Koch, C., et al., 2006. Characterization of the tuberculous granuloma in murine and human lungs: cellular composition and relative tissue oxygen tension. Cell Microbiol. 8, 218–232. Available from: https://doi.org/10.1111/j.1462-5822.2005.00612.x. PMID: 16441433.

Tullius, M.V., Harmston, C.A., Owens, C.P., Chim, N., Morse, R.P., McMath, L.M., et al., 2011. Discovery and characterization of a unique mycobacterial heme acquisition system. Proc. Natl. Acad. Sci. U. S. A. 108 (12), 5051–5056.

van der Wel, N., Hava, D., Houben, D., Fluitsma, D., van Zon, M., Pierson, J., et al., 2007. *M. tuberculosis and M. leprae* translocate from the phagolysosome to the cytosol in myeloidcells. Cell 129, 1287–1298.

Vandal, O.H., Pierini, L.M., Schnappinger, D., Nathan, C.F., Ehrt, S., 2008. A membrane protein preserves intrabacterial pH in intraphagosomal *Mycobacterium tuberculosis*. Nat. Med. 14, 849–854.

Via, L.E., Fratti, R.A., McFalone, M., Pagan-Ramos, E., Deretic, D., Deretic, V., 1998. Effects of cytokines on mycobacterial phagosome maturation. J. Cell Sci. 111, 897–905.

Voskuil, M.I., Visconti, K.C., Schoolnik, G.K., 2004. *Mycobacterium tuberculosis* gene expressionduring adaptation to stationary phase and low-oxygen dormancy. Tuberculosis (Edinb) 84, 218–227.

Ward, S.K., Hoye, E.A., Talaat, A.M., 2008. The global responses of *Mycobacterium tuberculosis* to physiological levels of copper. J. Bacteriol. 190, 2939e46.

Watanabe, S., Zimmermann, M., Goodwin, M.B., Sauer, U., Barry 3rd, C.E., Boshoff, H.I., 2011. Fumaratereductase activity maintains an energized membrane in anaerobic *Mycobacterium tuberculosis*. PLoS Pathog 7, e1002287. Available from: https://doi.org/10.1371/journal.ppat.1002287. PMID: 21998585.

Watson, R.O., Manzanillo, P.S., Cox, J.S., 2012. Extracellular *M. tuberculosis* DNA targets bacteria for autophagy by activating the host DNA-sensing pathway. Cell 150, 803–815. Available from: https://doi.org/10.1016/j.cell.2012.06.040.

Wells, R.M., Jones, C.M., Xi, Z., Speer, A., Danilchanka, O., Doornbos, K.S., et al., 2013. Discovery of a siderophore export system essential for virulence of *Mycobacterium tuberculosis*. PLoS Pathog. 9 (1), e1003120.

Wolschendorf, F., Ackart, D., Shrestha, T.B., Hascall-Dove, L., Nolan, S., Lamichhane, G., et al., 2011. Copper resistance is essential for virulence of *Mycobacterium tuberculosis*. Proc. Natl. Acad. Sci. U. S. A. 108 (4), 1621–1626.

Zhang, Z., Hillas, P.J., Ortiz de Montellano, P.R., 1999. Reduction of peroxides and dinitrobenzenes by *Mycobacterium tuberculosis* thioredoxin and thioredoxin reductase. Arch. Biochem. Biophys. 363, 19–26.

CHAPTER 2

Modulation of host pathways by *Mycobacterium tuberculosis* for survival

Shatabdi Sengupta[1], Aveepsa Sengupta[1], Arif Hussain[2], Jayshree Sarma[1], Aniruddha Banerjee[1], Saurabh Pandey[3], Deeksha Tripathi[4], Vidyullatha Peddireddy[5] and Ashutosh Kumar[1]

[1]Department of Microbiology, Tripura University (A Central University), Agartala, Tripura, India
[2]Laboratory Sciences and Services Division, International Centre for Diarrhoeal Disease Research (icddr,b), Dhaka, Bangladesh
[3]Department of Biochemistry, School of Chemical and Life Sciences, Jamia Hamdard, New Delhi, Delhi, India
[4]Microbial Pathogenesis and Microbiome Lab, Department of Microbiology, School of Life Sciences, Central University of Rajasthan, Ajmer, Rajasthan, India
[5]Department of Nutrition Biology, School of Interdisciplinary and Applied Sciences, Central University of Haryana, Mahendragarh, Haryana, India

Introduction

Mycobacterium tuberculosis (*Mtb*) is the causative agent of the disease tuberculosis taking the lives of millions of people around the world. The bacterial biofilms induces the development of antibiotic tolerance. Cyclophilin peptidyl-prolyl isomerase (PpiB), is a vital gene that has a major role in the formation of biofilm and drug tolerance (Kumar et al., 2019a) Cyclophilins are the enzymes that mediates folding of proteins by means of isomerization of peptide bonds of the subsequent propyl residues. *Mtb* peptidyl-prolyl isomerases are invoved inaltering cytokine secretion and may paly chaperone like activity (Pandey et. al. 2016; Pandey et al. 2017). *Mtb* infects the lungs of the host via aerosol route. After engulfment by macrophages, it can survive harsh physiological conditions for long time in a latent phase. To survive and cause pathogenesis under such conditions, *Mtb* has adopted various ways to modulate host defense mechanisms. The common host pathways modulated by *Mtb* are glycolytic flux, endoplasmic reticulum (EPR) stress, modulation in host mitochondria, programmed death, phagosome maturation, mitochondrial membrane disruption (Chen et al., 2006), necrosis induction (Mehrotra et al. 2014), host signalling pathway suppression autophagy etc. Although, *Mtb* can modulate itself in such a way that both the macrophage and mycobacteria can survive in several stress conditions like limited amount of nutrients, inside the granuloma. They survive in stress conditions by preserving their energy consumption (Kumar et al. 2017). The toxin-antitoxin systems are associated with the adaptation to stress, tolerance and resistance to antimicrobial agents wards other microbes, alteration of the physiological state, formation of biofilm, regulation of growth for survival, maintenance of plasmid, virulence, apoptosis etc.

Bacterial Survival in the Hostile Environment
DOI: https://doi.org/10.1016/B978-0-323-91806-0.00003-5

(Kumar et al., 2019b). Although, metabolic reprogramming of *Mtb*-infected macrophages aids in its survival within the host. Particularly *Mtb* regulates the glycolytic pathway to control the immune responses. Glycolysis is essential to generate adenosine triphosphate (ATP) and other precursors required to produce inflammatory mediators. The glycolytic pathway is usually regulated at four stages that include glucose import, hexokinase, phosphofructokinase (PFK), and lactate export. *Mtb* exploits these steps to repress the glycolytic pathway to facilitate their survival and growth in the macrophages. This disruption induces the production of proinflammatory molecules, mainly interleukin-1b (IL-1b). Regulation at the level of PFK is required for glycolysis and the suppression of lactate accumulation in macrophages infected by *Mtb* (Cumming et al., 2018).

Mtb has developed themselves in such a way that it can withstand different types of environmental stress conditions. There are various regulons which regulates espression of various genes to withstand various stresses. One such regulons is the DosR regulon which takes part in the survival in the latent stage of *Mtb*. For the maintenance of the granuloma, a balanced secretion of different cytokines is necessory. Dormancy associated translation inhibitor (DATIN) protein can trigger peripheral blood mononuclear cells (PBMC) and macrophages through the interaction with TLR2 (Kumar et al. 2013). The DATIN is also responsible to regulate translation by interacting with ribosomal subunits of bacteria (Kumar et al. 2012). *Mtb* modulates the host EPR stress. The EPR stress is resolved through the unfolded protein response (UPR). EPR stress is induced by the accumulation of lipids in the caseum region of TB granuloma. These lipids are triacylglycerols, unesterified cholesterol, and cholesterol esters. Accumulation of modified aggregated and remnant lipoproteins, that is, unesterified cholesterol, 7-ketocholesterol, and saturated fatty acids, induces EPR stress and causes macrophage apoptosis. EPR stress marker, C/EBP homologous protein (CHOP), is involved as proapoptotic stress factor. *Mtb* modulates the host mitochondria and further it induces the death of cell via an intrinsic apoptotic pathway. During the regulation of cell death via apoptotic pathway, B-cell lymphoma-2 (Bcl-2) protein family plays an important role. Bcl-2 homologous antagonist/killer (Bak) and Bcl-2-associated X protein (Bax) are the main apoptotic factors that are required for apoptosis pathway. These proteins are then bound to the outer mitochondrial membrane (OMM) to induce its permeabilization to produce MOMP (mitochondrial outer membrane permeabilization). The formation of MOMP helps in the delivery of proapoptotic factors such as cytochrome c and second mitochondria-derived activator of caspases (SMAC/DIABLO) into the cytoplasm. Some virulent strains of *Mtb* can induce translocation of proapoptotic factor Bax to mitochondria stimulating the permeabilization of OMM allowing the discharge of cytochrome c into the cytoplasm which ultimately leads to cell death through apoptosis. *Mtb*-infected macrophages exhibit MOMP phenotype.

During active *Mtb* infection, intraalveolar neutrophil infiltration and the extreme production of inflammatory response stimulates necrosis, which is characterized by the initial caseation as well as liquefaction of granulomas in lungs (Marzo et al., 2014; Niazi et al., 2015).

Mtb induce this neutrophil-mediated necrosis by inhibiting ROS production (Dallenga et al., 2017). *Mtb* can promote necrosis in macrophages by regulating the production of eicosanoids in host cells (Chen et al., 2008). Several mycobacterial proteins are involved in infected cell necrosis. Also, the secretion system of Esx-1 in *Mtb* have an important role in virulence, this may induce lysis of infected cells in a contact-dependent manner (van der Wel et al., 2007). Furthermore, cell death of host necrotic cell is also mediated by the CpnT, an NAD^+ glycohydrolase. It form pores in outer cell membrane, implicated in the depletion of cellular NAD^+ pools (Danilchanka et al., 2014; Sun et al., 2015). PPE11 (Rv0453) protein of *Mtb* is also involved in macrophage necrosis (Peng et al., 2019). Also, PE17 (Rv1646) protein of *Mtb* downregulates the synthesis of some proinflammatory cyto-kines (IL-6, IL-12, and tumor necrosis factor alpha (TNF-α)) and promotes necrosis in infected macrophages.

In *Mtb*-infected cells, the phagosome maturation is inhibited by antiinflammatory cyto-kine IL-10, which performs a key role in their survival. IL-10 suppresses the functions of some phagocytic cells, that is, macrophages and dendritic cells which block the initiation of the innate immunity in retort to the invading pathogens. IL-10 with the help of mitogen-activated protein kinase p38 ($MAPK^{p38}$), extracellular regulated kinase 1/2 (ERK1/2), and signal transducer and activator of transcription 3 (STAT3) are responsible to prevent phago-some development (Murray et al., 1997; Williams et al., 2004). Autophagy serves as an immune effector that results in the destruction of intracellular pathogens like *Mtb* (Gutierrez et al., 2004). To deal with such a situation, *Mtb* synthesizes the phenolic glycolipid (PGL) with the help of polyketide synthase, encoded by intact pks 1−15. *Mtb* also regulates autop-hagy by manipulating Ca^{2+} signaling, which results in the intracellular survival of *Mtb*.

Modulation of glycolytic flux

The glycolytic flux describes the exchange of metabolites of glycolysis; for instance, the alter-ation of fructose 6-phosphate to fructose 1,6-bisphosphate. This flux is regulated by the demand for ATP. The organisms produces enzymes that catalyze each of the 12 steps of gly-colysis, which are regulated by the presence of extracellular glucose and excreted lactate accumulation, thereby controlling the glycolytic flux. The glycolytic flux is regulated using four rate-limiting stages that include import of glucose, interconversion of fructose 6-phosphate to fructose 1,6-bisphosphate by PFK, glucose to glucose 6-phosphate by hexoki-nase, and lactate export.

Mtb modulates the glycolytic pathway of the infected host cells for their survival. The proliferating mycobacterium induces the production of anti-inflammatory microRNA-21 (miR-21) in host cells that in turn suppresses the glycolytic pathway of the host cells (Fig. 2.1). The suppression of the glycolytic pathway by miR-21 is achieved by targeting the key regulator of the glycolysis, the phosphofructokinase muscle (PFK-M) isoform. This suppression facilitates mycobacterial growth as the

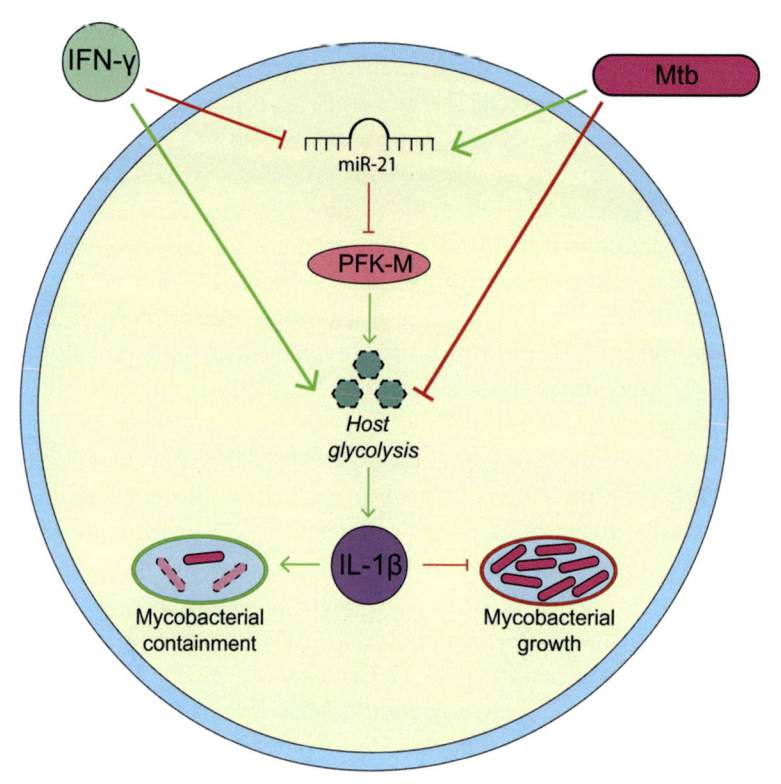

Figure 2.1 Modulation of glycolysis pathway by *Mtb* (Hackett et al., 2020). *Adapted from Hackett, E.E., Charles-Messance, H., O'Leary, S.M., Gleeson, L.E., Muñoz-Wolf, N., Case, S., et al., 2020. Mycobacterium tuberculosis limits host glycolysis and IL-1β by restriction of PFK-M via MicroRNA-21. Cell Rep. 30 (1), 124–136.*

production of proinflammatory cytokines dampens, mainly the IL–1β. Interferon–γ (INF–γ) plays a vital role in protecting the host from *Mtb* infection. INF–γ inhibits the miR–21 expression in the infected cells causing a switch in the isoenzyme PFK complex, leading to domination of active isoenzymes (PFK–M) and increased macrophage glycolysis. The miRNA, miR–21, acts as the regulator of the macrophage immuno–metabolic response. On the one hand, miR–21 represses the expression and activity of PFK–M and glycolysis. Also, miR–21 modulates the host's innate immunity by negatively regulating the anti–inflammatory cytokine secretion. Therefore, inhibition of miR–21 can be a potential therapeutic intervention strategy against defective immune responses, including cancer treatment, boosting mycobacterial vaccine effectiveness, or treating Th2–associated allergic inflammation.

Infection of *Mtb* will block fatty–acid oxidation in macrophages which results in mobilization of the intracellular cholesterol and fatty acids of *Mtb* that favors their growth. In the case of *Mtb* infection, the reprogramming of energy metabolism in macrophages is not precisely the "Warburg effect" that is observed in cancer cells or lipopolysaccharide (LPS)-stimulated macrophages wherein the glycolytic rate is increased in the infected macrophages (Koppenol et al., 2011). Whereas during *Mtb* infection, the proglycolytic isoenzyme form of PFK-1 limits the rate of glycolysis. This results in restriction of glycolytic flux through the glycolytic pathway with an overall decrease in lactate accumulation and glycolysis in macrophages infected by macrophages (Cumming et al., 2018). This helps in the better survival of *Mtb* in macrophages as decreased glucose metabolism through glycolysis suppression impedes macrophage activation due to the short supply of rapid sources of ATP and reducing the number of biosynthetic precursors responsible for inflammatory responses. In the case of *Mtb* infection, the activity of PFK-1 is also regulated by cytokines produced by the host, particularly the production of IFN-γ. As IFN-γ can relieve the inhibition of PFK-M by controlling the mature miR-21, it reprograms the cellular metabolism shifting the balance in support of the proinflammatory response. The posttranscriptional regulation can be brought by IFN-γ using the downregulation of mature miR-21. The SMAD proteins (Su et al., 2015) aid in signaling of transforming growth factor-β (TGF-β) which in turn elicits IFN-γ signaling (Ghosh et al., 2001). This negative regulation of miR-21 results in an increase in expression of PFK-M and glycolytic flux in stimulated macrophages, thereby enabling signaling of proinflammatory pathways. The lipid mobilization in *Mtb*-infected cells could subsidize to activity of IFN-γ (Knight et al., 2018). This IFN-γ aiming miR-21 could drive glycolysis for promoting glyceraldehyde 3-phosphate production. Newer antimycobacterial approaches are being developed, which will encourage IFN-γ activity and modulate miRNA expression. The original mRNA target and other cellular processes are mediated by miR-21. This justifies the defective boosting immunity against chronic pulmonary diseases like tuberculosis and cancer.

Endoplasmic reticulum stress

Mtb infection can evade macrophage apoptosis . EPR stress is mainly caused by misfolding or unfolding structure of proteins in the EPR that is overcome by unfolded protein response (UPR). The UPR is found to be activated in various disease conditions like diabetes, atherosclerosis, cancer, infectious diseases, and obesity. It is an adaptive strategy that gets started by the assembly of misfolded proteins in retort to various cellular abuses resulting in EPR stress. If the EPR stress is excessive and is accompanied by additional stimuli, such stress may eventually lead to apoptosis of the cell. Apoptosis occurs when EPR-folding capacity is exceeded, which is mainly depending on the upregulation of the UPR-inducible transcription factor CHOP. CHOP deficiency protects macrophages from EPR-induced apoptosis. In tuberculosis granulomas, apoptotic macrophages and foamy macrophages accumulate. The foamy appearance of macrophages is due to the presence of esterified cholesterol. Both *Mtb* and *Mycobacterium bovis* help in the development of foam cells (Seimon et al., 2010).

In TB granuloma (caseum), the primary lipid accumulation is found in the form of unesterified cholesterol, cholesterol ester, and triacylglycerols in the center of the granuloma of TB. Lipids like saturated fatty acids, 7-ketocholesterol, and unesterified cholesterol, when accumulated from the uptake of altered, accumulated and leftover host lipoproteins, would cause high and prolonged EPR stress; this induces apoptosis in the macrophages. During the accumulation of lipids like saturated fatty acid, 7-ketocholesterol, and unesterified cholesterol can result in high and prolonged EPR stress, which induces apoptosis in the macrophages. Macrophage apoptosis portrays an essential role in the protection of host from infection. *Mtb*-loaded apoptotic macrophages can be identified by TdT-mediated dUTP nick-end labelling (TUNEL) assay by costaining the Mac-3 (macrophage marker) and CHOP(Seimon et al., 2010).

There are two stages of mycobacterial infection. In early infection, apoptosis in macrophages prevents spreading and promotes clearance of the infection in the host. It was observed that when infected apoptotic macrophages are uptaken by dendritic cells, resulting in the collapse and further doling out of *Mtb* antigen and thus, it stimulates T cells and then induces adaptive immunity(Lee et al., 2007). At late stage of tuberculosis, granuloma has a restricted supply of both lymphocytes and macrophages and has little vascularization. This results in the accumulation of caseum and progression of the disease pathology. *Mtb* releases the infectious bacilli by liquefaction and cavitation of infected and foamy macrophages. Apoptosis is advantageous in the case of low bacterial load, but it can be detrimental in the late stage when fibrocaseous granulomas are already formed. Thus, the induction of ER stress and apoptosis are linked to mycobacterial infection and its progression effects the infection level in patients (Russell, 2007; Lee et al., 2009; Russell et al., 2009).

Modulation in host mitochondria

Mitochondria regulate and maintain the cellular ATP level, ROS production, host immune and calcium signaling responses upon infection. Viruses and bacteria have evolved molecular tools to modulate mitochondrial dynamics (Fig. 2.2). This disturbance of mitochondrial dynamics increases the intracellular survival of pathogen or bypassing of host immunity (Tiku et al., 2020). Mitochondria regulate the metabolism of carbohydrates, fatty acids, and amino acids in macrophages. In this case, pyruvate can be formed from glucose in the cytoplasm via glycolysis, shuttling the pyruvate into mitochondria. The oxidation of pyruvate takes place with the help of the electron transport chain and tricarboxylic acid (TCA) cycle to generate ATPs as a source of energy. Complete oxidation of each glucose molecule generates 36 molecules of ATP. As *Mtb* hijacks metabolism of the host by augmenting glycolysis, immune cells infected with *Mtb* show a deterioration in

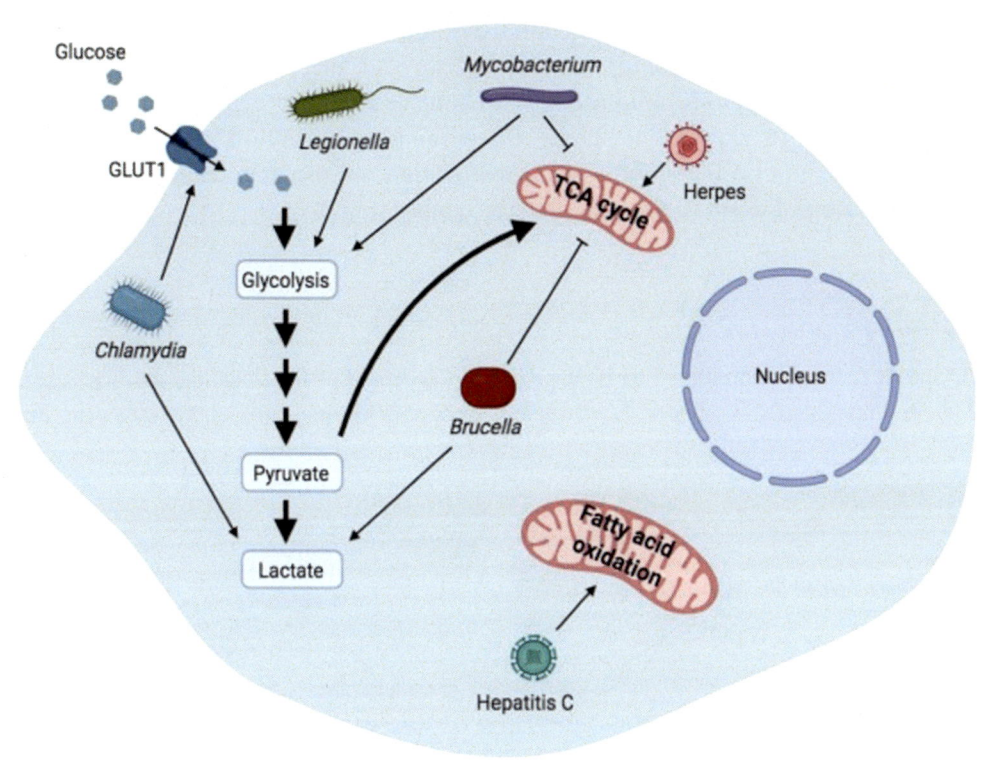

Figure 2.2 Modulation of mitochondrial metabolism. *Adapted from Tiku, V., Tan, M.W., Dikic, I., 2020. Mitochondrial functions in infection and immunity. Trends Cell Biol. 30, 263–275. https://doi.org/10.1016/j.tcb.2020.01.006.*

the TCA cycle, causing an increase in glycolytic flux. It was observed that in the lungs of TB patients, the induction of aerobic glycolysis takes place in granulomas infected by *Mycobacterium* having active tuberculosis (Tiku et al., 2020).

Death of cell can be regulated by the process of intrinsic apoptotic pathway with the help of Bcl-2 protein family. Both Bax and Bak are vital apoptotic factors. These two proteins bind to the OMM to induce MOMP (Kale et al., 2018). SMAC/DIABLO and cytochrome c are few proapoptotic factors which are released upon formation of MOMP. A protein complex known as apoptosome is formed with the interaction of apoptotic protease activating factor 1 with cytochrome c. This apoptosome helps in the formation of active caspase from procaspase-9 which subsequently activates executioner caspase-3, caspase-6, and caspase-7 that results in cell death/apoptosis (Boice and Bouchier-Hayes, 2020; Kale et al., 2018). This pathway is exploited by many pathogens to their favor by inducing death of the host cell to obtain nutrients or to further disseminate. By suppressing the cell death, some pathogens allow the proliferation of infected cells. Some virulent strains of *Mtb* induce translocation of proapoptotic factor, Bax, to mitochondria-stimulating permeabilization of OMM allowing to release cytochrome c into cytoplasm and eventually leading to death of the cell through apoptosis. *Mtb*-infected macrophages seen to exhibit a MOMP phenotype (Chen et al., 2006). Thus, mitochondrial disruption is a hallmark of *Mycobacterium* infection, leading to the stimulation of the mitochondrial apoptotic machinery for cell death to promote its survivability (Fig. 2.3).

Apoptosis

Apoptosis is usually considered to be a protection mechanism for shielding the host against *Mtb* in initial stages of illness. Nevertheless, apoptosis during subsequent stages in lung

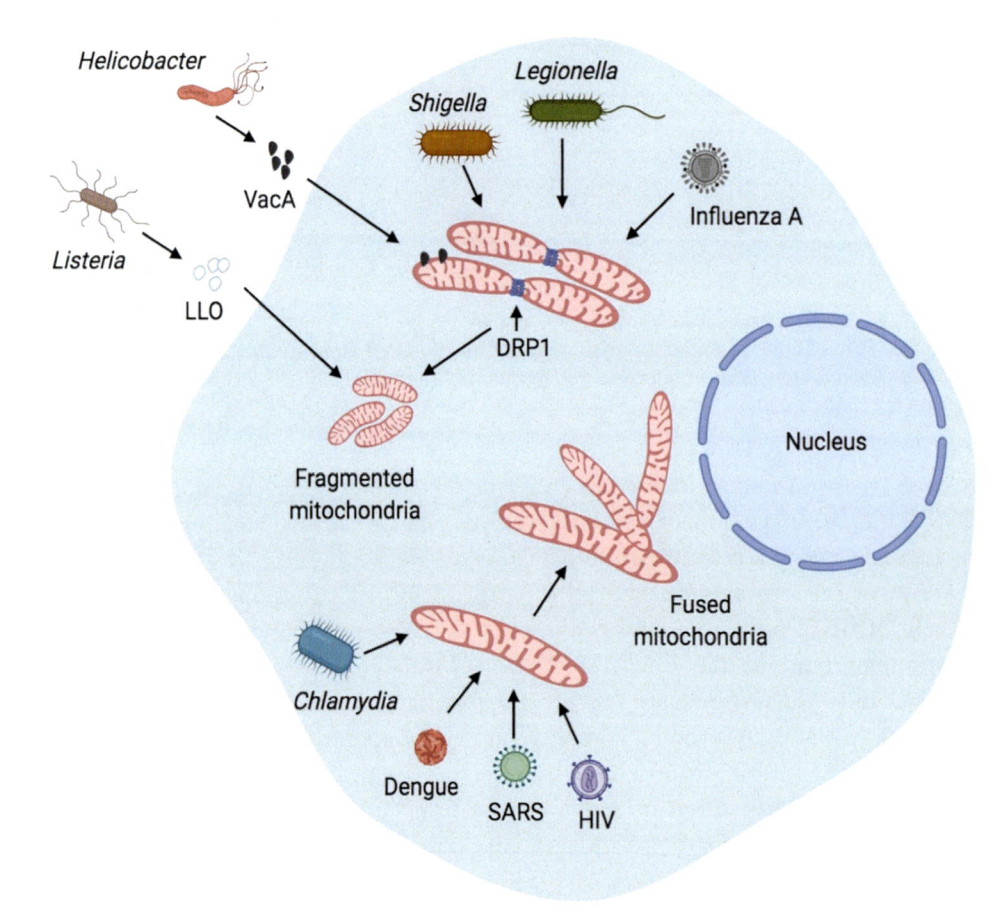

Figure 2.3 Mitochondrial apoptotic pathway modulated by pathogens (*Mtb*) during infection. *Adapted from Tiku, V., Tan, M.W., Dikic, I., 2020. Mitochondrial functions in infection and immunity. Trends Cell Biol. 30, 263−275. https://doi.org/10.1016/j.tcb.2020.01.006.*

granulomas might encourage the bacterium in spreading the disease. *Mtb* has adapted for its survival and replication in the phagosomes of macrophage. Apoptosis is a host defense against the invading pathogens. *Mtb*-infected alveolar macrophages in humans undergo apoptosis using a TNF-α-dependent mechanism (Keane et al., 1997). Among the strains of *Mtb*, the isogenic avirulent strain *Mtb* H37Ra induces more alveolar macrophages as compared to virulent *Mtb* H37Rv. However, the stimulation of IL-10 results in the release of soluble TNFR2 (sTNFR2) by the alveolar macrophages and subsequently it can antagonize the bioactivity of TNF-α (Balcewicz-Sablinska et al., 1998). For a successful host defense against tuberculosis, TNF-α expression is crucial (Flynn et al., 1995). A novel mechanism of *Mtb* to escape the host defense constitutes the induction of IL-10 that results in the inhibition of TNF-α activity. Macrophage apoptosis limits *Mtb* propagation and replication in the lungs (Molloy et al., 1994; Oddo et al., 1998). *Mtb*-induced apoptosis of alveolar macrophages can be considered as a phenotype, which is virulence-associated (Keane et al., 2000).

The initiation of infected macrophage apoptosis performs a vital role in the containment of disease pathology. The maintenance of a supportive intracellular environment for the growth of bacteria is linked to the capacity of the virulent mycobacterial strains in modulating alveolar macrophage apoptosis. *Mtb* can also avoid secondary phagocytosis by newly recruited mononuclear cells and inhibit the host macrophage apoptosis by inducing IL-10 (Zhang et al., 1994; Flynn et al., 1995), which indicates that the induction of cytokines after *Mtb* infection can influence alveolar macrophage apoptosis (Balcewicz-Sablinska et al., 1998). Therefore, other mechanisms might be involved in the modulation of mycobacterial-infected alveolar macrophage apoptosis, beyond IL-10 and TNF-α responses of *Mtb*-loaded alveolar macrophages. Two isogenic strains H37Ra (avirulent) and H37Rv (virulent) showed contrasting phenotypes with respect to apoptosis induction. Investigations on murine macrophage cells on analysis of apoptosis conclude that in tuberculosis infection, the genetic factor of the host may be responsible to the regulation of the fate on the cell (Rojas et al., 1997).

Numerous *Mtb* proteins/Antigens engender apoptosis. ROS signaling, intrinsic pathway, and suppressor of cytokine signaling1 (SOCS1) pathway are triggered by early secreted antigenic target 6-kDa protein (ESAT-6) for stimulating apoptosis in macrophages (Yang et al., 2015). Further ESAT-6, with phthiocerol dimycocerosate, stimulates the fissure of phagosomal membranes leading to apoptosis of the host cells, consequently promoting virulence (Augenstreich et al., 2017). The PE-PGRS domain of Rv0297 (Rv0297 PGRS), thought to be implicated in the localization of endoplasmic reticulum, was demonstrated to be involved in EPR stress-induced cell mortality and caspase-8-facilitated apoptosis (Grover et al., 2018). Toll-like receptor 2 (TLR2)/p38 MAPK/cyclooxygenase (COX)-2 signaling pathway was reported to be involved in the macrophage apoptosis induced by secreted lipoprotein of *Mtb* Ag MPT83 (Rv2872) (Wang et al., 2017). Rv1016c lipoprotein, a TLR2 ligand of *Mtb*, promotes macrophage apoptosis and prevents IFN-γ stimulated MHC-

II expression, leading to dodging from immune supervision thus fostering chronic infection (Su et al., 2016). The 38-kDa Ag, an *Mtb* lipoprotein provokes monocyte chemoattractant protein-1 (MCP-1), which boosts the fabrication of EPR-stress and ROS-induced apoptosis. A1 protein, a family member of antiapoptotic B-cell CLL/lymphoma-2 is stimulated by Mannose-capped lipoarabinomannan to prevent apoptosis of the host cell (Halder et al., 2015). Intracellular *Mtb* endurance is promoted by chaperone protein Cpn60.2 by repressing apoptosis via mitochondrial mortalin interaction (Joseph et al., 2017). PE_PGRS18, the *Mtb* PE/PPE family protein fosters pathogenesis by mitigating macrophage and TLR2-reliant apoptosis and inflammatory signaling is prevented by LpqT. An acyl carrier protein (Rv2244) of *Mtb*, responsible for the biosynthesis of mycolic acid, prevents apoptosis of macrophages by repressing the JNK/ROS signaling trail, thus promoting *Mtb* virulence (Paik et al., 2019). Thus, the pattern of apoptotic responses is very consistent though there is a noticeable variability in the production of cytokines by alveolar macrophages during the *Mtb* infection.

Necrosis

Mtb pathogenesis relies on its ability to regulate the cell cycle of the infected macrophages. Though apoptosis of infected cells is bactericidal, whereas necrosis would facilitate bacterial dissemination and transmission. In phagocytic cells, the microbicidal nature of the bacillary replication niche is influenced by adopting the apoptotic route. In contrast, cell lysis and facilitation of bacilli release and cell-to-cell spread are achieved by necrotic cell death. The control and progression of *Mtb* infection in the host cell depends on the fate of the infected primary host cell. During active *Mtb* infection, intra-alveolar neutrophil intrusion and the undue inflammatory reactions perform a significant role in necrosis which is distinguished primarily by the caseous and liquefaction of granulomas in the lungs (Marzo et al., 2014; Niazi et al., 2015). *Mtb* can induce neutrophil necrosis by inhibiting ROS production (Dallenga et al., 2017). *Mtb* can promote necrosis in macrophages by adjusting eicosanoid fabrication in host cells (Chen et al., 2008). Cells are protected from necrosis via the process of stimulating the production of the prostanoid i.e. prostaglandin E_2 (PGE$_2$) by avirulent strain of H37ra (Divangahi et al., 2009). In contrast, H37rv, a virulent strain that stimulates the lipoxin production A4 (LXA4) that results in downregulation of the cyclooxygenases, eventually reducing PGE2 production and hence inducing necrotic cell death (Chen et al., 2008; Divangahi et al., 2009).

Several mycobacterial proteins are involved in infected cell necrosis. For instance, the Esx-1 secretion system of *Mtb* can stimulate lysis of infected cells in a contact-dependent approach (van der Wel et al., 2007). Additional protein implicated in the host cell necrosis is the CpnT (channel protein with necrosis inducing toxin) an

NAD^+ glycohydrolase implicated in the reduction of cellular NAD^+ pools. CpnT can helps in transportation of nutrients as well as responsible for host cell death with the help of *Mtb*. (Danilchanka et al., 2014; Sun et al., 2015). PPE11 (Rv0453) was found to be implicated in macrophage necrosis (Peng et al., 2019). PE17 (Rv1646) downregulates the production of proinflammatory cytokines (IL-6, TNF-α, and IL-12) and promotes cell injury in infected macrophages. *Mtb* can evade host macrophage apoptosis by modulating the delivery of TNF-α and antiapoptotic proteins like Bcl 2. *Mtb* obstructs the extrinsic apoptotic pathway and subsequently initiates the intrinsic apoptotic pathway, resulting in the damage of mitochondrial transmembrane and induction of macrophage death. Both the virulent and the attenuated strains H37Rv and H37Ra stains demonstrated to induce characteristic damage of the mitochondrial outer membrane whereas the induction of mitochondrial transmembrane potential loss can be seen only for the virulent H37Rv strain. This results in macrophage necrosis and spread of *Mtb* bacilli (Chen et al., 2006). *Mtb* uses an intrinsic apoptotic pathway which is correlated directly with the increased levels of macrophages necrosis.

There is a link between PPE68 and the ESX-1 secretion system (Demangel et al., 2004). Inactivation of PPE68 results in the enhanced ESAT-6 secretion, which implies that PPE68 plays active participation in regulation and secretion by the type VII apparatus (Teutschbein et al., 2009). The PPE68 protein can dimerize and serves as a segregation gate prevents from secreting prematurely. Overexpression of this protein in *Mtb* can induce necrosis in macrophages. Thus, the cell-to-cell spreading can be facilitated by several virulence factor of the *Mtb* which helps in the necrosis of the host cell and escape from the macrophages.

Phagosome maturation

Phagocytosis protects cells against invading pathogens. The microbial pathogens and apoptotic cells are subjected to degradation after fusion of phagosome with an acidic lysosome containing antimicrobicidal proteases and lipases. It also clears the apoptotic bodies. Bacteria induce phagocytosis by activating the TLR signaling pathway. However, apoptotic cells regulate phagocytosis by several pathways, which include internalization and phagosome maturation. The absence of TLR signaling will impair bacterial phagocytosis. Phagosome maturation is considered as an essential innate immune reaction for intracellular illness. *Mtb* modulates and prevents the host responses leading to its endurance in the host cell. *Mtb* can also inhibit the phagosome maturation in macrophages and can inactivate the microbicidal properties of macrophages. This is achieved by the induction of inhibitory cytokine production and interfering with efficient antigen presentation.

After *Mtb* infection, the macrophages and monocytes rapidly secrete TNF-α and IL-10. The IL-10 which is an antiinflammatory cytokine inhibits the maturation of phagosome in *Mtb*-infected cells. IL-10 represses the functions of phagocytic cells, that is, dendritic cells and macrophages blocking the initiation of the innate immune reaction against the invading pathogen. IL-10-mediated inhibition of phagosome maturation is observed in THP-1 cells treated with phorbol myristate acetate, monocytes-derived macrophages (MDMs), and in cells infected with killed *Mtb*. Moreover, the IL-10 and MDMs addition results to enhanced survival and growth of mycobacterial infection. IL-10 exerts its effects partly using the STAT3 in the stage of early macrophage response. These functions are brought about independent of ERK1/2 and MAPK^{p38} activity. This way IL-10 favor pathogen survivability by arresting the *Mtb*-phagosome development in human macrophages. The reticence of the maturation pathway blocks the phagosome lysosome fusion prevents the formation of phagolysosome that contains acid hydrolases for microbicidal degradation and in turn, creates a suitable environment for its survival and replication (Clemens and Horwitz, 1995; Rohde et al., 2007; de Chastellier, 2009).

Cytokine and chemokine secretions after mycobacterial infections activate macrophages, in turn, employ inflammatory cells consist of T cells, neutrophils, and natural killer (NK) cells (Berrington and Hawn, 2007). TNF-α have major function in controlling the intracellularly replicating *Mtb* by activating macrophage (Raja, 2004; Berrington and Hawn, 2007). Whereas the production of proinflammatory cytokines are inhibited by IL-10 which results in deactivating the macrophages. It prevents the release of reactive oxygen/nitrogen intermediates within macrophages for killing of intracellularly replicating *Mtb* (Fiorentino et al., 1991; Gong et al., 1996). In the case of *Mtb* infection, the fabrication of IL-10 gets increased and is coupled with the increase in susceptibility to infection. MAPK^{p38} is responsible for the excretion of IL-10 in monocytes and macrophages. The MAPK^{p38} and ERK1/2 determine the level of IL-10 induction (Murray, 2006; Souza et al., 2006; Williams et al., 2004). IL-10 exerts its effects on the innate immune reactions through STAT3 and thereby helps in blocking phagosome maturation. IL-10 binds to two chains of receptors, IL-10R1 and CRF2−4/IL-10R2, which transduces the signal through Janus kinase 1 or tyrosine kinase 2 pathways resulting in the initiation of STAT3, STAT1, and STAT5. Further, the activation and signaling mainly through STAT3 mediate the IL-10 inhibitory signal on macrophages (Williams et al., 2004; Bajaj et al., 2006).

Other cytokines can also activate STAT3 but the antiinflammatory activities can only be transduced via IL-10 signaling. In human macrophages, STAT3 has a significant responsibility in the reticence of *Mtb*-phagosome development. STAT3 gets phosphorylated upon contamination with *Mtb*. Also, the inhibition of STAT3 activity could enhance phagosome maturation in *Mtb*-infected macrophages. This shows the potential mechanism of the STAT3 signaling pathway that is activated by IL-10. So, in the management of *Mtb*-phagosome maturation in humans, a major role is carried out by IL-10 which enables the long-term infection in the lungs. It implies that for treating *Mtb* infection, IL-10 can be a possible focus for its therapy.

Autophagy Regulation

Autophagy is an intracellular catabolic process that consumes body tissues during starvation. It maintains homeostasis and removes invading pathogens using the lysosomal degradation process. Using drugs to activate autophagy can be used for the treatment of *Mtb* infection. Induction of autophagy is mediated by the vitamin D receptor signaling, sirtuin-1 activation, AMP-activated protein kinase pathway, and nuclear receptors. Autophagy portrays an essential part in the management of innate and adaptive immunity. Autophagy is an essential component of innate immune defense mechanism employed by the host cells to remove intracellular bacteria, viruses, and parasites (Deretic et al., 2013; Huang and Brumell, 2014).

Autophagy plays regulatory roles through antigenic presentation, B cell development, T cell selection, and effector roles through digestion of intracellular microbes, Th1 and Th2 polarization and effector output of TLRs during host responses against intracellular infections. The autophagic pathways include microautophagy, macroautophagy, and chaperone-mediated autophagy. It is the macroautophagy that is activated during *Mtb* infection. Macroautophagy produces a double-membraned vesicle around intracytoplasmic cargos called phagophore which gets enlarged and closes to create the autophagosome which combines together with the lysosome to produce autolysosome. These autolysosomes then disintegrate the cytoplasmic cargos. Many bacteria that cause intracellular infections are eliminated by the autophagy of infected macrophages and nonphagocytic cells (Amer and Swanson, 2005; Birmingham et al., 2006; Huang and Brumell, 2014).

Mycobacterium activates innate immune signaling that includes receptors like TLRs (TLR2, TLR4, and TLR9) and some non-TLRs (NOD2, Dectin-1, and Mincle) receptors. Autophagy serves as an immune effector for phagosome maturation and is also responsible for the clearing of *Mtb* (Rajaram et al., 2014). The processing and demonstration of antigen by MHC class II molecule is facilitated by the process of autophagy and it mounts the adaptive immune response. Here, IFN-γ induces host defensive immune reactions by connecting autophagy with innate immune activation for limiting the *Mtb* infection. During starvation, induction of autophagy can kill *Mtb*, due to which it has evolved mechanisms to survive in host cells by synthesizing the phenolic glycolipid (PGL) due to the presence of intact pks 1−15. This helps in the downregulation of Th-1 type cytokines production (Reed et al., 2004; Hanekom et al., 2011). However, there is no major association with the intact pks 1−15 presence and the mycobacterial capability to resist autophagic exclusion stimulated by starvation, indicating that PGL is not always responsible for mycobacterial bypass from autophagic restriction.

TANK-binding kinase 1 (TBK1) and Rab8b are two host proteins that play main function in autophagy arbitrated phagosome maturation in *Mtb*. The mycobacterial phagosome acidification and cathepsin D are decreased by siRNA-mediated depletion of Rab8b or pharmacological inhibition of TBK1. Starvation of host cells also induces autophagy. Autophagosomes engulf cytosolic substrates and deliver them to acidic lysosomes. These are then transferred to acidic lysosomes which subsequently effect in the deprivation of sequestered contents (Yim and Mizushima, 2020). The position of the peripheral lysosomes along three microtubules toward the perinuclear region for the fusion with autophagosome is also induced by starvation. Kxd1 and Plekhm2 show properties in lysosomal transportation along microtubules toward the cell periphery (Cabukusta and Neefjes, 2018). The expression of these genes is induced in macrophages during starvation.

Mtb has progressed to endure and reproduce even in the acidified atmosphere of autophagosomes/autolysosomes. It has achieved this by manipulating Ca^{2+} signaling pathway and escaping autophagy, which results in the intracellular persistence of *Mtb*. During *Mtb* infection, the expression of miR-27a is increased in the host, the miR-27a, causes reduced expression of Ca^{2+} carrier subunit CACNA2D3, which diminishes the ER Ca^{2+} signaling. The manipulation of Ca^{2+} signaling by *Mtb* results in preventing the phagosome maturation and phagosome—lysosome fusion ultimately avoiding autophagy (Malik et al., 2000; Jayachandran et al., 2007; Sharma and Meena, 2017). Thus, modulation of autophagy is an important strategy to promote innate immune defense against *Mtb*. Improved understanding of the antibacterial autophagic mechanisms is still required to progress in the avenues of advancement of new medicinal strategies to cure tuberculosis.

Conclusion

Mtb modulates various pathways in the host for its survival and exerts long-term pathogenesis. *Mtb* contains various mechanisms, proteins, and enzymes to bypass or to modulate host defense responses to evade them and establish a chronic infection. There is continued struggle among the host and pathogen and both, mycobacterium and humans are evolving their molecular arsenals. The pathogen can hide within macrophages and escape the immune attack and latent stage of life cycle provide a vast reservoir for future clinical manifestation at a suitable time. A better understanding of host pathways may open avenues to develop host-directed therapies. They can triggered the immune system and are selectively tweak the inflammation therefore they are better protected from mutations.

References

Amer, A.O., Swanson, M.S., 2005. Autophagy is an immediate macrophage response to Legionella pneumophila. Cell Microbiol. 7, 765−778. Available from: https://doi.org/10.1111/j.1462-5822.2005.00509.x.

Augenstreich, J., Arbues, A., Simeone, R., Haanappel, E., Wegener, A., Sayes, F., et al., 2017. ESX-1 and phthiocerol dimycocerosates of Mycobacterium tuberculosis act in concert to cause phagosomal rupture and host cell apoptosis. Cell Microbiol. 19, e12726.

Bajaj, B.G., Verma, S.C., Lan, K., Cotter, M.A., Woodman, Z.L., Robertson, E.S., 2006. KSHV encoded LANA upregulates Pim-1 and is a substrate for its kinase activity. Virology 351, 18−28. Available from: https://doi.org/10.1016/j.virol.2006.03.037.

Balcewicz-Sablinska, M.K., Keane, J., Kornfeld, H., Remold, H.G., 1998. Pathogenic *Mycobacterium tuberculosis* evades apoptosis of host macrophages by release of TNF-R2, resulting in inactivation of TNF-alpha. J. Immunol. 161, 2636−2641.

Berrington, W.R., Hawn, T.R., 2007. *Mycobacterium tuberculosis*, macrophages, and the innate immune response: Does common variation matter? Immunol. Rev. 219, 167−186. Available from: https://doi.org/10.1111/j.1600-065X.2007.00545.x.

Birmingham, C.L., Smith, A.C., Bakowski, M.A., Yoshimori, T., Brumell, J.H., 2006. Autophagy controls Salmonella infection in response to damage to the Salmonella-containing vacuole. J. Biol. Chem. 281, 11374−11383. Available from: https://doi.org/10.1074/jbc.M509157200.

Boice, A., Bouchier-Hayes, L., 2020. Targeting apoptotic caspases in cancer. Biochim Biophys Acta Mol Cell Res 1867 (6), 118688. Available from: https://doi.org/10.1016/j.bbamcr.2020.118688. Epub 2020 Feb 19. PMID: 32087180; PMCID: PMC7155770.

Cabukusta, B., Neefjes, J., 2018. Mechanisms of lysosomal positioning and movement. Traffic. 19, 761−769. Available from: https://doi.org/10.1111/tra.12587.

Chen, M., Gan, H., Remold, H.G., 2006. A mechanism of virulence: Virulent *Mycobacterium tuberculosis* strain H37Rv, but not attenuated H37Ra, causes significant mitochondrial inner membrane disruption in macrophages leading to necrosis. J. Immunol. 176, 3707−3716. Available from: https://doi.org/10.4049/jimmunol.176.6.3707.

Chen, M., Divangahi, M., Gan, H., Shin, D.S.J., Hong, S., Lee, D.M., et al., 2008. Lipid mediators in innate immunity against tuberculosis: opposing roles of PGE 2 and LXA 4 in the induction of macrophage death. J. Exp. Med. 205, 2791−2801. Available from: https://doi.org/10.1084/jem.20080767.

Clemens, D.L., Horwitz, M.A., 1995. Characterization of the *Mycobacterium tuberculosis* phagosome and evidence that phagosomal maturation is inhibited. J. Exp. Med. 181, 257−270. Available from: https://doi.org/10.1084/jem.181.1.257.

Cumming, B.M., Addicott, K.W., Adamson, J.H., Steyn, A.J.C., 2018. *Mycobacterium tuberculosis* induces decelerated bioenergetic metabolism in human macrophages. Elife . Available from: https://doi.org/10.7554/eLife.39169.

Dallenga, T., Repnik, U., Corleis, B., Eich, J., Reimer, R., Griffiths, G.W., et al., 2017. *M. tuberculosis*-induced necrosis of infected neutrophils promotes bacterial growth following phagocytosis by macrophages. Cell Host Microbe. 22, 519−530. Available from: https://doi.org/10.1016/j.chom.2017.09.003. e3.

Danilchanka, O., Sun, J., Pavlenok, M., Maueröder, C., Speer, A., Siroy, A., et al., 2014. An outer membrane channel protein of *Mycobacterium tuberculosis* with exotoxin activity. Proc. Natl. Acad. Sci. U.S.A. 111, 6750−6755. Available from: https://doi.org/10.1073/pnas.1400136111.

de Chastellier, C., 2009. The many niches and strategies used by pathogenic mycobacteria for survival within host macrophages. Immunobiology. 214, 526−542. Available from: https://doi.org/10.1016/j.imbio.2008.12.005.

Demangel, C., Brodin, P., Cockle, P.J., Brosch, R., Majlessi, L., Leclerc, C., et al., 2004. Cell envelope protein PPE68 contributes to *Mycobacterium tuberculosis* RD1 immunogenicity independently of A 10-kilodalton culture filtrate protein and ESAT-6. Infect. Immun. 72, 2170−2176. Available from: https://doi.org/10.1128/IAI.72.4.2170-2176.2004.

Deretic, V., Saitoh, T., Akira, S., 2013. Autophagy in infection, inflammation and immunity. Nat. Rev. Immunol. 13, 722−737. Available from: https://doi.org/10.1038/nri3532.

Divangahi, M., Chen, M., Gan, H., Desjardins, D., Hickman, T.T., Lee, D.M., et al., 2009. *Mycobacterium tuberculosis* evades macrophage defenses by inhibiting plasma membrane repair. Nat. Immunol. 10, 899–906. Available from: https://doi.org/10.1038/ni.1758.

Fiorentino, D.F., Zlotnik, A., Mosmann, T.R., Howard, M., O'Garra, A., 1991. IL-10 inhibits cytokine production by activated macrophages. J. Immunol. 147, 3815–3822.

Flynn, J.A.L., Goldstein, M.M., Chan, J., Triebold, K.J., Pfeffer, K., Lowenstein, C.J., et al., 1995. Tumor necrosis factor-α is required in the protective immune response against *Mycobacterium tuberculosis* in mice. Immunity 2, 561–572. Available from: https://doi.org/10.1016/1074-7613(95)90001-2.

Ghosh, A.K., Yuan, W., Mori, Y., Chen, S.J., Varga, J., 2001. Antagonistic regulation of type I collagen gene expression by interferon-γ and transforming growth factor-β: Integration at the level of p300/CBP transcriptional coactivators. J. Biol. Chem. 276, 11041–11048. Available from: https://doi.org/10.1074/jbc.M004709200.

Gong, J.H., Zhang, M., Modlin, R.L., Linsley, P.S., Iyer, D., Lin, Y., et al., 1996. Interleukin-10 down-regulates *Mycobacterium tuberculosis*-induced Th1 responses and CTLA-4 expression. Infect. Immun. 64, 913–918. Available from: https://doi.org/10.1128/iai.64.3.913-918.1996.

Grover, S., Sharma, T., Singh, Y., Kohli, S., Manjunath, P., Singh, A., et al., 2018. The PGRS domain of *Mycobacterium tuberculosis* PE_PGRS protein Rv0297 is involved in endoplasmic reticulum stress-mediated apoptosis through toll-like receptor 4. MBio. 9, e01017–e01018.

Gutierrez, M.G., Master, S.S., Singh, S.B., Taylor, G.A., Colombo, M.I., Deretic, V., 2004. Autophagy is a defense mechanism inhibiting BCG and *Mycobacterium tuberculosis* survival in infected macrophages. Cell. 119, 753–766. Available from: https://doi.org/10.1016/j.cell.2004.11.038.

Hackett, E.E., Charles-Messance, H., O'Leary, S.M., Gleeson, L.E., Muñoz-Wolf, N., Case, S., et al., 2020. *Mycobacterium tuberculosis* limits host glycolysis and IL-1β by restriction of PFK-M via MicroRNA-21. Cell Rep. 30 (1), 124–136.

Halder, P., Kumar, R., Jana, K., Chakraborty, S., Ghosh, Z., Kundu, M., et al., 2015. Gene expression profiling of *Mycobacterium tuberculosis* Lipoarabinomannan-treated macrophages: A role of the Bcl-2 family member A1 in inhibition of apoptosis in mycobacteria-infected macrophages. IUBMB Life. 67, 726–736.

Hanekom, M., Gey Van Pittius, N.C., McEvoy, C., Victor, T.C., Van Helden, P.D., Warren, R.M., 2011. *Mycobacterium tuberculosis* Beijing genotype: A template for success. Tuberculosis. 91, 510–523. Available from: https://doi.org/10.1016/j.tube.2011.07.005.

Hinchey, J., Lee, S., Jeon, B.Y., Basaraba, R.J., Venkataswamy, M.M., et al., 2007. Enhanced priming of adaptive immunity by a proapoptotic mutant of Mycobacterium tuberculosis. J Clin Invest 117, 2279–2288.

Huang, J., Brumell, J.H., 2014. Bacteria-autophagy interplay: A battle for survival. Nat. Rev. Microbiol. 12, 101–114. Available from: https://doi.org/10.1038/nrmicro3160.

Jayachandran, R., Sundaramurthy, V., Combaluzier, B., Mueller, P., Korf, H., Huygen, K., et al., 2007. Survival of mycobacteria in macrophages is mediated by coronin 1-dependent activation of calcineurin. Cell. 130, 37–50. Available from: https://doi.org/10.1016/j.cell.2007.04.043.

Joseph, S., Yuen, A., Singh, V., Hmama, Z., 2017. *Mycobacterium tuberculosis* Cpn60.2 (GroEL2) blocks macrophage apoptosis via interaction with mitochondrial mortalin. Biol. Open. 6, 481–488.

Kale, J., Osterlund, E.J., Andrews, D.W., 2018. BCL-2 family proteins: changing partners in the dance towards death. Cell Death Differ. 25, 65–80. Available from: https://doi.org/10.1038/cdd.2017.186.

Keane, J., Balcewicz-Sablinska, M.K., Remold, H.G., Chupp, G.L., Meek, B.B., Fenton, M.J., et al., 1997. Infection by *Mycobacterium tuberculosis* promotes human alveolar macrophage apoptosis. Infect. Immun. 65, 298–304. Available from: https://doi.org/10.1128/iai.65.1.298-304.1997.

Keane, J., Remold, H.G., Kornfeld, H., 2000. Virulent *Mycobacterium tuberculosis* strains evade apoptosis of infected alveolar macrophages. J. Immunol. 164, 2016–2020. Available from: https://doi.org/10.4049/jimmunol.164.4.2016.

Knight, M., Braverman, J., Asfaha, K., Gronert, K., Stanley, S., 2018. Lipid droplet formation in *Mycobacterium tuberculosis* infected macrophages requires IFN-γ/HIF-1α signaling and supports host defense. PLoS Pathog. Available from: https://doi.org/10.1371/journal.ppat.1006874.

Koppenol, W.H., Bounds, P.L., Dang, C.V., 2011. Otto Warburg's contributions to current concepts of cancer metabolism. Nat. Rev. Cancer 11 (5), 325−337.

Kumar, A., Alam, A., Bharadwaj, P., Tapadar, S., Rani, M., Hasnain, S.E., 2019b. Toxin-antitoxin (TA) systems in stress survival and pathogenesis. Mycobacterium tuberculosis: molecular infection biology, pathogenesis, diagnostics and new interventions. Springer, Singapore, pp. 257−274.

Kumar, A., Alam, A., Grover, S., Pandey, S., Tripathi, D., Kumari, M., et al., 2019a. Peptidyl-prolyl isomerase-B is involved in Mycobacterium tuberculosis biofilm formation and a generic target for drug repurposing-based intervention. NPJ Biofilms Microbiomes 5 (1), 1−11.

Kumar, A., Lewin, A., Rani, P.S., Qureshi, I.A., Devi, S., Majid, M., Kamal, E., Marek, S., Hasnain, S. E., Ahmed, N., 2013. Dormancy associated translation inhibitor (DATIN/Rv0079) of Mycobacterium tuberculosis interacts with TLR2 and induces proinflammatory cytokine expression. Cytokine 64 (1), 258−264.

Kumar, A., Majid, M., Kunisch, R., Rani, P.S., Qureshi, I.A., Lewin, A., Hasnain, S.E., Ahmed, N., 2012. Mycobacterium tuberculosis DosR regulon gene Rv0079 encodes a putative,'dormancy associated translation inhibitor (DATIN)'. PloS one 7 (6), e38709.

Kumar, A., Rani, M., Ehtesham, N.Z., Hasnain, S.E., 2017. Commentary: modification of Host responses by mycobacteria. Frontiers in Immunology 8, 466.

Lee, J., Hartman, M., Hornfeld, H., 2009. Macrophage apoptosis in Tuberculosis. Yonsai Med J. 50, 1−11.

Malik, Z.A., Denning, G.M., Kusner, D.J., 2000. Inhibition of Ca^{2+} signaling by *Mycobacterium tuberculosis* is associated with reduced phagosome-lysosome fusion and increased survival within human macrophages. J. Exp. Med. 191, 287−302. Available from: https://doi.org/10.1084/jem.191.2.287.

Marzo, E., Vilaplana, C., Tapia, G., Diaz, J., Garcia, V., Cardona, P.J., 2014. Damaging role of neutrophilic infiltration in a mouse model of progressive tuberculosis. Tuberculosis. 94, 55−64. Available from: https://doi.org/10.1016/j.tube.2013.09.004.

Mehrotra, P., Jamwal, S.V., Saquib, N., Sinha, N., Siddiqui, Z., Manivel, V., et al., 2014. Pathogenicity of Mycobacterium tuberculosis is expressed by regulating metabolic thresholds of the host macrophage. PLoS Pathog 10 (7), e1004265. Available from: https://doi.org/10.1371/journal.ppat.1004265.

Molloy, A., Laochumroonvorapong, P., Kaplan, G., 1994. Apoptosis, but not necrosis, of infected monocytes is coupled with killing of intracellular bacillus Calmette−Guérin. J. Exp. Med. 180, 1499−1509. Available from: https://doi.org/10.1084/jem.180.4.1499.

Murray, P.J., 2006. Understanding and exploiting the endogenous interleukin-10/STAT3-mediated anti-inflammatory response. Curr. Opin. Pharmacol. 6, 379−386. Available from: https://doi.org/10.1016/j.coph.2006.01.010.

Murray, P.J., Wang, L., Onufryk, C., Tepper, R.I., Young, R.A., 1997. T cell-derived IL-10 antagonizes macrophage function in mycobacterial infection. J. Immunol. 158, 315−321.

Niazi, M.K.K., Dhulekar, N., Schmidt, D., Major, S., Cooper, R., Abeijon, C., et al., 2015. Lung necrosis and neutrophils reflect common pathways of susceptibility to *Mycobacterium tuberculosis* in genetically diverse, immune-competent mice. DMM Dis. Model. Mech. 8, 1141−1153. Available from: https://doi.org/10.1242/dmm.020867.

Oddo, M., Renno, T., Attinger, A., Bakker, T., MacDonald, H.R., Meylan, P.R., 1998. Fas ligand-induced apoptosis of infected human macrophages reduces the viability of intracellular *Mycobacterium tuberculosis*. J. Immunol. 160, 5448−5454.

Paik, S., Choi, S., Lee, K.I., Back, Y.W., Son, Y.J., Jo, E.K., et al., 2019. *Mycobacterium tuberculosis* acyl carrier protein inhibits macrophage apoptotic death by modulating the reactive oxygen species/c-Jun N-terminal kinase pathway. Microbes Infect. 21, 40−49.

Pandey, S., Sharma, A., Tripathi, D., Kumar, A., Khubaib, M., Bhuwan, M., Chaudhuri, T.K., Hasnain, S.E., Ehtesham, N.Z., 2016. Mycobacterium tuberculosis peptidyl-prolyl isomerases also exhibit chaperone like activity in-vitro and in-vivo. PLoS One 11 (3), e0150288.

Pandey, S., Tripathi, D., Khubaib, M., Kumar, A., Sheikh, J.A., Sumanlatha, G., Ehtesham, N.Z., Hasnain, S.E., 2017. Mycobacterium tuberculosis peptidyl-prolyl isomerases are immunogenic, alter cytokine profile and aid in intracellular survival. Frontiers in cellular and infection microbiology 7, 38.

Peng, X., Luo, T., Zhai, X., Zhang, C., Suo, J., Ma, P., et al., 2019. PPE11 of *Mycobacterium tuberculosis* can alter host inflammatory response and trigger cell death. Microb. Pathog. 126, 45−55. Available from: https://doi.org/10.1016/j.micpath.2018.10.031.

Raja, A., 2004. Immunology of tuberculosis. Indian J. Med. Res. 120, 213−232. Available from: https://doi.org/10.5005/jp/books/10992_7.

Rajaram, MV, Ni, B., Dodd, C.E., Schlesinger, L.S., 2014. Macrophage immunoregulatory pathways in tuberculosis. Semin Immunol 26 (6), 471−485. Available from: https://doi.org/10.1016/j.smim.2014.09.010. Epub 2014 Oct 30. PMID: 25453226; PMCID: PMC4314327.

Reed, M.B., Domenech, P., Manca, C., Su, H., Barczak, A.K., Kreiswirth, B.N., et al., 2004. A glycolipid of hypervirulent tuberculosis strains that inhibits the innate immune response. Nature. 431, 84−87. Available from: https://doi.org/10.1038/nature02837.

Rohde, K., Yates, R.M., Purdy, G.E., Russell, D.G., 2007. *Mycobacterium tuberculosis* and the environment within the phagosome. Immunol. Rev. 219, 37−54. Available from: https://doi.org/10.1111/j.1600-065X.2007.00547.x.

Rojas, M., Barrera, L.F., Puzo, G., Garcia, L.F., 1997. Differential induction of apoptosis by virulent *Mycobacterium tuberculosis* in resistant and susceptible murine macrophages: Role of nitric oxide and mycobacterial products. J. Immunol. 159, 1352−1361.

Russell, D.G., Cardona, P.J., Kim, M.J., Allain, S., Altare, F., 2009. Foamy macrophages and the progression of the human tuberculosis granuloma. Nat Immunol 10, 943−948.

Russell, D.G., 2007. Who puts the tubercle in tuberculosis? Nat Rev Micro 5, 39−47.

Seimon, T.A., Kim, M.J., Blumenthal, A., Koo, J., Ehrt, S., Wainwright, H., et al., 2010. Induction of ER stress in macrophages of tuberculosis granulomas. PLoS One. 5, 1−12. Available from: https://doi.org/10.1371/journal.pone.0012772.

Sharma, S., Meena, L.S., 2017. Potential of Ca^{2+} in *Mycobacterium tuberculosis* H37Rv pathogenesis and survival. Appl. Biochem. Biotechnol. 181, 762−771. Available from: https://doi.org/10.1007/s12010-016-2247-9.

Souza, C.D., Evanson, O.A., Weiss, D.J., 2006. Mitogen activated protein kinasep38 pathway is an important component of the anti-inflammatory response in Mycobacterium avium subsp. paratuberculosis-infected bovine monocytes. Microbial pathogenesis 41 (2−3), 59−66. Available from: https://doi.org/10.1016/j.micpath.2006.04.002.

Su, Z., Xu, S., Chen, T., Chen, J., 2015. Dexmedetomidine protects spatial learning and memory ability in rats. J. Renin Angiotensin Aldosterone Syst. 16, 995−1000. Available from: https://doi.org/10.1177/1470320314562059.

Su, H., Zhu, S., Zhu, L., Huang, W., Wang, H., Zhang, Z., et al., 2016. Recombinant lipoprotein rv1016c derived from *Mycobacterium tuberculosis* is a TLR-2 ligand that induces macrophages apoptosis and inhibits MHC II antigen processing. Front. Cell Infect. Microbiol. 6, 147.

Sun, J., Siroy, A., Lokareddy, R.K., Speer, A., Doornbos, K.S., Cingolani, G., et al., 2015. The tuberculosis necrotizing toxin kills macrophages by hydrolyzing NAD. Nat. Struct. Mol. Biol. 22, 672−678. Available from: https://doi.org/10.1038/nsmb.3064.

Teutschbein, J., Schumann, G., Möllmann, U., Grabley, S., Cole, S.T., Munder, T., 2009. A protein linkage map of the ESAT-6 secretion system 1 (ESX-1) of *Mycobacterium tuberculosis*. Microbiol. Res. 164, 253−259. Available from: https://doi.org/10.1016/j.micres.2006.11.016.

Tiku, V., Tan, M.W., Dikic, I., 2020. Mitochondrial functions in infection and immunity. Trends Cell Biol. 30, 263−275. Available from: https://doi.org/10.1016/j.tcb.2020.01.006.

van der Wel, N., Hava, D., Houben, D., Fluitsma, D., van Zon, M., Pierson, J., et al., 2007. *M. tuberculosis* and *M. leprae* translocate from the phagolysosome to the cytosol in myeloid cells. Cell. 129, 1287−1298. Available from: https://doi.org/10.1016/j.cell.2007.05.059.

Wang, L., Zuo, M., Chen, H., Liu, S., Wu, X., Cui, Z., et al., 2017. *Mycobacterium tuberculosis* lipoprotein MPT83 induces apoptosis of infected macrophages by activating the TLR2/p38/COX-2 signaling pathway. J. Immunol. 198, 4772−4780.

Williams, L., Bradley, L., Smith, A., Foxwell, B., 2004. Signal transducer and activator of transcription 3 is the dominant mediator of the anti-inflammatory effects of IL-10 in human macrophages. J. Immunol. 172, 567−576. Available from: https://doi.org/10.4049/jimmunol.172.1.567.

Yang, S., Li, F., Jia, S., Zhang, K., Jiang, W., Shang, Y., et al., 2015. Early secreted antigen ESAT-6 of *Mycobacterium tuberculosis* promotes apoptosis of macrophages via targeting the microRNA155-SOCS1 interaction. Cell Physiol. Biochem. 35, 1276—1288.

Yim, W.W.Y., Mizushima, N., 2020. Lysosome biology in autophagy. Cell Discov. Available from: https://doi.org/10.1038/s41421-020-0141-7.

Zhang, M., Gong, J., Iyer, D.V., Jones, B.E., Modlin, R.L., Barnes, P.F., 1994. T cell cytokine responses in persons with tuberculosis and human immunodeficiency virus infection. J. Clin. Invest. 94, 2435—2442. Available from: https://doi.org/10.1172/JCI117611.

CHAPTER 3

Signaling nucleotides in bacteria

Kuldeepkumar Ramnaresh Gupta[1], Gunjan Arora[2] and Andaleeb Sajid[2]
[1]Department of Microbial Pathogenesis, Yale University School of Medicine, New Haven, CT, United States
[2]Section of Infectious Diseases, Department of Internal Medicine, Yale University School of Medicine, New Haven, CT, United States

Introduction

Second messengers are small signaling molecules produced by a cell or an organism to bring about changes in gene expression in response to external stimuli, which act as first messengers (Pesavento and Hengge, 2009). A second-messenger signaling module usually comprises of four components: an enzyme that synthesizes the second messenger in response to external stimuli; an effector molecule that binds the second messenger; the target gene or operon that is either expressed or repressed; and finally, an enzyme that degrades the second messenger once its function is fulfilled (Fig. 3.1) (Hengge, 2009). Bacteria need to adapt continuously to survive in a constantly changing environment. These ceaseless adaptations necessitate a need for rapid changes in the gene expression pattern. The synthesis of second messengers is one of the swift ways to bring about such adaptive changes in gene expression. Bacteria generally utilize nucleotide-based second messengers to sense the changes in their surroundings and accordingly modulate the gene expression. These nucleotides also regulate several important processes such as virulence, pathogenicity, biofilm formation, cell wall homeostasis, and alternative carbon source utilization (Fig. 3.2). Since the discovery of first nucleotide-based second messenger, cyclic-AMP (cAMP) in 1965, several nucleotide second messengers have been discovered (Makman and Sutherland, 1965). Apart from the archetypal cAMP, bacteria also use guanosine tetra-or pentaphosphate ((p)ppGpp), cyclic guanosine monophosphate (c-di-GMP), cyclic adenosine monophosphate (c-di-AMP), and recently discovered cyclic-GMP (cGMP) and cyclic-GMP-AMP (c-GAMP) (Kalia et al., 2013).

In 1958, Earl Sutherland discovered cAMP as a first nucleotide second-messenger mediating response to glucagon and epinephrine in liver homogenates (Sutherland and Rall, 1958; Rall and Sutherland, 1958). In 1965, the existence of cAMP in *Escherichia coli* was also demonstrated in his laboratory (Makman and Sutherland, 1965). The signaling nucleotide cAMP binds to the transcription factor cAMP receptor protein (CRP) and regulates the utilization of alternative carbon sources (Kolb et al., 1993). Over the years, its role in the regulation of virulence, biofilm formation, and secretion

Bacterial Survival in the Hostile Environment
DOI: https://doi.org/10.1016/B978-0-323-91806-0.00013-8

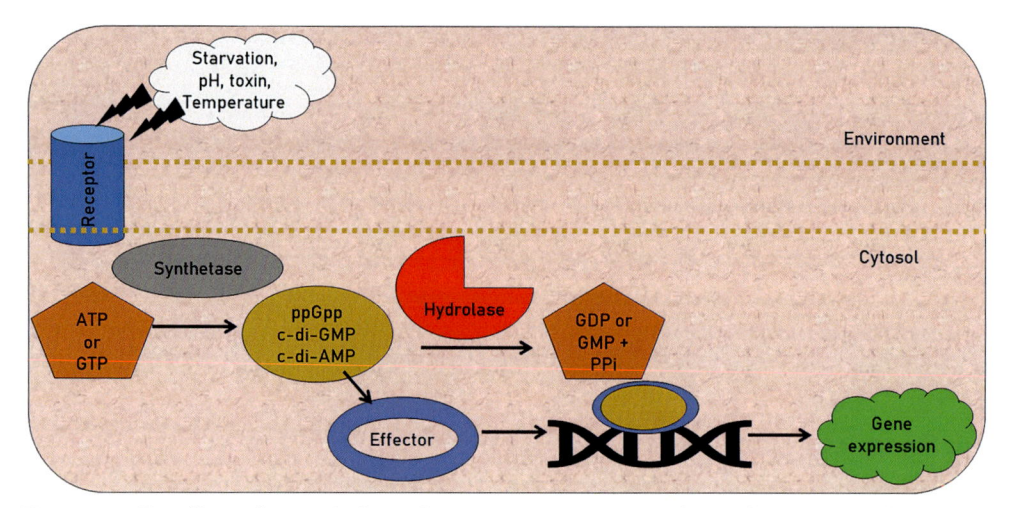

Figure 3.1 *Signaling schemes in bacteria.* Bacteria sense external signals in terms of changing nutrients, or some stress, which is sensed by a surface-receptor and transduced inside the cytosol, leading to synthesis of a second messenger. These in turn respond to external stimuli through an effector molecule that binds the second messenger; the target gene or operon, which is then either expressed or repressed.

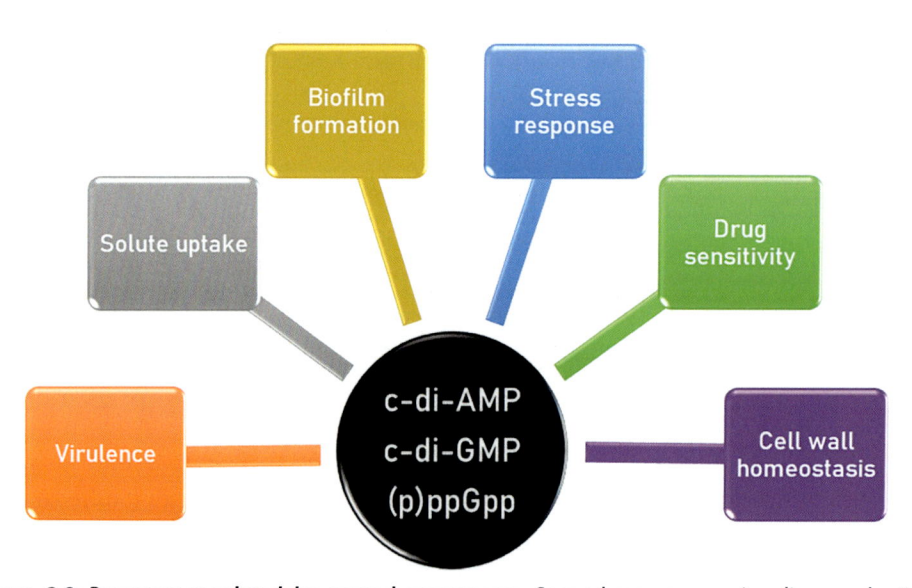

Figure 3.2 *Processes regulated by second messengers.* Second-messenger signaling nucleotides are involved in regulation of several important processes including stress, virulence, pathogenicity, biofilm formation, drug resistance, cell wall homeostasis, and solute uptake or alternative carbon source utilization.

systems in several bacteria has been studied (McDonough and Rodriguez, 2012). The second bacterial signaling nucleotide second messengers (p)ppGpp were discovered in *E. coli* in 1969. The locus responsible for its synthesis was named "*relA*." The associated stress response in this *E. coli* strain was termed stringent response and the molecules were called stringent factors (Cashel and Gallant, 1969). In 1987, a novel guanosine-based cyclic nucleotide was discovered which acted as an allosteric activator of cellulose biosynthesis in *Gluconacetobacter xylinus* (Ross et al., 1987). This molecule was termed c-di-GMP. The subsequent studies established c-di-GMP as the principal molecule governing the transition from motile (planktonic) to sessile (biofilm) forms of lifestyle in several bacterial species. In 2008, structural and biochemical analysis of DNA integrity scanning protein from *Thermotoga maritima* and *Bacillus subtilis* showed that protein exhibits c-di-AMP synthesis activity (Witte et al., 2008). Though cGMP was discovered in the extracts of *E. coli* and *Bacillus licheniformis* as early as 1974, its role as a bonafide bacterial second messenger was established in 2011 (Marden et al., 2011). In 2012 the existence of hybrid cyclic dinucleotide, c-GMP-AMP, was demonstrated (Davies et al., 2012). However, the fields of cGMP and c-GMP-AMP are still nascent and need to be investigated in detail. On the other hand, there is enough literature available for (p)ppGpp, c-di-GMP and c-di-AMP. Hence, in this chapter, we have provided a general overview of (p)ppGpp, c-di-GMP, and c-di-AMP signaling in bacteria.

cyclic-di-AMP

c-di-AMP signaling

c-di-AMP was discovered serendipitously while studying the crystal structure of *B. subtilis* sporulation associated protein called DNA integrity scanning protein, DisA, responsible for checking DNA damage before the onset of sporulation (Witte et al., 2008). The DisA protein monomer possesses one nucleotide-binding and DNA-binding domain. The octameric crystal structure of DisA had an unexplained electron density which corresponded to the c-di-AMP. Subsequent biochemical analysis established that the nucleotide-binding domain of DisA possessed diadenylate cyclase (DAC) activity and the domain was designated as the DAC domain (Witte et al., 2008). Based on the analysis of DAC domain-containing proteins or the structure of operons encoding DACs, a comprehensive classification of bacterial DACs has been proposed by Angelika Grundling and colleagues (Corrigan and Grundling, 2013). There are several families of DACs, but the three most abundant families—DacA (69.5%), DisA (24.1%), and DacB (5.5%)—together constitute 98.7% of all the DAC domain-containing proteins (Corrigan and Grundling, 2013). DacA families DGCs have three transmembrane domains at their N-terminus region. The *dacA* operon usually also encodes the *glmM* gene which encodes phosphoglucosamine mutase (GlmM), an important cell wall biosynthesis enzyme that is consistent with the observation that c-di-AMP is involved in the regulation of cell wall synthesis in bacteria where DacA proteins are main DACs. DisA domain proteins are the second most prevalent DACs. Indeed, *B. subtilis* DisA was the first protein to show DAC activity. As described above, the general architecture of DisA protein comprises a DAC domain linked to a DNA-binding HhH motif through a linker domain or region. The DisA DACs are involved in the maintenance of genomic integrity through sensing branched DNA structures such as branched replication forks and Holliday junction (Witte et al., 2008). DacB, the third most prevalent DAC domain-containing protein, is almost exclusively found in spore-forming bacteria belonging to order Bacillales with the exception of one belonging to *Clostridium* species (Corrigan and Grundling, 2013). There are two classes of c-di-AMP phosphodiesterases: DHH/DHH1 domain-containing PDEs and HD domain-containing PDEs (Fig. 3.3). The former has a catalytic aspartate-histidine-histidine motif while the latter contains a histidine-aspartate catalytic motif (Rao et al., 2010; Huynh et al., 2015). The DHH/DHH1 domain-containing PDEs are widespread in bacterial genomes and are found in many important pathogenic genera such as bacilli, mycobacteria, streptococci, borrelia, and staphylococci (Commichau et al., 2015). The second type of c-di-AMP PDE to be discovered was HD-domain phosphodiesterase, PgpH, in *Listeria monocytogenes*. It belongs to 7TM-HD class of phosphodiesterases which contain a C-terminal cytosolic HD domain and N-terminal 7 transmembrane helices (Huynh et al., 2015). The HD-domain c-di-AMP-specific PDEs are found in a large number of bacteria and seem to be the major PDEs degrading c-di-AMP in cyanobacteria, bacteroidetes, fusobacteria, and Thermotoga. In firmicutes, these enzymes cooccur with DHH/DHH1 type PDEs;

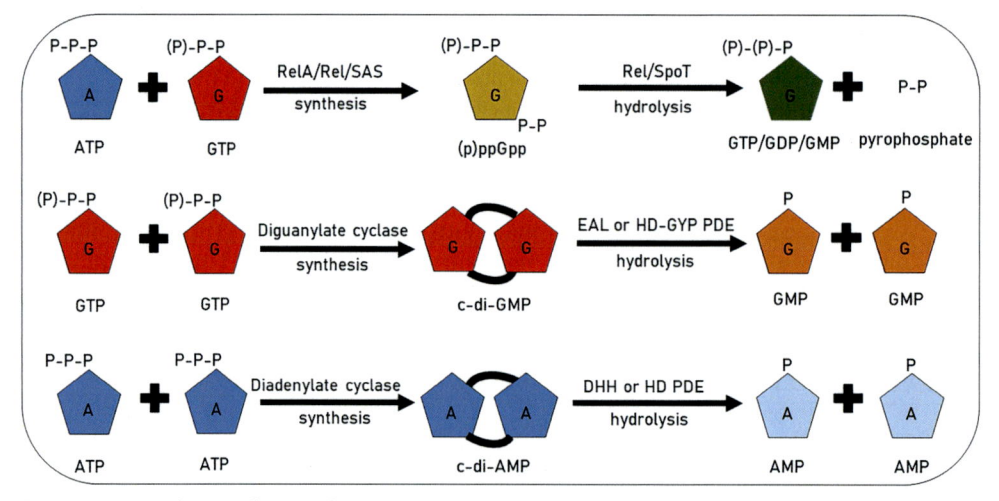

Figure 3.3 *Synthesis of second messengers.* Schematic showing synthesis of different second messengers.

while in groups such as staphylococci and streptococci they are not found, and in actinobacteria their occurrence is very rare (Commichau et al., 2015). c-di-AMP has been primarily found to regulate osmolyte uptake in bacteria and also in the maintenance of cell wall homeostasis. We will discuss how c-di-AMP regulates potassium uptake in *Staphylococcus aureus* and *Streptococcus pneumoniae*.

Regulation of potassium uptake by c-di-AMP

S. aureus contains two potassium (K^+) ion uptake systems: Ktr and Kdp systems. The Ktr system is a constitutively expressed, high K^+ ion uptake system. The Ktr system in bacteria comprises one membrane and one cytoplasmic gating component. The K^+ ion transport is believed to take place with the symport of Na^+ ion. High osmolarity and low K^+ ion conditions can only be tolerated through the expression of Ktr systems. The second K^+ ion transporter, the Kdp system, unlike Ktr system, is an active K^+ ion transporter and composed of four proteins: KdpF, KdpA, KdpB and KdpC. KdpA is K^+ ion transporter; KdpB, an ATPase; KdpC, a chaperone for KdpB; KdpF, helps in the assembly and stability of the complex. The expression of Kdp system is regulated by a two-component system, KdpDE with KdpD being membrane-bound sensor kinase and KdpE being response regulator. K^+ ion limiting condition upregulated activates KdpD which then phosphorylates KdpE. The activated transcription factor, KdpE, upregulates the expression of *kdpFABC* and *kdpDE* operon. Both K^+ ion transporters are regulated by c-di-AMP. KtrA (or KtrC), the cytoplasmic gating component of Ktr system and KdpD sensor kinase of KdpDE two-component signal transduction system bind c-di-AMP. During salt stress, the *gdpD* mutant with high levels of c-di-AMP and *ktrA* mutant show reduced growth indicating c-di-AMP inhibiting K^+ ion uptake. The growth defects of *gdpD* and *ktrA* mutants could be rescued by the exogenous supply of K^+ ion or by complementation with a functional copy or *gdpD* and *ktrA* genes, respectively (Corrigan et al., 2013). The levels of *kdpA* transcript during 1 M NaCl salt stress were significantly higher in wild-type (WT) strain compared with a *gdpP* deletion strain with a high level of c-di-AMP. Thus increased concentration of c-di-AMP inhibits the expression of potassium transporter KdpA (Moscoso et al., 2016). Recently, the role of c-di-AMP in potassium uptake has been demonstrated (Bai et al., 2014). c-di-AMP affinity chromatography of *S. pneumoniae* cell lysate showed enrichment of SPD_0077, which was annotated as Trk family potassium transporter. The protein was found to bind c-di-AMP specifically and named as c-di-AMP-binding protein or CabP. The *cabP* gene was clustered with SPD_0076 which is predicted to be a TrkH protein with 10 transmembrane helices and an ortholog of KtrB from *B. subtilis*. Both Trk and Ktr transporters are involved in potassium uptake. The knockout mutants Δ*cabP*, Δ*SPD_0076* and Δ*cabP*Δ*SPD_0076* failed to grow in media containing a low concentration of potassium, indicating that both CabP and SPD_0076 are needed for potassium uptake.

The operonic structure of *cabP* and *SPD_0076* implied that they might interact with each other. Bacterial two-hybrid assay confirmed that, indeed, CabP and SPD_0076 interacted physically with each other and high levels of c-di-AMP reduced the interaction between two proteins. *S. pneumoniae* encodes two c-di-AMP phosphodiesterases—Pde1 and Pde2 encoded by *pde1* and *pde2*, respectively. The double knockout strain *Δpde1Δpde2* showed significantly reduced growth in a medium containing low potassium which suggests that increased c-di-GMP concentration reduces potassium uptake by the pneumococci (Bai et al., 2014).

(p)ppGpp

(p)ppGpp signaling

During starvation, bacteria initiate a stress response by downregulating the synthesis of rRNA and ribosomal proteins and concomitantly upregulating the synthesis of amino acids. This phenomenon is called stringent response which is brought about by two linear nucleotides—ppGpp and ppGpp—collectively referred to as alarmones (p)ppGpp (Cashel and Gallant, 1969). The stringent factors are synthesized by the transfer of a pyrophosphate moiety to the 3′-end of GDP/GTP. Enzymes that metabolize (p)ppGpp belong to the RSH (**R**el **S**poT **H**omolog) superfamily of proteins (Potrykus and Cashel, 2008). The RSH proteins are long polypeptides comprising at least 750 amino acid residues. The N-terminal region of these proteins harbors both the (p)ppGpp synthase and hydrolase activities, while the C-terminal domain regulates N-terminal catalytic activities (Potrykus and Cashel, 2008). In gram-negative bacteria, RelA and SpoT enzymes are responsible for the metabolism of (p)ppGpp (Fig. 3.3). The monofunctional RelA enzyme synthesizes (p)ppGpp, while the bifunctional SpoT enzyme degrades (p)ppGpp. The bifunctional SpoT synthesizes (p)ppGpp in response to stressors that do not activate RelA. In gram-positive bacteria, a single bifunctional protein, Rel, both synthesizes and degrades (p)ppGpp (Atkinson et al., 2011). Recently, the genomes of various gram-positive bacteria were found to code for single domain (p)ppGpp synthases. These proteins were named Small Alarmone Synthetases (SASs) as their size is smaller as compared with "long" RSH proteins. These SASs do not have a C-terminal regulatory domain and N- terminal hydrolase domain found in the long RSH proteins. (Atkinson et al., 2011). SASs have been found in *B. subtilis*, *Enterococcus faecalis*, *Vibrio cholerae*, *S. aureus*, *Mycobacterium smegmatis* and *Mycobacterium tuberculosis*. SASs are implicated in maintaining the basal level of (p)ppGpp within the cell (Krishnan and Chatterji, 2020). Thus the metabolism of stringent factor, (p)ppGpp, is controlled by both long RSH proteins and short SAS proteins.

Not only the landscape of enzymes metabolizing the stringent factor but also the meaning of the term "stringent response" has changed over the years. Initially, it signified an amino acid starvation-induced increase in the level of (p)ppGpp with the concomitant decrease in the stable RNAs. Nowadays, the term stringent response applies to any phenomenon leading to an increase in the (p)ppGpp concentration in the bacterial cell (Potrykus and Cashel, 2008). The alarmone once synthesized brings about significant changes in bacterial physiology (Srivatsan and Wang, 2008). (p)ppGpp also regulates sigma factor competition, stress tolerance, long-term survival and virulence (Ramnaresh Gupta and Chatterji, 2016; Dalebroux and Swanson, 2012). In *E. coli*, (p)ppGpp binds to RNA polymerase and regulates transcription, translation, and replication (Srivatsan and Wang, 2008). In other bacteria, it can regulate transcription by modulating the levels of GTP without physically interacting with RNA polymerase (Dalebroux and Swanson, 2012).

Regulation of antibiotic resistance by (p)ppGpp

It has been observed that *E. coli* having higher levels of (p)ppGpp exhibit increased resistance to β-lactam antibiotics, while a strain defective in (p)ppGpp synthesis shows sensitivity to multiple antibiotics (Vinella et al., 1992; Rodionov and Ishiguro, 1995). Similarly, a mutant *Salmonella Typhimurium* with reduced levels of (p)ppGpp showed hypersusceptibility to multiple classes of antibiotics (Macvanin and Hughes, 2005). Recently, it was shown that the (p)ppGpp null strain of *P. aeruginosa* is more sensitive to multiple classes of antibiotics and is defective in biofilm formation (Nguyen et al., 2011). In the absence of (p)ppGpp, the mutant strain failed to express enough catalases and superoxide dismutases which are needed to neutralize the reactive oxygen and nitrogen species formed due to antibiotic treatment. The biofilm formed by (p)ppGpp null strain of *P. aeruginosa* was also sensitive to several antibiotics. Thus the antibiotic tolerance shown by nutrient-starved and biofilm cultures of *P. aeruginosa* is not because of passive growth arrest due to high levels of (p)ppGpp, but due to active starvation response (Nguyen et al., 2011). The gram-positive *E. faecalis* encodes a bifunctional synthetase/hydrolase RelA and a SAS, RelQ. The Δ*relA* strain of *E. faecalis* is tolerant to vancomycin. However, the ppGpp null strain, Δ*relAQ* which lacks both RelA and RelQ, is sensitive vancomycin (Abranches et al., 2009). A similar result was observed in gram-positive firmicute *B. subtilis*, which harbors a bifunctional Rel and two monofunctional SASs. The Δ*rel* strain of *B. subtilis* showed higher resistance to antimicrobial compared with the wild-type or (p)ppGpp null strain. However, the Δ*rel* strain became sensitive to antibiotics when the (p)ppGpp levels were artificially reduced when treated with stringent response inhibitor relacin (Tabone et al., 2014). *M. smegmatis* encodes one bifunctional Rel and one SAS, RelZ. The Δ*rel* strain of *M. smegmatis* displayed not only tolerance to multiple antibiotics but also the differential expression genes categories which included cell wall metabolism, virulence, reactive oxygen species and detoxification. The double deletion strain, Δ*rel*Δ*relZ*, was more sensitive to multiple antibiotics compared with Δ*rel* strain (Gupta et al., 2015, 2016; Petchiappan et al., 2020). Thus, it is clear that (p)ppGpp is involved in the regulation of antibiotic resistance, but the mode of action is not similar in different bacteria. In beta- and gamma-proteobacteria, both the Δ*rel* and pppGpp null strains display antibiotic sensitivity and the antimicrobial sensitivity is inversely proportional to (p)ppGpp concentration. However, in enterococci, bacilli, and mycobacteria, the Δ*rel* strain shows tolerance to antibiotics, while the (p)ppGpp null strain is antibiotic sensitive. Thus there appears to be a species-specific regulation of antibiotic resistance by stringent response. Hence, a complete understanding of stringent response in clinically important bacterial species will help in designing new inhibitors that target stringent response.

cyclic-di-GMP

c-di-GMP signaling

Signaling nucleotide cyclic-di-GMP was discovered as a cofactor and activator of cellulose synthase enzyme in alpha-proteobacterium, *Gluconacetobacter xylinus*, then known as *Acetobacter xylinum* (Ross et al., 1987). Diguanylate cyclases (DGCs) synthesize c-di-GMP from two molecules of GTP. The catalytic active site of DGCs contains a conserved sequence motif of Glycine (G), glycine (G), aspartic acid (D), glutamic acid (E), and phenylalanine (F), and hence they are called GGDEF domain proteins (Ausmees et al., 1999). There are two kinds of phosphodiesterases (PDEs) that break down c-di-GMP—EAL domain PDEs and HD-GYP domain PDEs. The active site of EAL domain PDEs is lined by glutamic acid (E), alanine (A), and leucine (L) (Fig. 3.3) (Tischler and Camilli, 2004; Simm et al., 2004). These PDEs break down c-di-GMP into two molecules of GMP. HD superfamily of enzymes contains metal-dependent phosphohydrolases. These phosphohydrolases are involved in nucleic acid metabolism and signal transduction in bacteria, archea, and eukarya (Aravind and Koonin, 1998). The term HD represents histidine (H) and aspartic acid (D), which are conserved in this superfamily and are involved in metal coordination during phosphohydrolytic reactions (Aravind and Koonin, 1998). In HD-GYP PDEs—a subset of HD superfamily, apart from conserved histidine and aspartic acid, three other amino acids viz. glycine (G), tyrosine (Y) and, proline (P) were found to be conserved. HD-GYP domain-containing PDEs convert c-di-GMP into two molecules of GMP instead of converting them to 5'pGpG (Ryan et al., 2006). We will discuss how c-di-GMP modulates the synthesis of extracellular matrix, fimbriae and pili in several bacterial species and promotes biofilm formation.

Regulation of biofilm formation by c-di-GMP

The synthesis and degradation of various biofilm components such as extracellular polysaccharides, pili and fimbriae are regulated by c-di-GMP (Valentini and Filloux, 2016).

Cellulose biosynthesis was the first physiological process discovered to be affected by c-di-GMP (Ross et al., 1987). Cellulose has been reported to be one of the exopolysaccharides found in environmental biofilms (Ogawa and Maki, 2003). Bacterial cellulose synthases, which bind to c-di-GMP, contain PilZ domains that bring about the activation of the enzyme. This ultimately enhances cellulose production. Conversely, impairing c-di-GMP signaling also inhibits cellulose biosynthesis. (Ryjenkov et al., 2006). Poly-β-1, 6-*N*-acetylglucosamine (PAG) is one of the polysaccharides found in biofilms formed by various species of bacteria (Itoh et al., 2005). In *E. coli*, PAG is synthesized by *pgaABCD* operon (Wang et al., 2004). Overexpression of the DGC protein YddV enhances the production of PAG (Tagliabue et al., 2010). c-di-GMP binds to both PgaC and PgaD, the inner membrane components of the PAG synthesis module, and stimulates their glycosyltransferase activity (Steiner et al., 2013). Thus enhancing the extracellular matrix production. Pel polysaccharides are part of non-mucoid biofilms formed by *P. aeruginosa* (Romling et al., 2013). The synthesis of Pel polysaccharides is enhanced by the response regulator WspR, which also harbors a GGDEF domain. Upon phosphorylation, the activity of WspR, which makes c-di-GMP, was enhanced and leads to increased biofilm formation (Hickman et al., 2005). In the absence of c-di-GMP, transcription factor FleQ makes a complex with another protein FleN and binds to the promoter of *pel* operon leading to its repression. However, in the presence of c-di-GMP, the FleQ-FleN complex fails to exclude RNA polymerase binding to the promoter of pel operon, which results in the expression of *pel* operon (Hickman and Harwood, 2008). *Klebsiella pneumoniae* is well-known for causing nosocomial infections. It gets attached to various surfaces using type III fimbriae (Johnson and Clegg, 2010). The type III fimbriae in *K. pneumoniae* are synthesized by *mrk*ABCDF operon. Immediately, next to the type III fimbrial gene cluster, *mrkJ* gene is located which encodes an EAL domain c-di-GMP phosphodiesterase. Deletion of *mrkJ* resulted in high levels of c-di-GMP which modulates the upregulation of type III fimbriae. Subsequent work found that MrkH, a PilZ domain receptor protein upon binding c-di-GMP interacts strongly with the promoter region of *mrkA* gene and influences the synthesis of type III fimbriae (Wilksch et al., 2011). Type IV pili are the most diverse and abundant pili in bacteria (Pelicic, 2008). The phenomenon of twitching motility has been attributed to these pili because of their unique ability to retract (Merz et al., 2000). Type IV pili mutant of *P. aeruginosa* failed to form mature biofilms (O'Toole and Kolter, 1998). FimX receptor protein, with degenerate GGDEF and EAL domain, is required for the synthesis of type IV pili. FimX is present at one of the poles of the bacterium, which is important for the correct assembly of type IV pili. c-di-GMP binds to the C-terminal EAL domain and brings about a long-range conformational change in the N-terminal domain of FimX which harbors the localization sequence for type IV proteins. If the degenerate C-terminal domain is mutated or deleted, then long-range conformational change is abolished. This results in the inhibition of type IV pili synthesis (Qi et al., 2011). Curli fimbriae are one of the extracellular components of the biofilms produced by Enterobacteriaceae. In addition, Curli fibers are needed for adhesion to surfaces and cell aggregation. Curli fimbriae also mediate host cell adhesion and invasion. They are strong inducers of the host inflammatory responses (Barnhart and Chapman, 2006). The transcription factor CsgD is a master regulator of genes involved in the curli fimbriae synthesis. Upon c-di-GMP binding the CsgD, it upregulates the transcription of curli structural operon *csgBAC* (Pesavento et al., 2008; Sommerfeldt et al., 2009). Adhesins are also one of the components of biofilm matrix and their synthesis is regulated by the second-messenger c-di-GMP. Adhesins are biofilm components of environmental *Pseudomonas* species such as *Pseudomonas fluorescens* and *Pseudomonas putida* (Hinsa et al., 2003; Gjermansen et al., 2005). The adhesin LapA is translocated to the outside surface by ABC transporter encoded by *lapECB* genes. When c-di-GMP concentrations are low, LapA is proteolytically cleaved by LapG, a periplasmic protease. When c-di-GMP concentrations are high, the transmembrane protein LapD binds c-di-GMP at its cytosolic site and brings about conformation change in its periplasmic domain, which sequesters protease LapG. The sequestration of LapG prevents the degradation of LapA and allows for biofilm formation (Newell et al., 2009). Taken together, high levels of c-di-GMP promote biofilm formation in several bacterial species by regulating the synthesis of various biofilm matrix components.

Conclusions and perspectives

When survival is threatened or the nutrients are scarce, bacteria must modulate their gene expression to either survive or adapt to new conditions. The survival can be ensured by utilizing a universal second messenger like (p)ppGpp which brings about large-scale changes in gene expression. Alternatively, specialized second messengers, like c-di-GMP or c-di-AMP, can also be used to bring about changes in s specific sets of genes. Irrespective of the identity of the second messenger, all signaling nucleotides follow the same module: cue, synthesis, binding to effector and eventual degradation. Given the presence of more than one nucleotide signaling system in the same bacterial species, it is possible that there might be an interplay between these systems, which makes the signaling landscape even more complex and exciting. There have been some reports of crosstalk between bacterial second messengers. In *S. aureus*, increased c-di-AMP levels enhance the synthesis of ppGpp; while increased (p)ppGpp levels inhibit c-di-AMP PDE GdpP (Corrigan et al., 2015). In *L. monocytogenes*, c-di-AMP PDE PgpH is inhibited by elevated (p)ppGpp levels (Huynh et al., 2015). In the last decade, apart from cGMP and c-GMP-AMP, several new bacterial signaling nucleotides have been discovered. These include c-UMP-AMP, c-AMP-AMP-GMP as well as pyrimidine-based second messengers cUMP and cCTP (Cohen et al., 2019; Whiteley et al., 2019; Tal et al., 2021). These newly discovered signaling nucleotides protect bacteria against phage infection. However, if these nucleotides have any other functions remain to be discovered. Thus the repertoire of signaling nucleotides that bacteria can deploy has expanded significantly. It is likely that this repertoire will add more nucleotide-based second messengers. The crosstalk between various second messenger signaling pathways adds a layer of complexity in biological regulation. Hence, a complete understanding of how second messengers' crosstalk and regulate various physiological processes such as virulence, pathogenesis, and antibiotic resistance becomes important to devise novel therapeutic approaches.

References

Abranches, J., Martinez, A.R., Kajfasz, J.K., Chavez, V., Garsin, D.A., Lemos, J.A., 2009. The molecular alarmone (p)ppGpp mediates stress responses, vancomycin tolerance, and virulence in *Enterococcus faecalis*. J. Bacteriol. 191, 2248−2256.

Aravind, L., Koonin, E.V., 1998. The HD domain defines a new superfamily of metal-dependent phosphohydrolases. Trends Biochem. Sci. 23, 469−472.

Atkinson, G.C., Tenson, T., Hauryliuk, V., 2011. The RelA/SpoT homolog (RSH) superfamily: distribution and functional evolution of ppGpp synthetases and hydrolases across the tree of life. PLoS One 6, e23479.

Ausmees, N., Jonsson, H., Hoglund, S., Ljunggren, H., Lindberg, M., 1999. Structural and putative regulatory genes involved in cellulose synthesis in *Rhizobium leguminosarum* bv. trifolii. Microbiology 145 (Pt 5), 1253−1262.

Bai, Y., Yang, J., Zarrella, T.M., Zhang, Y., Metzger, D.W., Bai, G., 2014. Cyclic di-AMP impairs potassium uptake mediated by a cyclic di-AMP binding protein in *Streptococcus pneumoniae*. J. Bacteriol. 196, 614−623.

Barnhart, M.M., Chapman, M.R., 2006. Curli biogenesis and function. Annu. Rev. Microbiol. 60, 131−147.

Cashel, M., Gallant, J., 1969. Two compounds implicated in the function of the RC gene of *Escherichia coli*. Nature 221, 838−841.

Cohen, D., Melamed, S., Millman, A., Shulman, G., Oppenheimer-Shaanan, Y., Kacen, A., et al., 2019. Cyclic GMP-AMP signalling protects bacteria against viral infection. Nature 574, 691−695.

Commichau, F.M., Dickmanns, A., Gundlach, J., Ficner, R., Stulke, J., 2015. A jack of all trades: the multiple roles of the unique essential second messenger cyclic di-AMP. Mol. Microbiol. 97, 189−204.

Corrigan, R.M., Grundling, A., 2013. Cyclic di-AMP: another second messenger enters the fray. Nat. Rev. Microbiol. 11, 513−524.

Corrigan, R.M., Campeotto, I., Jeganathan, T., Roelofs, K.G., Lee, V.T., Grundling, A., 2013. Systematic identification of conserved bacterial c-di-AMP receptor proteins. Proc. Natl. Acad. Sci. USA 110, 9084−9089.

Corrigan, R.M., Bowman, L., Willis, A.R., Kaever, V., Grundling, A., 2015. Cross-talk between two nucleotide-signaling pathways in *Staphylococcus aureus*. J. Biol. Chem. 290, 5826−5839.

Dalebroux, Z.D., Swanson, M.S., 2012. ppGpp: magic beyond RNA polymerase. Nat. Rev. Microbiol 10, 203−212.

Davies, B.W., Bogard, R.W., Young, T.S., Mekalanos, J.J., 2012. Coordinated regulation of accessory genetic elements produces cyclic di-nucleotides for *V. cholerae* virulence. Cell 149, 358−370.

Gjermansen, M., Ragas, P., Sternberg, C., Molin, S., Tolker-Nielsen, T., 2005. Characterization of starvation-induced dispersion in *Pseudomonas putida* biofilms. Environ. Microbiol 7, 894−906.

Gupta, K.R., Kasetty, S., Chatterji, D., 2015. Novel functions of (p)ppGpp and cyclic di-GMP in mycobacterial physiology revealed by phenotype microarray analysis of wild-type and isogenic strains of *Mycobacterium smegmatis*. Appl. Environ. Microbiol. 81, 2571−2578.

Gupta, K.R., Baloni, P., Indi, S.S., Chatterji, D., 2016. Regulation of growth, cell shape, cell division, and gene expression by second messengers (p)ppGpp and cyclic Di-GMP in *Mycobacterium smegmatis*. J. Bacteriol. 198, 1414−1422.

Hengge, R., 2009. Principles of c-di-GMP signalling in bacteria. Nat. Rev. Microbiol. 7, 263−273.

Hickman, J.W., Harwood, C.S., 2008. Identification of FleQ from *Pseudomonas aeruginosa* as a c-di-GMP-responsive transcription factor. Mol. Microbiol. 69, 376−389.

Hickman, J.W., Tifrea, D.F., Harwood, C.S., 2005. A chemosensory system that regulates biofilm formation through modulation of cyclic diguanylate levels. Proc. Natl. Acad. Sci. USA 102, 14422−14427.

Hinsa, S.M., Espinosa-Urgel, M., Ramos, J.L., O'toole, G.A., 2003. Transition from reversible to irreversible attachment during biofilm formation by *Pseudomonas fluorescens* WCS365 requires an ABC transporter and a large secreted protein. Mol. Microbiol. 49, 905−918.

Huynh, T.N., Luo, S., Pensinger, D., Sauer, J.D., Tong, L., Woodward, J.J., 2015. An HD-domain phosphodiesterase mediates cooperative hydrolysis of c-di-AMP to affect bacterial growth and virulence. Proc. Natl. Acad. Sci. USA 112, E747−E756.

Itoh, Y., Wang, X., Hinnebusch, B.J., Preston 3rd, J.F., Romeo, T., 2005. Depolymerization of beta-1,6-N-acetyl-D-glucosamine disrupts the integrity of diverse bacterial biofilms. J. Bacteriol. 187, 382−387.

Johnson, J.G., Clegg, S., 2010. Role of MrkJ, a phosphodiesterase, in type 3 fimbrial expression and biofilm formation in Klebsiella pneumoniae. J. Bacteriol. 192, 3944−3950.

Kalia, D., Merey, G., Nakayama, S., Zheng, Y., Zhou, J., Luo, Y., et al., 2013. Nucleotide, c-di-GMP, c-di-AMP, cGMP, cAMP, (p)ppGpp signaling in bacteria and implications in pathogenesis. Chem. Soc. Rev. 42, 305−341.

Kolb, A., Busby, S., Buc, H., Garges, S., Adhya, S., 1993. Transcriptional regulation by cAMP and its receptor protein. Annu. Rev. Biochem. 62, 749−795.

Krishnan, S., Chatterji, D., 2020. Pleiotropic effects of bacterial small alarmone synthetases: underscoring the dual-domain small alarmone synthetases in *Mycobacterium smegmatis*. Front. Microbiol. 11, 594024.

Macvanin, M., Hughes, D., 2005. Hyper-susceptibility of a fusidic acid-resistant mutant of Salmonella to different classes of antibiotics. FEMS Microbiol. Lett. 247, 215−220.

Makman, R.S., Sutherland, E.W., 1965. Adenosine 3',5'-phosphate in *Escherichia coli*. J. Biol. Chem. 240, 1309−1314.

Marden, J.N., Dong, Q., Roychowdhury, S., Berleman, J.E., Bauer, C.E., 2011. Cyclic GMP controls *Rhodospirillum centenum* cyst development. Mol. Microbiol. 79, 600−615.

McDonough, K.A., Rodriguez, A., 2012. The myriad roles of cyclic AMP in microbial pathogens: from signal to sword. Nat. Rev. Microbiol. 10, 27−38.

Merz, A.J., So, M., Sheetz, M.P., 2000. Pilus retraction powers bacterial twitching motility. Nature 407, 98−102.

Moscoso, J.A., Schramke, H., Zhang, Y., Tosi, T., Dehbi, A., Jung, K., et al., 2016. Binding of cyclic Di-AMP to the *Staphylococcus aureus* sensor kinase KdpD occurs via the universal stress protein domain and downregulates the expression of the Kdp potassium transporter. J. Bacteriol. 198, 98−110.

Newell, P.D., Monds, R.D., O'toole, G.A., 2009. LapD is a bis-(3',5')-cyclic dimeric GMP-binding protein that regulates surface attachment by *Pseudomonas fluorescens* Pf0−1. Proc. Natl. Acad. Sci. USA 106, 3461−3466.

Nguyen, D., Joshi-Datar, A., Lepine, F., Bauerle, E., Olakanmi, O., Beer, K., et al., 2011. Active starvation responses mediate antibiotic tolerance in biofilms and nutrient-limited bacteria. Science 334, 982−986.

Ogawa, K., Maki, Y., 2003. Cellulose as extracellular polysaccharide of hot spring sulfur-turf bacterial mat. Biosci. Biotechnol. Biochem. 67, 2652−2654.

O'Toole, G.A., Kolter, R., 1998. Flagellar and twitching motility are necessary for Pseudomonas aeruginosa biofilm development. Mol. Microbiol. 30, 295−304.

Pelicic, V., 2008. Type IV pili: e pluribus unum? Mol. Microbiol. 68, 827−837.

Pesavento, C., Hengge, R., 2009. Bacterial nucleotide-based second messengers. Curr. Opin. Microbiol. 12, 170−176.

Pesavento, C., Becker, G., Sommerfeldt, N., Possling, A., Tschowri, N., Mehlis, A., et al., 2008. Inverse regulatory coordination of motility and curli-mediated adhesion in *Escherichia coli*. Genes. Dev. 22, 2434–2446.

Petchiappan, A., Naik, S.Y., Chatterji, D., 2020. RelZ-mediated stress response in *Mycobacterium smegmatis*: pGpp synthesis and its regulation. J. Bacteriol. 202.

Potrykus, K., Cashel, M., 2008. (p)ppGpp: still magical? Annu. Rev. Microbiol. 62, 35–51.

Qi, Y., Chuah, M.L., Dong, X., Xie, K., Luo, Z., Tang, K., et al., 2011. Binding of cyclic diguanylate in the non-catalytic EAL domain of FimX induces a long-range conformational change. J. Biol. Chem. 286, 2910–2917.

Rall, T.W., Sutherland, E.W., 1958. Formation of a cyclic adenine ribonucleotide by tissue particles. J. Biol. Chem. 232, 1065–1076.

Ramnaresh Gupta, K., Chatterji, D., 2016. Sigma factor competition in *Escherichia coli*: kinetic and thermodynamic perspectives. Stress. Environ. Regul. Gene Expr. Adapt. Bact. .

Rao, F., See, R.Y., Zhang, D., Toh, D.C., Ji, Q., Liang, Z.X., 2010. YybT is a signaling protein that contains a cyclic dinucleotide phosphodiesterase domain and a GGDEF domain with ATPase activity. J. Biol. Chem. 285, 473–482.

Rodionov, D.G., Ishiguro, E.E., 1995. Direct correlation between overproduction of guanosine 3',5'-bis-pyrophosphate (ppGpp) and penicillin tolerance in *Escherichia coli*. J. Bacteriol. 177, 4224–4229.

Romling, U., Galperin, M.Y., Gomelsky, M., 2013. Cyclic di-GMP: the first 25 years of a universal bacterial second messenger. Microbiol. Mol. Biol. Rev. 77, 1–52.

Ross, P., Weinhouse, H., Aloni, Y., Michaeli, D., Weinberger-Ohana, P., Mayer, R., et al., 1987. Regulation of cellulose synthesis in *Acetobacter xylinum* by cyclic diguanylic acid. Nature 325, 279–281.

Ryan, R.P., Fouhy, Y., Lucey, J.F., Crossman, L.C., Spiro, S., He, Y.W., et al., 2006. Cell-cell signaling in *Xanthomonas campestris* involves an HD-GYP domain protein that functions in cyclic di-GMP turnover. Proc. Natl. Acad. Sci. USA 103, 6712–6717.

Ryjenkov, D.A., Simm, R., Romling, U., Gomelsky, M., 2006. The PilZ domain is a receptor for the second messenger c-di-GMP: the PilZ domain protein YcgR controls motility in enterobacteria. J. Biol. Chem. 281, 30310–30314.

Simm, R., Morr, M., Kader, A., Nimtz, M., Romling, U., 2004. GGDEF and EAL domains inversely regulate cyclic di-GMP levels and transition from sessility to motility. Mol. Microbiol. 53, 1123–1134.

Sommerfeldt, N., Possling, A., Becker, G., Pesavento, C., Tschowri, N., Hengge, R., 2009. Gene expression patterns and differential input into curli fimbriae regulation of all GGDEF/EAL domain proteins in *Escherichia coli*. Microbiology 155, 1318–1331.

Srivatsan, A., Wang, J.D., 2008. Control of bacterial transcription, translation and replication by (p)ppGpp. Curr. Opin. Microbiol. 11, 100–105.

Steiner, S., Lori, C., Boehm, A., Jenal, U., 2013. Allosteric activation of exopolysaccharide synthesis through cyclic di-GMP-stimulated protein-protein interaction. EMBO J. 32, 354–368.

Sutherland, E.W., Rall, T.W., 1958. Fractionation and characterization of a cyclic adenine ribonucleotide formed by tissue particles. J. Biol. Chem. 232, 1077–1091.

Tabone, M., Lioy, V.S., Ayora, S., Machon, C., Alonso, J.C., 2014. Role of toxin zeta and starvation responses in the sensitivity to antimicrobials. PLoS One 9, e86615.

Tagliabue, L., Antoniani, D., Maciag, A., Bocci, P., Raffaelli, N., Landini, P., 2010. The diguanylate cyclase YddV controls production of the exopolysaccharide poly-N-acetylglucosamine (PNAG) through regulation of the PNAG biosynthetic pgaABCD operon. Microbiology 156, 2901–2911.

Tal, N., Morehouse, B.R., Millman, A., Stokar-Avihail, A., Avraham, C., Fedorenko, T., et al., 2021. Cyclic CMP and cyclic UMP mediate bacterial immunity against phages. Cell 184, 5728–5739. e16.

Tischler, A.D., Camilli, A., 2004. Cyclic diguanylate (c-di-GMP) regulates *Vibrio cholerae* biofilm formation. Mol. Microbiol. 53, 857–869.

Valentini, M., Filloux, A., 2016. Biofilms and Cyclic di-GMP (c-di-GMP) Signaling: Lessons from Pseudomonas aeruginosa and Other Bacteria. J. Biol. Chem. 291, 12547–12555.

Vinella, D., D'ari, R., Jaffe, A., Bouloc, P., 1992. Penicillin binding protein 2 is dispensable in *Escherichia coli* when ppGpp synthesis is induced. EMBO J. 11, 1493–1501.

Wang, X., Preston 3rd, J.F., Romeo, T., 2004. The pgaABCD locus of *Escherichia coli* promotes the synthesis of a polysaccharide adhesin required for biofilm formation. J. Bacteriol. 186, 2724–2734.

Whiteley, A.T., Eaglesham, J.B., De Oliveira Mann, C.C., Morehouse, B.R., Lowey, B., Nieminen, E.A., et al., 2019. Bacterial cGAS-like enzymes synthesize diverse nucleotide signals. Nature 567, 194–199.

Wilksch, J.J., Yang, J., Clements, A., Gabbe, J.L., Short, K.R., Cao, H., et al., 2011. MrkH, a novel c-di-GMP-dependent transcriptional activator, controls *Klebsiella pneumoniae* biofilm formation by regulating type 3 fimbriae expression. PLoS Pathog. 7, e1002204.

Witte, G., Hartung, S., Buttner, K., Hopfner, K.P., 2008. Structural biochemistry of a bacterial checkpoint protein reveals diadenylate cyclase activity regulated by DNA recombination intermediates. Mol. Cell 30, 167–178.

CHAPTER 4

The fish immune armaments in response to pathogen invasion—a tour inside the macrophages

Chaitali Banerjee
Vidyasagar College for Women (Affiliated to University of Calcutta), Kolkata, West Bengal, India

Introduction

The vertebrates' immune system comprises of different immune organs which are categorized into primary and secondary. Primary lymphoid organs are those that generate special immune system cells called lymphocytes and allow further development and maturation. Organs such as bone marrow, thymus being the examples. Secondary lymphoid organs allow the interaction of mature lymphocytes with antigen. Examples being lymph nodes, spleen, Peyer's patches (Anderson and Anderson, 1975). Immunity provided by innate arm works foremost against intruding pathogen but generally doesn't show immunologic memory or the capability to memorize the same pathogen. Adaptive immunity is considered to be antigen specific; thus involves a time lag between antigen exposure and maximal response. The characteristics of adaptive immunity are to elicit memory response which enables the host to mount a profound immune response upon subsequent exposure to the similar antigen (Hoebe et al., 2004). Both the arms of immune system are complementary to each other, consequent to pathogen intrusion and other insults (Marshall et al., 2018). Here in this review, a focus on teleost immune system with a particular emphasis on macrophages has been attempted.

Immune organs and cell types of teleosts

Teleost fishes exhibit the greatest representatives of extant aquatic vertebrates. Another fascinating aspect is the diversified environment that they have colonized (Volff, 2005). A key factor responsible is the well-defined immune system that supports their free-living habitat from very early stage of their life cycle. Primarily, innate immune system protects against microbial infections. In fish lineage, bony fishes show greater extent of similarity to tetrapod's comparative to jawless and cartilaginous fishes (Takezaki et al., 2004).

Bacterial Survival in the Hostile Environment
DOI: https://doi.org/10.1016/B978-0-323-91806-0.00002-3

Hematopoiesis is the process of production of cellular components of different lineages such as myeloid, erythroid from the apical cell type of the hematopoietic hierarchy, that is, hematopoietic stem cells (HSCs). This process is localized into hematopoietic system such as bone marrow, spleen, liver. In teleosts, primarily hematopoiesis is reported to be initiated in yolk sac and intermediate cell mass, an intraembryonic structure (Zapata et al., 2006). Gradually as we attempt to understand the histogenesis of lymphoid organ; thymus, kidney (pro- and mesonephros), spleen, gut-associated lymphoid tissue (GALT) are the organs majorly noted. Each of these organs can yield varieties of immune cells—of which patrolling macrophage cells are among the earliest to respond and hence are indispensable to innate immune system for providing the first line of defense. The macrophages are the major cell types which have been largely implicated in triggering innate immune response and present antigens to the lymphocytes to activate adaptive immune response (Patel et al., 2021). Hence their role is exceptionally recognized. In this context it is worth mentioning that head-kidney or anterior portion of the kidney are important and rich source of macrophages (Joerink et al., 2006). Fig. 4.1 shows the different immune organs of teleost fish (Dixon et al., 2016).

Characteristics of teleosts' macrophages

The glycoproteins specific for stimulating macrophages, that is, M-CSF is important along with its dedicated receptor. They essentially influence the commitment of cell

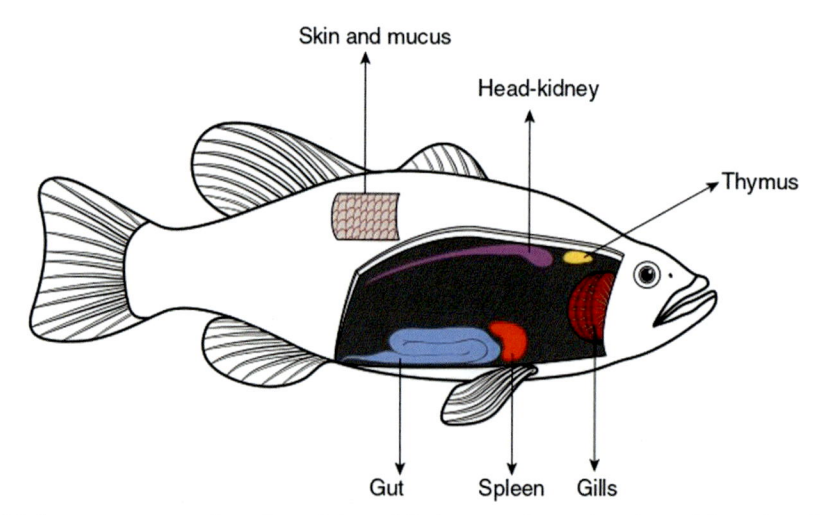

Figure 4.1 Immune organs of a teleost fish and their approximate locations. *Adapted, with permission, from Dixon, B., Heath, B., Semple, S.L., 2016. The immune system of bony fish. Encycl. Immunobiol., 481–485. https://doi.org/10.1016/B978-0-12-374279-7.12010-7. © 2016 Elsevier Ltd.*

development pathway to macrophage lineage. This process is central to monopoiesis across vertebrates. In teleosts, a little variation is seen owing to the presence of two distinct csf-1 genes, namely, 1.1 and 1.2. In higher vertebrates such as birds and mammals, csf-1 gene is reported to be present as a single alternatively spliced gene (Grayfer et al., 2018).

Macrophages are ubiquitous in most tissue structures; hence as the microbes invade it is inevitably noticed by them. The macrophages have inherent mechanism to engulf pathogens through a process called as phagocytosis, neutralization and present antigens to lymphocytes to activate adaptive immunity response (Hirayama et al., 2018). Similarly, the pathogens have several strategies to evade or counteract the host response and utilize its niche for survival and further replication (Leseigneur et al., 2020). An array of signal transduction pathways gets activated as soon as the microbe is sensed by the macrophages. Like higher vertebrates the pathogen—internalization mediated by cytoskeletal network or receptor-mediated sensing of cellular components are well elucidated (Hodgkinson et al., 2015). In this process of phagocytosis and in subsequent cell signaling events, intermediate signaling components are produced which lead to secretion of cytokines (Barreda et al., 2000). This process involves various cellular enzymes and caspases (Grayfer et al., 2018). Also, crosstalk between these different signaling molecules instructs the participation of various cellular organelles in response to microbial infections (Banerjee et al., 2015). This response by macrophages creates hostile environment to the invading pathogens. Next section describes different cellular events opted by macrophages in response to bacterial infection.

Phagocytosis by macrophages

Macrophages are the established professional phagocytes in teleost (Esteban et al., 2015). The finer steps of phagocytosis as conceived and hypothesized originally by E. Metchnikoff are—detection and recognition, attachment, encircling to internalize inside phagosome, phagolysosome formation, intracellular killing, and/or egestion or subsequent antigen presentation. This is the well-known process in fish macrophages in response to bacterial infection (Hodgkinson et al., 2015).

Pathogenic infiltration is sensed by the macrophages *via* the receptors, pathogen recognition receptors (PRRs) present both extracellular and intracellular (Grayfer et al., 2018). The different PRRs have been found in the fish macrophages against a range of pathogens—*Aeromonas hydrophila*, *A. salmonicida*, *Mycobacterium* sp. to name a few (Srivastava et al., 2017). Besides many, as seen in higher vertebrates, few toll like receptors (TLRs) are unique to fish immune response such TLRs 20, 21, 22, 23 (Gao et al., 2021). The complementary pathogen-associated molecular pattern (PAMP) such as LPS for TLR-4 and recognition of lipoproteins derived from bacteria, viruses, fungi, and parasites by TLR-2 are the dynamic examples of PRR—PAMP interaction in bony fishes (Sahoo, 2020). C-type lectins are the PRRs that principally recognize glucans (carbohydrates) on cell surface. In marine fish such as *Totoaba macdonaldi*, these lectins are intricately studied and found to be important against *Vibrio* sp infections (Angulo et al., 2019). Expressions of several nucleotide-binding and oligomerization domain (NODs)-like NLRs have been reported in the recognition of different bacterial or viral stimuli by fish macrophages (Li et al., 2020). These include the major NODs NOD1 and 2, NLR-C3, -C5, -X1. Retinoic acid-inducible gene-I like receptors, that is, RLRs and its different members—laboratory of genetics and physiology-2 (LGP2), melanoma differentiation-associated protein-5 (MDA5)are well-conserved across vertebrates and accordingly been identified in different teleost. Accordingly distinct downstream molecules such as MITA, TRAF3, and TANK-binding kinase-1 have been reported in piscine macrophages against bacteria such as *Edwadsiella* sp. (Chen et al., 2017). Pathogen uptake is greatly facilitated by cytoskeleton remodeling. In case of fish macrophages, the process is dynamically involved. The essential actin-filled protrusion is primary to phagocytic uptake and downstream orchestrated movements. The phenomenon has been visualized by several scientific evidences including microscopic studies (Ramirez et al., 2015) and pharmacological inhibitors (Banerjee et al., 2012) in macrophages infected with pathogen. The lysosomal activation, phagosome—lysosome fusion is equally important host strategies shown and executed by teleost macrophages in response to invading pathogen (Cortes et al., 2021). Hence, it's apparent that teleost macrophages are potent phagocyte and serve as essential player of innate immune system.

Antigen presentation by macrophages

The acquired branch of immune system relies on the presence of an antigen that is identified, internalized and further processed by different antigen presenting cells including macrophages. This enables microorganisms to be processed into molecular units/peptides. Consequently, it is presented in association with major histocompatibility complex (MHC) molecules to trigger immune response and generation of memory in immune cells (Drutman&Trombetta, 2010). Intracellular antigens are presented to T cytotoxic cells which are CD8 positive and with MHC class I molecules, while extracellular antigens are presented to T helper cells which are Class II restricted. Another important feature of APCs is the presence of costimulatory molecules, namely, B7−1, -2; alternatively, CD80 and CD86, respectively. They essentially prime naive T cells to induce their differentiation and release of cytokines (Newton et al., 2005). In this context, the fish professional APCs like macrophages have been reported to be antigen-specific, eliciting costimulatory signal and trigger cytokine release (Jin et al., 2022). The piscine macrophages in fugu, zebrafish, and others express MHC-II proving its role as APC and costimulatory molecules like B7 suggesting these to be antigen presenting (Yamaguchi and Dijkstra, 2019). Though all genes are not truly homologous as the evolution from teleosts to higher vertebrates progresses, many of the signaling cascades such as IFN-gamma is conserved and noted to be well functioning (Munoz-Flores et al., 2022). In trouts, IFNγ is reported to elicit the expression of diverse immune genes such as mhcIIb, il1ß, ifnγ, stat1, c-type lectin, tap1, ikb, junb, irf1 (Zou et al., 2015). Quite alike are the situation reported in zebrafish, black seabream, carps, and others where expression of several genes such as il12 subunits p35 and p40, il1ß isoforms 1&2, TNFα isoforms 1&2, ifng, ccl1, il8 (CXCL-8), viperin are evident (Aggad et al., 2010; Arts et al., 2010). The related antiinflammatory cytokines IL-4 and IL-13 attribute important role in teleosts such as trout, seabass, grass carp, and goldfish. They are implied in the upregulation of several immunosuppressive genes such as IL-10, TGF-β; and reducing proinflammatory cytokine's (IL-1β, TNFα, IFNγ) gene expressions. Specifically in the kidney phagocyte/macrophages, they induce alternative pathways of macrophages' activation as evident from gene expression of arginase and illustrative enzymatic activity of arginase (Junttila, 2018).

Subcellular crosstalk of teleost macrophages

Communication is important and is one of the main regulatory mechanisms. From the point of view of signal transduction, interorganelle crosstalk is indispensable and comprises both vesicular and nonvesicular mechanisms of transport (Namgaladze et al., 2019). Nonvesicular transfer occurs at the close proximity "membrane contact sites" and generally gets regulated by a set of specialized proteins. Categorically, these specialized proteins are involved in important phenomena such as calcium homeostasis, phospholipid biosynthesis, and cell death pathways. A fascinating aspect of interactive host and microbe session is the mode and mechanism of cell death. The host tries to clear the pathogen, restore the homeostasis probably arising from inflammation whereas the pathogen aims at rapid multiplication, neutralizing probable defense machineries, and creating a favorable niche for itself. The appropriate modulation of cell death is crucial determinant shaping the final outcome (FitzGerald et al., 2020). While attempting to understand these interactions and dynamics of manipulation, unraveling the intraorganelle chemistry is important.

A well-implicated communication network inside the cell is mitochondria—endoplasmic reticulum (Fig. 4.2). There are phenomena such as lipid metabolism, inflammasome activation, antiviral responses, calcium homeostasis, efferocytosis, phospholipid biosynthesis, redox signaling, and others that are critical to cell death mechanisms (Simmen et al., 2005). ER—mitochondrial interaction is such an aspect that shapes the fate of these mentioned processes. Mitochondria are the hub of important biochemical and physiological events. Structurally, mitochondria are organelle with unique compartmentalization, namely, presence of inner and outer membranes and a matrix (Iovine et al., 2021). Hence, mitochondria possess machineries that span

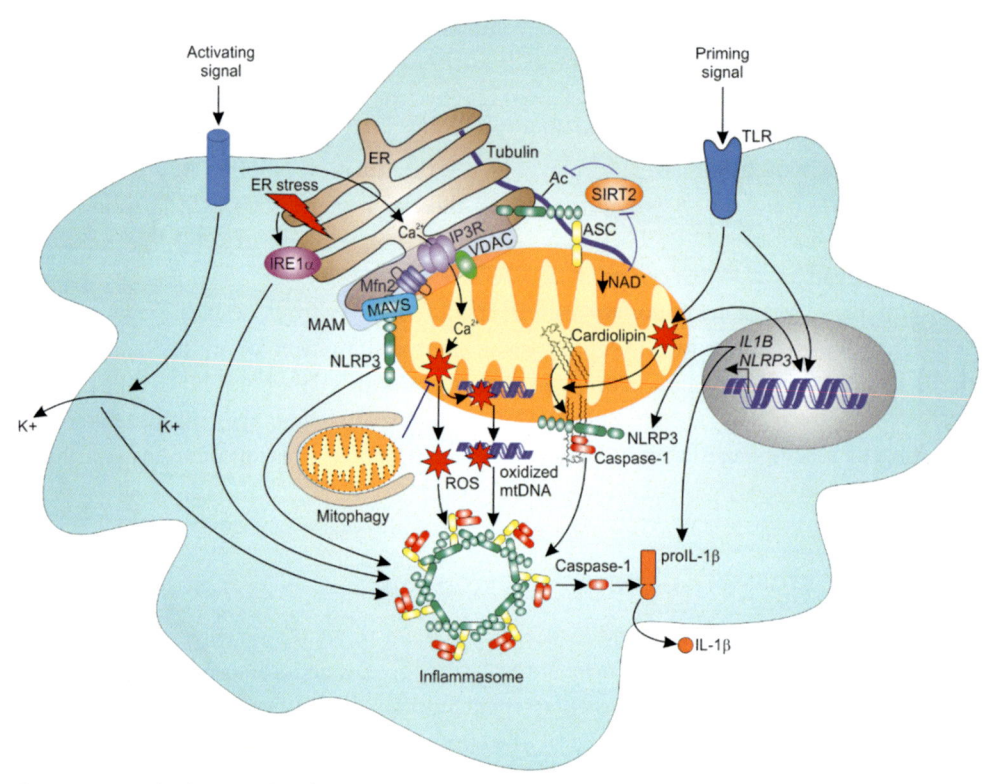

Figure 4.2 Endoplasmic reticulum—mitochondria interaction inside the macrophages in response to pathogen invasion. *Adapted from Namgaladze, D., Khodzhaeva, V., Brune, B., 2019. ER-mitochondria communication in cells of the innate immune system. Cell 8, 1088. doi:10.3390/cells8091088.*

the entire region of mitochondria and work in collaboration influencing several functioning. On the other hand, ER, a continuous membrane system severely involved in multiple functions is intricately spread in the cytoplasm apposing several other organelles (Gagnon et al., 2002). Interestingly, mitochondria undergo fission and fusion as a routine process of maintenance (Youle and van der Bliek, 2012). A similar situation has been reported during pathogen invasion as well. The mitochondria increase their number and moves near the ER (Hoppins et al., 2007). As a result, a plethora of mechanisms happen like calcium transfer and phospholipid transfer through mito—ER contact sites (Simmen et al., 2005).

Microbial pathogenesis in piscine macrophages has been reported to involve calcium dynamics between mitochondria and ER (Roca et al., 2019). It has been found that due to pathogen invasion, mitochondria and ER arrive in vicinity. This facilitates mitochondrial calcium uptake mediated by uniporters. This depletion of calcium is usually mediated by ER receptors such as IP3R and RyR, inositol triphosphate and ryanodine receptors, respectively. As a consequence, in the downstream signaling, proteins are unfolded in ER and typically involve accumulation of mis-folded proteins (Dahiya et al., 2021). Sensors dedicated to ER such as activating transcription factor 6 (ATF6), RNA-dependent protein kinase-like ER kinase (PERK), and inositol-requiring ER-to-nucleus signal kinase 1 (IRE1) try to establish the homeostasis. Important proteases such as calpains (proteolytic enzymes of calcium-dependent and nonlysosomal cysteine proteases family) get activated that activate caspase-12 in the ER. The latter protease is initially located facing the ER's outer membrane as inactive procaspase-12. It triggers several functioning and initiating the apoptotic cascade is one such process (Datta et al., 2018). Basically, calcium ion alterations inside the cell are a dynamic process and calcium plays an essential role as a second messenger for phagocytic process in teleost macrophages (Eijkenboom et al., 2019). The well-functioning mitochondrial uniporters triggering important signaling events as a result of microbial infection have been characterized in the phagocytes of many fish species (Wentzel et al., 2020, Banerjee et al., 2015).

Mitochondrial calcium uptake is further involved in cellular metabolism and cell death processes. The mitochondrial permeabilization is an important downstream consequence that channelizes plenty of events. As mentioned, mitochondrial calcium uptake and further efflux are important determinants of classical mitochondrial functioning and cell death (Murphy and Steenbergen, 2021). Besides, mitochondrion is a region of sequestration for calcium *via* chelation in the matrix with the help of amorphous phosphates (Chinopoulas and Adam-Vizi, 2010). This process of sequestration occurs upstream, with the collapse of mitochondrial potential, and the calcium is released leading to Ca^{2+} efflux from mitochondria into cytosol. The reverse phenomenon of calcium uptake by mitochondria from the cytosol is an equally important event that has been noticed in several cases (Rizzuto et al., 2000). The electron transport chain (ETC) is located at the inner mitochondrial membrane (IMM) and generates electrochemical potential due to ATP synthesis (Griffiths and Rutter, 2009). The mitochondrial membrane potential hence generated is crucial to cytosolic calcium uptake by mitochondria (Bhatti et al., 2017). Reports suggest that during microbial pathogenesis several mitochondrial processes act as central effector in shaping phagocytic activity by macrophages. In response to infections, mitochondria have shown both anterograde and retrograde signaling (do Amarel et al., 2021). During these infections, mitochondrial nitric oxide synthesis, intracellular reactive oxygen species are well documented in phagocytic teleost (do Amarel et al., 2021). Uniporters and mitochondrial calcium sequestration play critical in the process (Datta et al., 2018; Paredas et al., 2019; Datta et al., 2016) and are detrimental in several mitochondria-mediated biological processes. Thus mitochondria play a significant role in downstream signaling cascade and in maintaining homeostasis in macrophages in response to bacterial infections.

Modalities of cell death

The host and pathogen interactions are vivid, fascinating, and complex leading to killing of pathogen by host immune system. On the contrary, the situation may proceed for generating favorable environment for pathogen dissemination (Best, 2008). However, host cell death due to microbial invasion is a generalized phenomenon. Killing of the infected cells is crucial to pathogenesis and cell death serves as an effective way of reducing the pathogen load (Labbe and Saleh, 2008). The detailed mechanisms and mode of cell death following pathogen invasion may vary; and apoptosis, necrosis, pyroptosis, autophagy are to name some predominant pathways (Fig. 4.3) (Lemasters, 2018).

Apoptosis proceeds with distinct morphological changes such as cellular shrinkage, fragmentation into smaller membrane-bound bodies (designated apoptotic bodies), rapid phagocytosis by neighboring cells and involves distinct signaling cascade of different caspases. The integrity of plasma membrane surrounding the fragmented apoptotic bodies renders apoptosis a controlled death pathway (Lamkanfi and Dixit, 2010). Biochemically, apoptotic cells undergo declining potential of inner mitochondrial

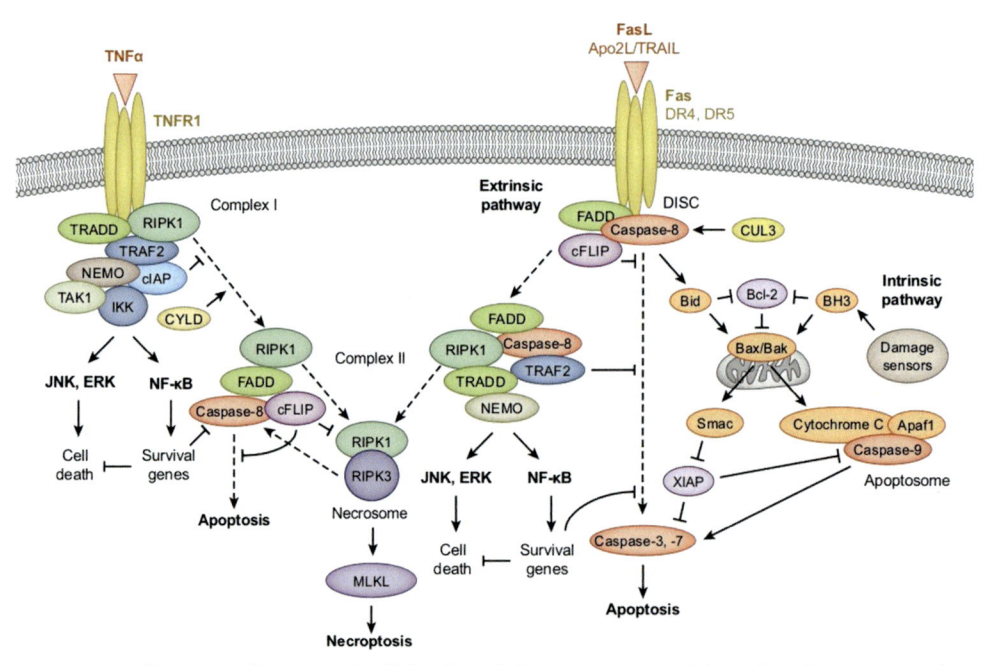

Figure 4.3 Different mechanisms of cell death and their accessories. *Adapted, with permission, from Lemasters, J.J., 2018. Molecular mechanisms of cell death, In: Coleman, W.B., Tsongalis, G.J. (Eds.), Molecular Pathology, second ed. Academic Press, pp. 1–24 (Chapter 1). https://doi.org/10.1016/B978-0-12-802761-5.00001-8. © 2018 Elsevier Inc.*

membrane, followed by the activation of proteases, fragmentation of chromosomal DNA into characteristics laddering. This further trigger activation of different proteases, and then the flipping of specific phospholipids such as phosphatidylserine from the inner to the outer side of the plasma membrane. It is a two-step cascade that starts with "initiator caspases" such as 2, 8, 9, and 10. They have larger prodomains and are the first to be recruited. The large protein complexes thus formed undergoes proximity-induced oligomerization and autoprocessing. The DISC signaling complex and apoptosome get activated. They activate different apoptotic signaling pathways which communicate with each other through activation of Bid protein. The intrinsic apoptotic pathway converges into activation of Bak and Bax. These proapoptotic Bcl-2 members induce release of proapoptogenic factors such as HtrA2/Omi, cytochrome C, and SMAC/Diablo from the mitochondrial intermembrane space. Cytochrome C is localized in the cytosol that assembles with Apaf-1 to form the apoptosome complex that activates caspase-9. These initiator caspases process the executioner players, caspases-3 being primary one to operate. The latter acts on substrates such as PolyADP ribose polymerase, process the inhibitor of CAD to promote oligonucleosomal DNA fragmentation by the endonuclease CAD, cellular cytoskeletons like actin to induce apoptosis once the inhibition by XIAP is lifted by HtrA2 and Diablo. Precisely it is fine balance of pro- and antiapoptotic factors that determines whether host cell would undergo induction or inhibition of apoptosis (Salvesen and Riedl, 2008).

Contrasting to it, necrosis [Greek word "nekros," = "dead body"] involves inflammation leading to release of proinflammatory cytokines. This promotes spilling out of cytosolic content into the outside space. Usually, a necrotic cell undergoes increased cell volume, disorganized hydrolysis of chromatin and DNA and lost membrane integrity, hence establishing tissue damage. Necrosis, a programmed signaling cascade meticulously involves destabilized organelle such as lysosomes, release of cathepsins, reactive species participation, and enzymatic activity like calpain (Scaffidi et al., 2002).

Pyroptosis, the classic caspase-1-mediated death cascade involves distinct proinflammatory cytokines. Interleukins (IL)-1β and -18 are the primary proinflammatory cytokines and are implicated to be important in host—pathogen interaction. Inflammasome formation is the characteristics feature in the pyroptosis. Cytosolicreceptors such as nucleotide-binding and oligomerization domain (NOD)-like receptors (NLRs) sense specific microbial PAMP to activate inflammasome complex formation to induce innate immunity. These receptors contain LRR (for ligand sensing), NACHT [nucleotide-binding domain (NBD) for oligomerization of NLRs] and PYRIN (effector) domains. NLR family of proteins generally contains N-terminal PYRIN domain as interaction domain, effector PYRIN domain contains caspase activation and recruitment domain (CARD), necessary for recruiting caspases in the downstream signaling in response to microbial infections, middle NACHT domain which contain NBD and NBD-associated domains which are crucial for nucleotide binding and self-oligomerization of NLRs upon activation, and leucine-rich repeats (LRRs) domain at C-terminal for sensing pathogen patterns. NLRs such as NLRP3 bind to adaptor protein ASC (apoptosis-associated speck-like protein containing CARD) with PYD-PYD domain interaction. ASC is also known as PYCARD which contains PYD and CARD domains. Subsequently, PYD-PYD interaction-dependent oligomerization of ASC promotes CARD domain-mediated interaction of ASC and procaspase-1 through CARD-CARD interaction and induces its oligomerization and secretion of effector caspase-1. The effector caspase-1 converts proIL-1βto active IL-1β and secretion of inflammasome-mediated cytokines (IL-1β and IL-18) (Salvesen and Riedl, 2008).

In teleosts macrophages, following pathogen sensing and/or invasion, the diverse mode of cell death has been noted and well elucidated (Li et al., 2020; Grayfer et al., 2018; Banerjee et al., 2015).

Immune evasion strategies of microbial pathogen

The macrophages are the professional phagocytes that trigger various cellular signal once it senses pathogen. However, the microbe also deploys its own mechanisms to modulate the immune system. Thus often it's seen that the pathogen evades or inhibits different defense strategies of the host. It has been reported that pathogens can effectively conceal the expression of molecular patterns such as lipopolysaccharide, lipoteichoic acids, glycolipids, phosphorylcholine, N-formylmethionine, and others; thus preventing their recognition by these patrolling cells. Lipopolysaccharide is a readily recognizable cell wall component of gram-negative bacteria. Pathogens such as *Yersinia* sp., *Shigella* sp., often modify the composition of lipid A chain, an important component of LPS. The other components being o-antigen and core of oligosaccharide. Several bacteria are endowed with effective secretory system, for example, type III, IV, and VI enabling to deliver "effector" proteins, directly into the host cell (Reddick and Alto, 2014). Apart from that there exist several other strategies and interruption of signal transduction pathways by microbial pathogen is worth mentioning. The sequential activating cascade like mitogen-activated protein kinase pathway and nuclear factor kappa B (NF-κB) are the targets which get modulated by several pathogens. In *Shigella flexneri*, OspF an effector type III protein effectively interferes with the phosphorylation of different MAPKs such as ERK, JNK, and p38 kinases (Mukherjee et al., 2006). Besides manipulating the different signaling cascade through, an inevitable approach is to neutralize the antimicrobial peptides. Often seen is the modulation of host's secretory system and disruptive crosstalk of the cellular organelles by the invading pathogen. These strategies are the probable explanation of pathogenic commandeer over various host maneuver, hence highlighting different immune mechanisms effectively modulated by the pathogen.

Conclusion

Teleost's immune system is endowed with essential immune organs and mediators alike higher vertebrates. Similarly, the sentinel armamentarium is functionally highly evolved and macrophages being worth mentioning. In this review, attempt has been made to elucidate the functional aspects of teleosts' macrophages during microbial pathogenesis. Macrophages are important immune cells and are one of initial cell types that respond as pathogen is sensed. Phagocytosis could be the major event and as discussed earlier it is initiated by the involvement of several receptors, involves signaling molecules, generation of free radicals and sequential activation of different cascades and activation of caspases. On the other hand, pathogen can also modulate the innate immune responses of macrophages through elegant strategies. They have the potential to disguise the recognition patterns, steal the classical interactive process by using specialized secretory needle or directly hamper the signaling cascades that may bring out deleterious effects for pathogen. Precisely there are numerous events that happen during the interaction between the host and pathogen. At times the cellular environment may turn hostile to the pathogen and mediate its clearance. Alternatively, the host immune response may also get modulated and allow pathogen to oppose clearance and facilitate multiplication. This fine balance of events is critical and influenced by multiple factors. Hence it would be really tempting to further shed light on understanding the complexity of the process and elucidate different triggers and factors shaping the outcome.

References

Aggad, D., Stein, C., Sieger, D., Mazel, M., Boudinot, P., Herbomel, P., et al., 2010. In vivo analysis of Ifn-γ1 and Ifn-γ2 signaling in zebrafish. J. Immunol. 185, 6774−6782. Available from: https://doi.org/10.4049/jimmunol.1000549.

Anderson, A.O., Anderson, N.D., 1975. Studies on the structure and permeability of the microvasculature in normal rat lymph nodes. Am. J. Pathol. 80, 387−418.

Angulo, C., Sanchez, V., Delgado, K., Reyes-Becerril, M., 2019. C-type lectin 17A and macrophage-expressed receptor genes are magnified by fungal β-glucan after Vibrio parahaemolyticus infection in *Totoaba macdonaldi* cells. Immunobiology 224, 102−109.

Arts, J.A.J., Tijhaar, E.J., Chadzinska, M., Savelkoul, H.F.Z., Verburg-van Kemenade, B.M.L., 2010. Functional analysis of carp interferon-gamma: evolutionary conservation of classical phagocyte activation. Fish Shellfish Immunol. 29, 793−802. Available from: https://doi.org/10.1016/j.fsi.2010.07.010.

Banerjee, C., Goswami, R., Verma, G., Datta, M., Mazumder, S., 2012. Aeromonas hydrophila induced head kidney macrophage apoptosis in *Clariasbatrachus* involves the activation of calpain and is caspase-3 mediated. Dev. Comp. Immunol. 37, 323−333.

Banerjee, C., Singh, A., Das, T.K., Raman, R., Shrivastava, A., Mazumder, S., 2015. Ameliorating ER-stress attenuates aeromonas hydrophila-induced mitochondrial dysfunctioning and caspase mediated HKM apoptosis in *Clariasbatrachus*. Sci. Rep. 4, 5820.

Barreda, D.R., Neumann, N.F., Belosevic, M., 2000. Flow cytometric analysis of PKH26-labeled goldfish kidney-derived macrophages. Dev. Comp. Immunol. 24, 395−406.

Best, S.M., 2008. Viral subversion of apoptotic enzymes: escape from death row. Annu. Rev. Microbiol. 62, 171–192.

Bhatti, J.S., Bhatti, G.K., Reddy, P.H., 2017. Mitochondrial dysfunction and oxidative stress in metabolic disorders - a step towards mitochondria based therapeutic strategies. Biochimica et. Biophysica Acta (BBA) — Mol. Basis Dis. 1863, 1066–1077.

Chen, S.N., Zou, P.F., Nie, P., 2017. Retinoic acid-inducible gene I (RIG-I)-like receptors (RLRs) in fish: current knowledge and future perspectives. Immunology . Available from: https://doi.org/10.1111/imm.12714.

Chinopoulas, C., Adam-Vizi, V., 2010. Mitochondrial Ca^{2+} sequestration and precipitation revisited. FEBS J. . Available from: https://doi.org/10.1111/j.1742-4658.2010.07755.x.

Cortes, H.D., Gomez, F.A., Marshall, S.H., 2021. The phagosome—lysosome fusion is the target of a purified quillaja saponin extract (PQSE) in reducing infection of fish macrophages by the bacterial pathogen *Piscirickettsia salmonis*. Antibiotics 10, 847. Available from: https://doi.org/10.3390/antibiotics10070847.

Dahiya, P., Hussain, M.A., Mazumder, S., 2021. mtROS induced via TLR-2-SOCE signaling plays proapoptotic and bactericidal role in Mycobacterium fortuitum-infected head kidney macrophages of *Clariasgariepinus*. Front. Immunol. 20, 748758. Available from: https://doi.org/10.3389/fimmu.2021.748758.

Datta, D., Khatri, P., Banerjee, C., Singh, A., Meena, R., Saha, D.R., et al., 2016. Calcium and superoxide-mediated pathways converge to induce nitric oxide-dependent apoptosis in mycobacterium fortuitum-infected fish macrophages. PLoS One 11, e0146554.

Datta, D., Khatri, P., Singh, A., Saha, D.R., Verma, G., Raman, R., et al., 2018. *Mycobacterium fortuitum*-induced ER-mitochondrial calcium dynamics promotes calpain/caspase-12/caspase-9 mediated apoptosis in fish macrophages. Cell Death Discov. 4, 30. Available from: https://doi.org/10.1038/s41420-018-0034-9.

Dixon, B., Heath, B., Semple, S.L., 2016. The immune system of bony fish. Encycl. Immunobiol. 481–485. Available from: https://doi.org/10.1016/B978-0-12-374279-7.12010-7.

do Amarel, M.A., Paredes, L.C., Padovani, B.N., Mendonca-Gomes, M.J., Montes, L.F., Camara, N.O.S., et al., 2021. Mitochondrial connections with immune system in Zebrafish. Fish Shellfish Immunol. Rep. 2, 100019.

Drutman, S.B., Trombetta, E.S., 2010. Dendritic cells continue to capture and present antigens after maturation *in vivo*. J. Immunol. 185, 2140–2146. Available from: https://doi.org/10.4049/jimmunol.1000642.

Eijkenboom, I., Vanoevelen, J.M., Hoeijmakers, J.G., Wijnen, I., Gerards, M., Faber, C.G., et al., 2019. A zebrafish model to study small-fiber neuropathy reveals a potential role for GDAP1. Mitochondrion 47, 273–281.

Esteban, M.A., Cuesta, A., Chaves-Pozo, E., Meseguer, J., 2015. Phagocytosis in teleosts. Implications of the new cells involved. Biology 4, 907–922. Available from: https://doi.org/10.3390/biology4040907.

FitzGerald, E.S., Luz, N.F., Jamieson, N.M., 2020. Competitive cell death interactions in pulmonary infection: host modulation vs pathogen manipulation. Front. Immunol. . Available from: https://doi.org/10.3389/fimmu.2020.00814.

Gagnon, E., Duclos, S., Rondeau, C., Chevet, E., Cameron, P.H., Steele-Mortimer, O., et al., 2002. Endoplasmic reticulum-mediated phagocytosis is a mechanism of entry into macrophages. Cell 12, 119–131. Available from: https://doi.org/10.1016/s0092-8674(02)00797-3.

Gao, F., Liu, J., Lu, M., Liu, Z., Wang, M., Ke, X., et al., 2021. Nile tilapia toll-like receptor 7 subfamily: intracellular TLRs that recruit MyD88 as an adaptor and activate the NF-κB pathway in the immune response. Dev. Comp. Immunol. 125, 104173. Available from: https://doi.org/10.1016/j.dci.2021.104173.

Grayfer, L., Kerimoglu, B., Yaparla, A., Hodgkinson, J.W., Xie, J., Belosevic, M., 2018. Mechanisms of fish macrophage antimicrobial immunity. Front. Immunol. . Available from: https://doi.org/10.3389/fimmu.2018.01105.

Griffith, E.J., Rutter, G.A., 2009. Mitochondrial calcium as a key regulator of mitochondrial ATP production in mammalian cells. Biochim. Biophys. Acta 1787, 1324–1333.

Hirayama, D., Iida, T., Nakase, H., 2018. The phagocytic function of macrophage-enforcing innate immunity and tissue homeostasis. Int. J. Mol. Sci. 19, 92.

Hodgkinson, J.W., Ge, J.Q., Katzenback, B.A., Havixbeck, J., Barreda, D.R., Stafford, J.L., et al., 2015. Development of an in vitro model system to study the interactions between Mycobacterium marinum and teleost neutrophils. Dev. Comp. Immunol. 53, 349−357.

Hoebe, K., Janssen, E., Beutler, B., 2004. The interface between innate and adaptive immunity. Nat. Immunol. 5, 971−974.

Hoppins, S., Lackner, L., Nunnari, J., 2007. The machines that divide and fuse mitochondria. Annu. Rev. Biochem. 76, 751−780.

Iovine, J.C., Claypool, S.M., Alder, N.N., 2021. Mitochondrial compartmentalization: emerging themes in structure and function. Trends Biochem. Sci. 46, 902−917. Available from: https://doi.org/10.1016/j.tibs.2021.06.003.

Jin, C.Y., Su, N., Hu, C.B., Shao, T., Ji, J.F., Qin, L.L., et al., 2022. Regulatory role of BTLA and HVEM checkpoint inhibitors in T cell activation in a perciform fish *Larimichthyscrocea*. Dev. Comp. Immunol. 128, 104312.

Joerink, M., Ribeiro, C.M.S., Stet, R.J.M., Hermsen, T., Savelkoul, H.F.Z., Wiegertjes, G.F., 2006. Head kidney-derived macrophages of common carp (*Cyprinus carpio* L.) show plasticity and functional polarization upon differential stimulation. J. Immunol. 177, 61−69.

Junttila, I.S., 2018. Tuning the cytokine responses: an update on interleukin (IL)-4 and IL-13 receptor complexes. Front. Immunol. 7, 888. Available from: https://doi.org/10.3389/fimmu.2018.00888.

Labbe, K., Saleh, M., 2008. Cell death in the host response to infection. Cell Death Differ. 15, 1339−1349.

Lamkanfi, M., Dixit, V.M., 2010. Manipulation of host cell death pathways during microbial infections. Cell Press. 8, 44−54.

Lemasters, J.J., 2018. -Molecular mechanisms of cell death. In: Coleman, W.B., Tsongalis, G.J. (Eds.), Molecular Pathology, second ed. Academic Press, pp. 1−24. (Chapter 1). Available from: https://doi.org/10.1016/B978-0-12-802761-5.00001-8.

Leseigneur, C., Le-Bury, P., Pizarro-Cerda, J., Dussurget, O., 2020. Emerging evasion mechanisms of macrophage defenses by pathogenic bacteria. Front. Cell. Infect. Microbiol. 10, n577559. Available from: https://doi.org/10.3389/fcimb.2020.577559.

Li, J.N., Zhao, Y.T., Cao, S.L., Wang, H., Zhang, J.J., 2020. Integrated transcriptomic and proteomic analyses of grass carp intestines after vaccination with a double-targeted DNA vaccine of Vibrio mimicus. Fish Shellfish Immunol. 98, 641−652.

Marshall, J.S., Warrington, R., Watson, W., Kim, H.L., 2018. An introduction to immunology and immunopathology. Allergy, Asthma Clin. Immunol. 14, 49.

Mukherjee, S., Keitany, G., Li, Y., Wang, Y., Ball, H.L., Goldsmith, E.J., et al., 2006. Yersinia YopJ acetylates and inhibits kinase activation by blocking phosphorylation. Science 312, 1211−1214.

Munoz-Flores, C., Astuya-Villalon, A., Romero, A., Toledo, J.R., 2022. Salmonid MyD88 is a key adapter protein that activates innate effector mechanisms through the TLR5M/TLR5S signaling pathway and protects against Piscirickettsiasalmonis infection. Fish Shellfish Immunol. . Available from: https://doi.org/10.1016/j.fsi.2021.12.030.

Murphy, E., Steenbergen, C., 2021. Regulation of mitochondrial Ca^{2+} uptake. Annu. Rev. Physiol. 10, 107−126. Available from: https://doi.org/10.1146/annurev-physiol-031920-092419.

Namgaladze, D., Khodzhaeva, V., Brune, B., 2019. ER-mitochondria communication in cells of the innate immune system. Cell 8, 1088. Available from: https://doi.org/10.3390/cells8091088.

Newton, S., Ding, Y., Chung, C.S., Chen, Y., Lomas-Neira, J.L., Ayala, A., 2005. Sepsis-induced changes in macrophage co-stimulatory molecule expression: CD86 as a regulator of anti-inflammatory IL-10 response. Surg. Infect. 5. Available from: https://doi.org/10.1089/sur.2004.5.375.

Paredas, L.C., Camara, N.O.S., Braga, T.T., 2019. Understanding the metabolic profile of macrophages during the regenerative process in zebrafish. Front. Physiol. . Available from: https://doi.org/10.3389/fphys.2019.00617.

Patel, A.A., Ginhoux, F., Yona, S., 2021. Monocytes, macrophages, dendritic cells and neutrophils: an update on lifespan kinetics in health and disease. Immunology . Available from: https://doi.org/10.1111/imm.13320.

Ramirez, R., Gomez, F.A., Marshall, S.H., 2015. The infection process of *Piscirickettsia salmonis* in fish macrophages is dependent upon interaction with host-cell clathrin and actin. FEMS Microbiol. Lett. Oxf. 362. Available from: https://doi.org/10.1093/femsle/fnu012.

Reddick, L.E., Alto, N.M., 2014. Bacteria fighting back: how pathogens target and subvert the host innate immune system. Mol. Cell 54, 321−328.

Rizzuto, R., Bernardi, P., Pozzan, T., 2000. Mitochondria as all-round players of the calcium game. J. Physiol. 15, 37−47. Available from: https://doi.org/10.1111/j.1469-7793.2000.00037.x.

Roca, F.J., Whitworth, L.J., Redmond, S., Jones, A.A., Ramakrishnan, L., 2019. TNF induces pathogenic programmed macrophage necrosis in tuberculosis through a mitochondrial-lysosomal-endoplasmic reticulum circuit. Cell 5, 1344−1361. Available from: https://doi.org/10.1016/j.cell.2019.08.004.

Sahoo, B.R., 2020. Structure of fish toll-like receptors (TLR) and NOD-like receptors (NLR). Int. J. Biol. Macromol. 161, 1602.

Salvesen, G.S., Riedl, S.J., 2008. Caspase mechanisms, Programmed Cell Death in Cancer Progression and Therapy. Advances in Experimental Medicine and Biology, vol. 615. Springer, Dordrecht. Available from: https://doi.org/10.1007/978-1-4020-6554-5_2.

Scaffidi, P., Misteli, T., Bianchi, M., 2002. Release of chromatin protein HMGB1 by necrotic cells triggers inflammation. Nature 418, 191−195.

Simmen, T., Aslan, J.E., Blagoveshchenskaya, A.D., Thomas, L., Wan, L., Xiang, Y., et al., 2005. PACS-2 controls endoplasmic reticulum-mitochondria communication and Bid-mediated apoptosis. EMBO J. 24 (4), 717−729. Available from: https://doi.org/10.1038/sj.emboj.7600559. 23.

Srivastava, N., Shelly, A., Kumar, M., Pant, A., Das, B., Majumdar, T., et al., 2017. Aeromonas hydrophila utilizes TLR4 topology for synchronous activation of MyD88 and TRIF to orchestrate anti-inflammatory responses in zebrafish. Cell Death Discov. 3, 17067.

Takezaki, N., Figueroa, F., Zaleska-Rutczynska, Z., Takahata, N., Klein, 2004. The phylogenetic relationship of tetrapod, coelacanth, and lungfish revealed by the sequences of forty-four nuclear genes. Mol. Biol. Evol. 21, 1512−1524.

Volff, J.N., 2005. Genome evolution and biodiversity in teleost fish. Heredity 94, 280−294.

Wentzel, A., Jannsen, J., De Boer, V.C.J., Van Veen, W.G., Forlenza, M., Wiegertjes, 2020. Fish macrophages show distinct metabolic signatures upon polarization. Front. Immunol. 11. Available from: https://doi.org/10.3389/fimmu.2020.00152.

Yamaguchi, T., Dijkstra, J.M., 2019. Major histocompatibility complex (MHC) genes and disease resistance in fish. Cells 8, 378. Available from: https://doi.org/10.3390/cells8040378.

Youle, R.J., van der Bliek, 2012. Mitochondrial fission, fusion, and stress. Science 31, 1062−1065. Available from: https://doi.org/10.1126/science.1219855.

Zapata, A., Diez, B., Cejalvo, T., Gutierrez-de Frias, C., Cortes, A., 2006. Ontogeny of the immune system of fish. Fish Shellfish Immunol. 20, 126−136.

Zou, J., Redmond, A.K., Qi, Z., Dooley, H., Secombes, C.J., 2015. The CXC chemokine receptors of fish: insights into CXCR evolution in the vertebrates. Gen. Comp. Endocrinol. 215, 117−131.

CHAPTER 5

Essential proteins for the survival of bacteria in hostile environment

Shivendra Tenguria* and Sana Ismaeel*

Department of Pathology and Laboratory Medicine, Cedars-Sinai Medical Center (UCLA), Los Angeles, CA, United States

Introduction

Sudden changes in surrounding environmental conditions or niche may be favorable or hostile. Bacteria sense the surrounding environments with their sensor kinase of two-component system (detailed in Chapter 6) and respond with differential gene expression through response regulator, resulting in expression of specific bacterial proteins/factors which facilitate bacteria to adapt to changed and/or stressful conditions. These proteins form enzymes specialized to work at extreme conditions within or outside the respective host and facilitate bacterial adaptation to extreme environmental conditions.

All the living organisms are divided into three domains of life: Bacteria, Archaea, and Eukarya. Bacteria and archaea are prokaryotes. Bacteria have capability to adapt to a wide range of environmental conditions. However, archaea have exceptional capability for adaptation in adverse environments. The bacteria and other organisms that exist at extreme environmental conditions are known as extremophiles. These extremophiles range from low to high temperatures (*Psychromonas ingrahamii* at $-12°C$ to *Geogemma barossii* at $121°C$, an hyperthermophile) (Kashefi & Lovley, 2003; Auman et al., 2006; Maciejewska et al., 2019), extreme pH (*Plectonema nostocorum* at below pH 0 and *Hydrogenophaga* sp. at pH 13), extreme pressure of more than 100 MPa (*Shewanella benthica*), extreme saline conditions (*Haloferax volcani*), extreme radiations of UV more than $100 \, \mathrm{J \, m^{-2}}$, and gamma radiatioin of more than 12 kGy outside the host organisms (*Deinococcus radiodurans* and *Halobacterium* sp. NRC1) (Auman et al., 2006). Bacterial pathogens also face challenging environmental conditions (low pH and immune response) within human host.

Hostile environment outside human host: extreme temperature

Temperature is one of the major factors that regulate efficiency of biochemical reactions at extreme temperatures in the environment. Reaction kinetics of these biochemical

* These authors contributed equally.

Bacterial Survival in the Hostile Environment
DOI: https://doi.org/10.1016/B978-0-323-91806-0.00008-4

reactions are greatly reduced at low temperatures and can be approximated by Arrhenius law (Carvalho-Silva et al., 2019). Proteins including enzymes facilitate these biochemical reactions necessary for bacterial growth. Therefore, proteins and enzymes must have adapted to carry out these biochemical reactions at low temperatures, and efficiency of psychrophilic enzymes must be high to their counterparts in mesophiles to carry out these biological processes. Bacteria have been isolated from different temperatures. *Psychromonas ingrahamii* (gas vacuole bacterium), a psychrophile isolated from Alaska, United States, in arctic region can divide and grow at $10°C-15°C$ and is able to survive at $-12°C$. Proteins of psychrophilic bacteria undergo molecular changes under low temperatures. Coil regions of the secondary structures of psychrophilic proteins are overrepresented with serine, threonine, aspartic acid, and alanine amino acids, which provide higher protein flexibility to these proteins. Amino acids with aliphatic, aromatic, basic, and hydrophilic side chains (glutamic acid and leucine) are underrepresented in helical regions of these proteins (Metpally & Reddy, 2009).

Geogemma barossii, a single-celled microbe of archaea domain, can reproduce at $121°C$ (also called strain 121). The biostatic temperature for this organism is $130°C$, which means it can remain viable at $130°C$ but growth is halted. Although it can survive under standard autoclave conditions ($121°C$) used to sterile surgical or medical tools in laboratory practice, this strain is not infectious to humans because of its inability to grow at temperatures around $37°C$ and does not pose any threat as pathogen to humans.

In early of 1960 it was believed that microbes do not grow and survive at temperatures surpassing $60°C$, but discovery of *Thermus aquaticus* by Thomas D. Brock and Hudson Freeze in 1969 from hot springs of Yellowstone national park (United States) (temperature reaching around $70°C$) sparked the light for microbial existence at extremely higher temperatures and caught the attention of biochemists for the source of thermostable enzymes. In 1976 DNA polymerase was isolated from *T. aquaticus* (called as Taq polymerase) and became the cornerstone of invention of polymerase chain reaction (PCR) by Kary Mullis in 1983 (Analytical Methods Committee Amctb, 2013). The optimum growing temperature for *T. aquaticus* ranges is $65°C-72°C$, and it can survive at temperature up to $50°C-80°C$. This remarkable thermostability of DNA polymerase provides adaptation potential to grow at these extremely high temperatures. Taq polymerase is stable in successive heating cycle of PCR reaching up to $95°C$. Today, the commercial use of Taq polymerase emerged as foundation of countless findings and innovations in molecular biology and medicine because of its use in DNA amplification and sequencing methods leading to emergence of multimillion dollar biotechnological industries (Engelke et al., 1990; Terpe, 2013).

Hostile environment within the human host

Bacteria are able to evade the immune defenses presented by the host through differential protein expression. There are three levels of defense mechanism placed by the host which bacteria has to overcome—physical barrier, innate immunity barrier, and adaptive immunity barrier. Here, in this chapter, we have used *Helicobacter pylori* and *Salmonella* sp. as models to explain evasion of hostile host environment.

Bacteria like *H. pylori* and *Salmonella* sp. invade the host through oral route by contaminated food or water. The first barrier which these two organisms must breach is the acid barrier of the stomach to establish niche in the host. *H. pylori* overcomes this barrier and creates the niche in the stomach itself, whereas the surviving bacteria of *Salmonella* sp. travel further and reach intestines. Here, they are taken up by the non-phagocytic intestinal epithelium cells by upregulating the expression of Type-3-secretion system (T3SS) encoded by the Salmonella pathogenicity island-1 (SPI-1) genes, and niche is created inside the acidic vacuole of host cells by the action of Salmonella pathogenicity island-2 (SPI-2) genes. We will explain in detail both the mechanisms next.

I. *Helicobacter pylori*

Urease activity for pH homeostasis as a survival strategy of H. pylori in highly acidic hostile environment

H. pylori presents a good example of acidophilic pathogen within the human stomach with pH of 1.5–2.0. The extreme acidic environment within the stomach creates hostile acidic environment for bacterial survival. However, some pathogens have evolved to subvert challenges posed by environmental conditions within the host and host-immune system. *H. pylori* uses its urease protein (enzyme) to neutralize and survive in the acidic environments.

Earlier, before discovery of *H. pylori*, there was a belief in scientific community that due to high acidic conditions, human stomach was sterile. After the discovery, researchers investigated the mechanisms by which *H. pylori* survives in highly acidic environment.

The role of urease in nitrogen metabolism by producing ammonia is well defined, but excess ammonia produces toxic effect on gastric epithelium, and *H. pylori* produces urease up to 10% of total proteins (Kusters et al., 2006). Then why does *H. pylori* produce this much urease? And how ammonia, a product of urea, is used by bacteria?

On the basis of several investigations, researchers concluded the role of urease enzyme in pH homeostasis inside the highly acidic gastric environment. This could be understood by the proposed hypothesis as shown in Fig. 5.1, suggesting the use of extracytoplasmic and cytoplasmic urease enzyme to maintain pH homeostasis (Stingl et al., 2002; Stingl & De Reuse, 2005). Extracytoplasmic urease is used to neutralize acidic microenvironment outside the bacterial cells by producing ammonia clouds around the bacterial cells. Cytoplasmic urease buffers protons by two ways: first, the ammonium produced by urease is transported to periplasmic space which buffers protons there, thus prevents acidification of the periplasm. Second, the ammonia produced by urease activity in the cytoplasm buffers protons leaking in through the

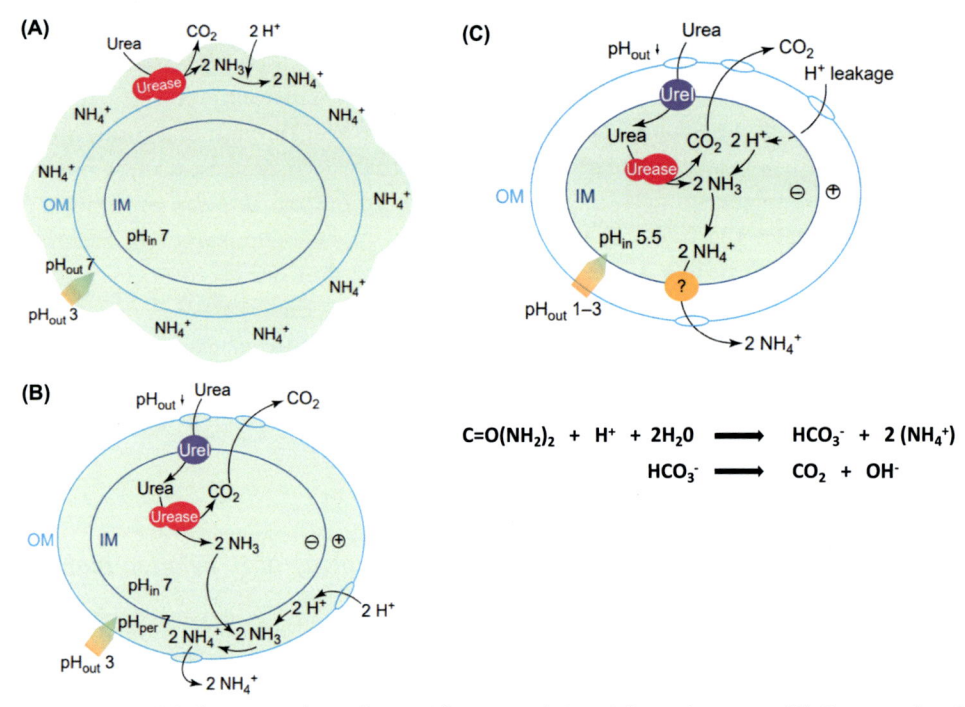

Figure 5.1 Models for urease-dependent pH homeostasis in acidic environment. (A) Extracytoplasmic urease activity buffers extracellular pH; (B) cytoplasmic urease activity buffers periplasmic H + ; (C) cytoplasmic urease activity buffers cytoplasmic H + . *Adapted with permission from Stingl, K., Altendorf, K., Bakker, E.P., 2002. Acid survival of* Helicobacter pylori: *how does urease activity trigger cytoplasmic pH homeostasis? Trends Microbiol. 10, 70–74.* © *2001 Elsevier Science Ltd.*

cytoplasmic membrane, and charged ammonium ions (NH_4) are extruded from cytoplasm probably through putative ammonium (NH_4) transporter. Thus urease activity maintains pH homeostasis inside and outside the bacterial cells (Stingl et al., 2002; Stingl & De Reuse, 2005).

Evasion of innate immune response with bacterial proteins (*H. pylori*)

Host-immune system creates challenging hostile environment for bacterial pathogens. Innate immune response is the first line of host defense. It employs various types of cells: monocytes, neutrophils, NK cells, and dendritic cells (DCs) to recognize and eliminate the bacteria from the infection sites. *H. pylori* uses several strategies to evade innate immune system of the human host. Unlike other Gram-negative bacteria, *H. pylori* avoids phagocytosis at submucosal sites. It is reported that a particular type of professional phagocytes engulfs *H. pylori*. Delayed phagocytosis by macrophages has been reported resulting in the accumulation of *H. pylori* in "megasomes" from phagosome fusion. In these phagocytes, *H. pylori* induces apoptosis thus enabling to escape the bacteria (Allen et al., 2000; Chaturvedi et al., 2004).

Bacteria use their enzymes, catalase and superoxide dismutase activity, to prevent the toxic effects of reactive oxygen species (ROS) generated by activated phagocytes. *H. pylori* prevents the toxic effect of ROS produced by activated neutrophils and monocytes by employing its catalase and superoxide dismutase activity (Ramarao et al., 2000; Seyler et al., 2001). *H. pylori* employs other important strategies for survival by producing arginase enzyme. This bacterial arginase utilizes L-arginine (at physiological concentration), a common substrate for arginase and iNOS of macrophages, and thus limits the L-arginine to be utilized by macrophages. The macrophages utilize L-arginine to produce nitric oxide (NO) which is an antimicrobial agent for killing bacterial pathogens (Gobert et al., 2001). Thus arginase of *H. pylori* inhibits production of NO by macrophages and helps evasion of *H. pylori* through macrophages, an essential component of innate immunity.

Toll-like receptors (TLRs), expressed on macrophage surface, are used to recognize the Gram-negative bacteria by macrophages but *H. pylori* has PAMPs that have evolved to subvert recognition by TLRs. LPS of *H. pylori*, with only four acyl chains instead of six, is 1000 times less active to be recognized by TLR4 (Moran et al., 1997; Moran, 1999). Also, flagellin of *H. pylori* is not recognized by TLR5 due to difference in amino acid sequence at TLR5 recognition site (Andersen-Nissen et al., 2005). DNA of *H. pylori* is not recognized by TLR9 due to high methylation. It is well known that TLR9 recognizes unmethylated CpGp islands of bacterial and viral DNA (Blaser & Atherton, 2004).

In stress conditions, when rod-shaped *H. pylori* is converted to coccoid form, the bacteria use other approaches from host detection. Rod-shaped *H. pylori* is sensed by interaction of bacterial peptidoglycan and intracellular NOD1 receptor in the epithelial cells. *H. pylori* escape its recognition from innate immune system in its coccoid form because peptidoglycan from coccoid form of *H. pylori* is not sensed by NOD1 receptors and is also unable to activate NFκB in gastric epithelial cells (Viala et al., 2004; Chaput et al., 2006).

DCs in the gastrointestinal mucosa are unable to recognize LPS of *H. pylori*. However, in response to CagA positive *H. pylori*, DCs produce IL-12 to induce Th1-specific response (Guiney et al., 2003). Simultaneously, DCs produce IL-10 in moderate amount inducing Th2 specific response and activate T-regulatory cells (T_{reg}) (Kranzer et al., 2004). This, in turn, suppresses T-cell memory response to *H. pylori*. Thus DCs act as bridge between innate and adaptive immune responses.

Evasion of adaptive immune response (*H. pylori*)

Induction of both local and systemic antibiotic responses by *H. pylori* has been reported through raised titers of IgA, IgM, and IgG antibodies (Rathbone et al., 1986; Crabtree et al., 1991). But elimination of the bacteria in response to these antibodies is ineffective (Thomas et al., 1993; Ermak et al., 1998). This may be due to phase variability in surface proteins of the bacteria and inaccessibility of the bacteria to the antibodies, since bacteria reside in the gastric mucosal protective layer.

Persistent colonization with *H. pylori* causes chronic active gastritis which increases accumulation of $CD4^+$ T_H cells, responsible for Th1 response by secreting proinflammatory cytokines, IL-12, IL-8, and TNF-α (Tummala et al., 2004). This response causes progression of chronic active gastritis to atrophic gastritis and gastric cancer (Houghton et al., 2002), subverting the adaptive immune response of the host. *H. pylori* overcomes the adaptive responses in many ways mainly by inducing apoptosis of T-cells or by suppressing the cellular responses. *H. pylori* has been reported to induce T-cells' apoptosis as one of the ways of immune evasion (Wang et al., 2001). Gamma-glutamyl transpeptidase has also reported to show immunosuppressive effect by inhibiting T-cell proliferation (Schmees et al., 2007). cagPAI-positive *H. pylori* has also shown DCs mediated increase in T_{reg} cell population, which, in turn, suppress the T_H17 and T_H1 cells and memory T-cell response to *H. pylori* infection (Lundgren et al., 2003).

II. *Salmonella* typhimurium

Acid shock proteins for pH homeostasis as a survival strategy of S. typhimurium

Salmonella enterica serovar *typhimurium* is an enteropathogenic neutrophilic bacteria thriving at neutral pH. It is an extremely efficient pathogen causing a variety of diseases including typhoid fever, gastric enteritidis, and iNTS. Its lifestyle from the outside host environment to the host's stomach, intestines and finally to cellular phagolysosomes, requires drastic alterations in its physiology to survive. Similar to *H. pylori*, it has a variety of adaptive mechanisms through which it is able to resist and survive in the acidic pH of the host. However, unlike *H. pylori* which uses urease for its survival, *Salmonella* sp. uses acid shock proteins (ASPs) for surviving in the acidic stomach environment. Growth phase, prior exposure to acidic environment, and availability of extracellular factors like glutamine play an important role in *S. typhimurium* acid stress response. Depending on these conditions, two types of low-pH induced acid tolerance response (ATR) is characterized in *S. typhimurium*: log-phase ATR and stationary-phase ATR. There is also a third type which is activated in stationary phase without any pH stress. These ATRs once induced, protect the bacteria from pH as low as 3 for an extended period of time.

The log-phase ATR is the most commonly studied and is induced by rapid pH changes in exponentially growing bacteria. When external environment pH drops between 4.5 and 5.8, several amino acid decarboxylases become activated and lead to internal pH maintenance.

More than 60 ASPs are reported to be produced, and these acid-adapted bacteria can tolerate a pH as low as 3 up to 1000 times better than unadapted bacteria. Three major regulatory genes take part in this pathway—*rpoS* which encodes an alternate sigma factor

(σ^s), *fur* which encodes for iron uptake regulator Fur, and *phoP/phoQ* which encodes for PhoP/Q, the two-component sensor regulatory system. Increased transcription of these genes upregulates the production of ASPs and thus lead to acid stress tolerance.

The stationary-phase ATR, as the name suggests, is induced in the stationary phase of bacteria. It is very different from the log-phase ATR in terms of regulatory proteins and ASPs involved. Essentially, there are two independently regulated subsets of stationary-phase response. When the bacteria enters stationary phase, a generalized stress response is initiated which is independent of pH but is regulated by the alternate sigma factor (σ^s). As stationary phase progresses, the level of σ^s increases, which activates a variety of genes called σ^s regulon. These genes then help the bacteria survive stationary-phase stress like low pH and high osmolarity. These bacteria (which are not acid-stressed) are 1000-times more resistant to acid than log-phase *S. typhimurium*. Additionally, when the pH is also lowered, the stationary-phase ATR gets activated, which is regulated by OmpR/EnvZ, a two-component response regulator and independent of σ^s. Mutations in *Fur* or *PhoP* also have no effect on the stationary-phase ATR. These bacteria can now withstand pH 3 conditions for an extended time. Acid-induced activation of OmpR requires its Env-Z-dependent phosphorylation.

OmpR in its activated phosphorylated state induces the expression of acid-dependent stationary-phase AMPs. Over 48 stationary-phase ASPs have been identified including OmpR which in itself is an ASP.

Apart from these regulatory proteins, repair mechanisms such as DNA repair mediated by Ada (encoded by *ada* gene) also take part in extreme pH tolerance. Low pH causes DNA methylation and depurination leading to DNA damage. *Ada* encodes for DNA methyl transferase which repairs damages like O^6-methyl guanine, methyl phosphotriesters, and O^4-methyl thymine, important for survival under pH stress.

Evasion of innate immune response with bacterial proteins (*S. typhimurium*)

The bacteria which are able to evade the acidic stress of the stomach, now have the task of reaching the intestines and cause their uptake by the host cells to create an intracellular niche. It uses T3SS-1 regulated by SPI-1 proteins to infect the host epithelial cells or cause its phagocytosis by monocytes or DCs. Once inside the cells, *S. typhimurium* resides inside an acidic vacuole called Salmonella-containing vacuole (SCV), where it proliferates using SPI-2 proteins and cause cell death. Cell death helps in systemic dissemination of the bacteria to internal organs like spleen, lymph nodes, and liver. During both these intracellular and transient extracellular phases, *S. typhimurium* is challenged with the innate immune components.

Salmonella poses several PAMPs for the identification by the innate immune system. This includes outer membrane lipid polysaccharide LPS (recognized by TLR-4) and protein components flagellin (recognized by TLR-5 and cytosolic inflammasome NLRC4).

Once Salmonella reaches intracellular vacuoles, it downregulates flagellin expression to prevent NLRC4 inflammasome activation and facilitates its systemic dissemination. *AcnB* controls level of citrate through the tricarboxylic acid cycle inside the bacteria and helps evade the NLRP3 inflammasome activation. T3SS effectors such as AvrA (of T3SS1) and SseL (of T3SS2) possess an antiinflammatory activity. They inhibit IkBa degradation by its deubiquitination. PipA, GtgA, and GogA are proteases that cleave RelA and RelB, members of the Nf-kB family, thus inhibiting Nf-kB signaling.

Another member of the Salmonella sp., *S. typhi*, which causes typhoid fever in humans possesses an outer capsular polysaccharide called Vi (for Virulence factor). Vi can inhibit complement deposition, thus inhibiting phagocytosis and killing. It also possesses an antiinflammatory activity and inhibits inflammatory cytokine secretion from epithelial and T-cells via interaction with membrane prohibitin.

PgtE, a surface protein of *S. typhimurium* cleave complement components C3b, C4b, C5, B, and H, thereby inhibiting complement activation.

Evasion of adaptive immune response (*S. typhimurium*)

Adaptive immunity starts by phagocytosis of bacteria and presentation by antigen-presenting cells to T-cells. The most effective mechanism of *Salmonella* to survive and replicate inside the cell is by inhibiting the host lysosomal degradation machinery in the SCV. It injects SPI-2 effectors from the SCV inside the host cytoplasm through T3SS2. T3SS2 mutants cannot survive inside DCs. For example, SopD2 binds and inhibits host GTPase Rab7, thereby inhibiting endosomal trafficking and phagosome——lysosome fusion. Another effector, SifA inhibit Rab9, central to retrograde trafficking and prevent delivery of hydrolytic enzymes to SCV. Several SPI-1 effectors such as SopE and SopB play a role in inhibiting lysosomal fusion. Thus *Salmonella* avoids antigen processing and presentation by inhibiting lysosomal degradation.

Downregulation of flagellin, major protein component of *Salmonella* against which adaptive immune response is generated, is another strategy. This limits availability of antigen for T-cell presentation and response. *Salmonella* also causes selective culling of activated CD4 T-cells in a SPI-2-dependent manner. It increases cellular expression of PD-L1 on activated CD4 cells, thus increasing their apoptosis during clonal expansion.

References

Allen, L.A., Schlesinger, L.S., Kang, B., 2000. Virulent strains of *Helicobacter pylori* demonstrate delayed phagocytosis and stimulate homotypic phagosome fusion in macrophages. J. Exp. Med. 191, 115—128.

Analytical Methods Committee Amctb, N., 2013. PCR — the polymerase chain reaction. Anal. Methods 6, 333—336.

Andersen-Nissen, E., Smith, K.D., Strobe, K.L., Barrett, S.L., Cookson, B.T., Logan, S.M., et al., 2005. Evasion of Toll-like receptor 5 by flagellated bacteria. Proc. Natl. Acad. Sci. USA 102, 9247—9252.

Auman, A.J., Breezee, J.L., Gosink, J.J., Kampfer, P., Staley, J.T., 2006. *Psychromonas ingrahamii* sp. nov., a novel gas vacuolate, psychrophilic bacterium isolated from Arctic polar sea ice. Int. J. Syst. Evol. Microbiol. 56, 1001—1007.

Blaser, M.J., Atherton, J.C., 2004. *Helicobacter pylori* persistence: biology and disease. J. Clin. Invest. 113, 321—333.

Carvalho-Silva, V.H., Coutinho, N.D., Aquilanti, V., 2019. Temperature dependence of rate processes beyond arrhenius and eyring: activation and transitivity. Front. Chem. 7, 380.

Chaput, C., Ecobichon, C., Cayet, N., Girardin, S.E., Werts, C., Guadagnini, S., et al., 2006. Role of AmiA in the morphological transition of *Helicobacter pylori* and in immune escape. PLoS Pathog. 2, e97.

Chaturvedi, R., Cheng, Y., Asim, M., Bussiere, F.I., Xu, H., Gobert, A.P., et al., 2004. Induction of polyamine oxidase 1 by *Helicobacter pylori* causes macrophage apoptosis by hydrogen peroxide release and mitochondrial membrane depolarization. J. Biol. Chem. 279, 40161—40173.

Crabtree, J.E., Taylor, J.D., Wyatt, J.I., Heatley, R.V., Shallcross, T.M., Tompkins, D.S., et al., 1991. Mucosal IgA recognition of *Helicobacter pylori* 120 kDa protein, peptic ulceration, and gastric pathology. Lancet 338, 332—335.

Engelke, D.R., Krikos, A., Bruck, M.E., Ginsburg, D., 1990. Purification of *Thermus aquaticus* DNA polymerase expressed in *Escherichia coli*. Anal. Biochem. 191, 396—400.

Ermak, T.H., Giannasca, P.J., Nichols, R., Myers, G.A., Nedrud, J., Weltzin, R., et al., 1998. Immunization of mice with urease vaccine affords protection against *Helicobacter pylori* infection in the absence of antibodies and is mediated by MHC class II-restricted responses. J. Exp. Med. 188, 2277—2288.

Gobert, A.P., McGee, D.J., Akhtar, M., Mendz, G.L., Newton, J.C., Cheng, Y., et al., 2001. *Helicobacter pylori* arginase inhibits nitric oxide production by eukaryotic cells: a strategy for bacterial survival. Proc. Natl. Acad. Sci. USA 98, 13844—13849.

Guiney, D.G., Hasegawa, P., Cole, S.P., 2003. *Helicobacter pylori* preferentially induces interleukin 12 (IL-12) rather than IL-6 or IL-10 in human dendritic cells. Infect. Immun. 71, 4163—4166.

Houghton, J., Fox, J.G., Wang, T.C., 2002. Gastric cancer: laboratory bench to clinic. J. Gastroenterol. Hepatol. 17, 495—502.

Kashefi, K., Lovley, D.R., 2003. Extending the upper temperature limit for life. Science 301, 934.

Kranzer, K., Eckhardt, A., Aigner, M., Knoll, G., Deml, L., Speth, C., et al., 2004. Induction of maturation and cytokine release of human dendritic cells by *Helicobacter pylori*. Infect. Immun. 72, 4416—4423.

Kusters, J.G., van Vliet, A.H., Kuipers, E.J., 2006. Pathogenesis of *Helicobacter pylori* infection. Clin. Microbiol. Rev. 19, 449—490.

Lundgren, A., Suri-Payer, E., Enarsson, K., Svennerholm, A.M., Lundin, B.S., 2003. *Helicobacter pylori*-specific CD4+ CD25 high regulatory T cells suppress memory T-cell responses to H. pylori in infected individuals. Infect. Immun. 71, 1755—1762.

Maciejewska, N., Walkusz, R., Olszewski, M., Szymanska, A., 2019. New nuclease from extremely psychrophilic microorganism *Psychromonas ingrahamii* 37: identification and characterization. Mol. Biotechnol. 61, 122—133.

Metpally, R.P., Reddy, B.V., 2009. Comparative proteome analysis of psychrophilic vs mesophilic bacterial species: insights into the molecular basis of cold adaptation of proteins. BMC Genomics 10, 11.

Moran, A.P., 1999. *Helicobacter pylori* lipopolysaccharide-mediated gastric and extragastric pathology. J. Physiol. Pharmacol. 50, 787—805.

Moran, A.P., Lindner, B., Walsh, E.J., 1997. Structural characterization of the lipid A component of *Helicobacter pylori* rough- and smooth-form lipopolysaccharides. J. Bacteriol. 179, 6453—6463.

Ramarao, N., Gray-Owen, S.D., Meyer, T.F., 2000. *Helicobacter pylori* induces but survives the extracellular release of oxygen radicals from professional phagocytes using its catalase activity. Mol. Microbiol. 38, 103—113.

Rathbone, B.J., Wyatt, J.I., Worsley, B.W., Shires, S.E., Trejdosiewicz, L.K., Heatley, R.V., et al., 1986. Systemic and local antibody responses to gastric *Campylobacter pyloridis* in non-ulcer dyspepsia. Gut 27, 642–647.

Schmees, C., Prinz, C., Treptau, T., Rad, R., Hengst, L., Voland, P., et al., 2007. Inhibition of T-cell proliferation by *Helicobacter pylori* gamma-glutamyl transpeptidase. Gastroenterology 132, 1820–1833.

Seyler Jr., R.W., Olson, J.W., Maier, R.J., 2001. Superoxide dismutase-deficient mutants of *Helicobacter pylori* are hypersensitive to oxidative stress and defective in host colonization. Infect. Immun. 69, 4034–4040.

Stingl, K., De Reuse, H., 2005. Staying alive overdosed: how does *Helicobacter pylori* control urease activity? Int. J. Med. Microbiol. 295, 307–315.

Stingl, K., Altendorf, K., Bakker, E.P., 2002. Acid survival of *Helicobacter pylori*: how does urease activity trigger cytoplasmic pH homeostasis? Trends Microbiol. 10, 70–74.

Terpe, K., 2013. Overview of thermostable DNA polymerases for classical PCR applications: from molecular and biochemical fundamentals to commercial systems. Appl. Microbiol. Biotechnol. 97, 10243–10254.

Thomas, J.E., Austin, S., Dale, A., McClean, P., Harding, M., Coward, W.A., et al., 1993. Protection by human milk IgA against *Helicobacter pylori* infection in infancy. Lancet 342, 121.

Tummala, S., Keates, S., Kelly, C.P., 2004. Update on the immunologic basis of *Helicobacter pylori* gastritis. Curr. Opin. Gastroenterol. 20, 592–597.

Viala, J., Chaput, C., Boneca, I.G., et al., 2004. Nod1 responds to peptidoglycan delivered by the *Helicobacter pylori* cag pathogenicity island. Nat. Immunol. 5, 1166–1174.

Wang, J., Brooks, E.G., Bamford, K.B., Denning, T.L., Pappo, J., Ernst, P.B., 2001. Negative selection of T cells by *Helicobacter pylori* as a model for bacterial strain selection by immune evasion. J. Immunol. 167, 926–934.

CHAPTER 6

Kinases and phosphatases in bacterial survival in hostile environment

Faizan Ahmed[1] and Shivendra Tenguria[2]
[1]Department of Medical Sciences, Cedars-Sinai Medical Center (UCLA), Los Angeles, CA, United States
[2]Department of Pathology and Laboratory Medicine, Cedars-Sinai Medical Center (UCLA), Los Angeles, CA, United States

Introduction

Bacteria are present in all spheres of life, ranging from milder to extreme environmental conditions. Extreme temperatures pose challenge to bacteria for their growth and survival. To survive, bacteria need to perceive and adapt quickly to ever changing environmental conditions. Their adaptation potential depends on their ability to sense and transduce environmental signals and respond to changing stress stimuli by expressing genes differentially to encode specific proteins needed for the adaptations in the hostile environment. Bacteria use protein kinases and phosphatases to sense and transduce various environmental signals to their genetic machinery which alters the gene expression according to environmental conditions. These kinases and phosphatases are the essential part of signal transduction not only in bacteria but in all three domains of life: bacteria, archaea, and eukarya (Kennelly, 2002, 2003). These protein kinases phosphorylate proteins (a posttranslational modification) mainly at conserved amino acid residues: Ser, Thr, Tyr, His, and Arg with reversible noncovalent phosphorylation. Protein phosphorylation constitutes more than half of the 30 types of posttranslational modification events for signal transduction (Engin, 2021). Protein phosphorylation regulates the activity of target proteins, either directly by inducing conformational changes in proteins or indirectly by involving protein−protein interaction in intermediate partner proteins.

In bacteria, protein kinases are classified in five major categories: (1) His kinases; (2) Arg kinases; (3) Tyr kinases; (4) Hanks-type Ser/Thr kinases (also known as eukaryotic-like Ser/Thr kinases); and (5) atypical Ser kinases (Mijakovic et al., 2016). Recently, a new family of protein kinases, Ser/Thr/Tyr kinases have been identified which have a unique adenosine triphosphate (ATP) binding fold, not present in other protein kinases (Nguyen et al., 2017). All the four types of protein kinases are present in bacteria but atypical Ser kinases are restricted to only some species. His kinases are involved in regulation of gene expression in response to changing environmental conditions while atypical Ser kinases control bacterial metabolism. Tyr and Hanks-type Ser/Thr kinases regulate the processes associated with bacterial physiology (Mijakovic et al., 2016). His kinases and phosphatases play very crucial role in bacteria for sensing and responding to hostile environmental conditions.

Both His kinases and phosphatases work together in bacteria to sense and respond to ever changing environmental conditions and forms two components of this sensory and responding machinery. Together, these His kinases and cognate phosphatases are known as two-component system (TCS). TCS is a widely distributed phenomenon in bacteria (both Gram-positive and Gram-negative) to respond to various kinds of stress stimuli, where bacteria modulate and change expression of different genes in response to these stress stimuli (van Hoek et al., 2019).

In this chapter, we discuss in detail the involvement of protein kinases and phosphatases in stressful abiotic and biotic conditions. This will help us understand the mechanisms involved in how bacteria have evolved distinct sensory system by involving their protein kinases and phosphatases to sense stressed environmental conditions, and how this system uses two different components for sensing stress signals from surrounding environments and transmits these signals to differentially express genes so as to adapt to surrounding hostile conditions.

Kinases and phosphatases in abiotic conditions

Several environmental stress factors activate the TCS in bacteria depending on the conditions involved. TCS consists of two components: (1) transmembrane histidine kinase (HK) (sensor kinase) and (2) cytosolic response regulator. The transmembrane sensor kinase (HK) gets activated upon sensing the stress stimuli from hostile environment and autophosphorylates at conserved histidine residue. Then the phosphoryl group is transferred to the aspartate residue of cytosolic response regulator to regulate bacterial gene expression (Fig. 6.1). Bacteria use phosphatases to dephosphorylate the response regulator once bacteria pass the activation signal to the corresponding gene to reset the TCS in default state. Moreover, Landry et al. showed using the TCS of *Bacillus subtilis* and *Escherichia coli*, phosphatase activity tunes the detection threshold of the TCS (Landry et al., 2018). Protein kinase of *Staphylococcus aureus* is a good example of it. *S. aureus* is an opportunistic human pathogen which can colonize in human intestine, skin, or other perineal regions of the host and causes acute and chronic diseases. *S. aureus* has developed a well-sophisticated system to sense surrounding changing environmental conditions. It has developed 16 pairs of TCS for sensing different hostile environmental conditions (Bleul et al., 2021). It has four transmembrane-sensing HKs (BraS, GraS, VraS, and SaeS), one HK as quorum sensor (AgrB) that enables *S. aureus* to sense and monitor cell density, two HKs (PhoR and WalK) containing Per-ARNT-Sim (PAS) domain for ligand-interacting site, and two HKs (AirS and NreB) containing iron—sulfur (Fe—S) protein for sensing various environmental signals.

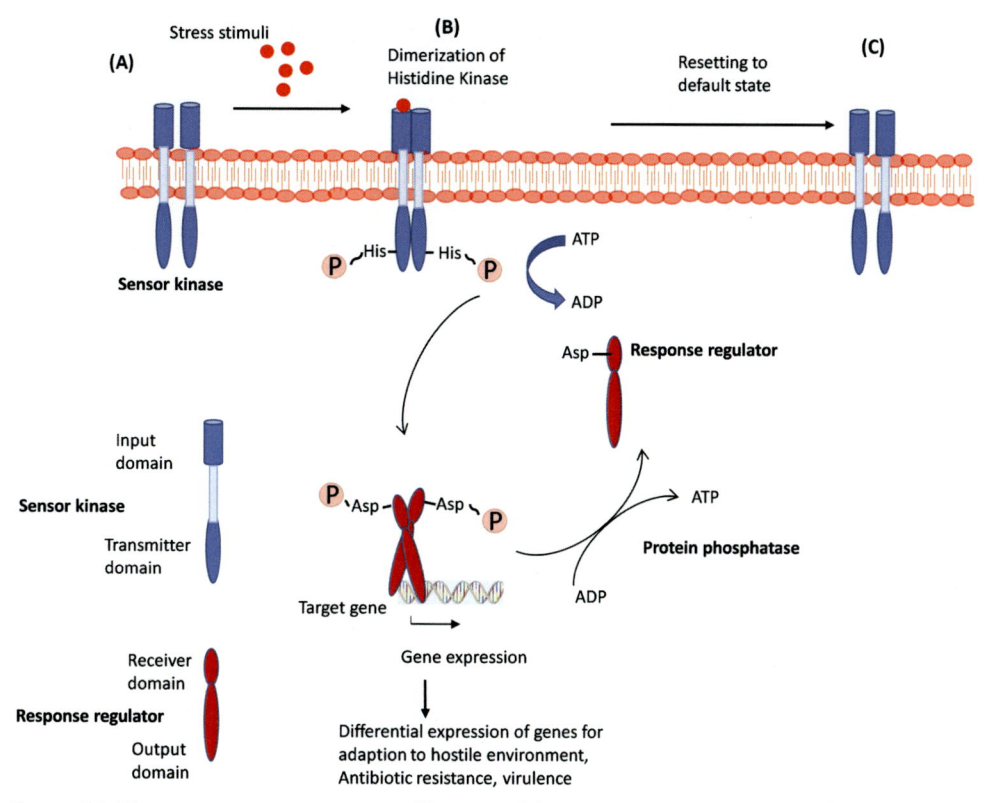

Figure 6.1 The two-component system of bacteria. (A) Bacteria sense stress stimuli from the hostile environment through sensor kinase, the histidine kinase and (B) histidine kinase autophosphory-lates on conserved His residue in the transmitter domain and transmits phosphoryl group to Asp residue of receiver domain within response regulator. Response regulator undergoes to conformational changes and output signals to express the corresponding gene(s) to adapt to hostile environment. (C) Response regulator after transmitting output signals gets dephosphorylated with bacterial protein phosphatase to reset the two-component system in default state.

Reversible and continuous phosphorylation is required by bacteria to respond to the altering environmental factors. Serine/threonine kinases (STKs) and serine/threonine phosphatases (STPs) are regulating the process of sensing the changes in these environmental factors (Pereira et al., 2011). The characterized number of STPs from STKs is much lower owing to less interest in researchers and presence of lower number of STPs than STKs in bacterial cells. There is substantially low number of STPs identified in the Gram-positive bacteria while in case of Gram-negative bacteria, the reports of these enzymes are identically significantly lower (Sajid et al., 2015). These enzymes have not been significantly explained since they were not considered necessary for dephosphorylation of phosphorylated threonine (Thr-P) and phosphorylated serine (Ser-P) regulatory proteins. His-P is more stable than the Ser-P and Thr-P residues, which are needed for quenching signaling cascades by phosphatases involving STKs

(Shi, 2009). Nutrients are essential for *Myxococcus xanthus* to grow in unicellular form growing into extensive swarms. But when the nutrients are depleted, these cells form multicellular bodies and can lead to the formation of myxospores until the cells get nutrients available again. MrpC is a transcription factor that is controlled by TCS in these bacteria. When the cells generate starvation signals, a His kinase MrpA activates and leads to phosphorylation of the regulator MrpB which finally leads to aggregation signals owing to the MrpC regulation (Inouye and Nariya, 2008). Similarly, during nutrient limiting conditions, desiccation- and heat-resistant spores are formed by *B. subtilis*. Specific amino acids and other nutrient sources stimulate and germinate the dormant spores. The transmembrane Ser−Thr kinase is important for the induction of this germination by growing cells due to the peptidoglycan release by these cells. The PASTA repeats bind to the peptidoglycan fragments in PrkC and activate kinases, leading to germination. This PrkC-dependent germination pathway is phosphorylated by utilizing the elongation factor EF-G (Shah et al., 2008).

B. subtilis responds to metabolic (limitations of glucose, phosphate, and oxygen) or environmental (heat, salt, ethanol, pH, and temperature) stress through two Sigma B-activating transduction pathways. Dephosphorylation of RsbV is the regulatory output of both of these pathways. This RsbV protein is essential for the environmental and metabolic stress pathways, where loss of this RsbV protein can halt the induction of formation of the SigB−RsbW complex (Hecker and Völker, 2001). The interactions of SigB with RsbV and RsbW is necessary during environmental stress conditions while there is a different pathway for the SigB during low temperature conditions (Brigulla et al., 2003).

WalK HK controls the expression of genes involved in cell wall metabolism and is responsible for biofilm formation, virulence, and autolysis. This HK is responsible for evolving vancomycin sensitive strain of *S. aureus* into vancomycin-intermediate resistant strain when the bacteria face hostile environment containing vancomycin. Deletion of extracellular PAS domain in WalK of *B. subtilis* results into constitutive signaling and suggests that WalK is inhibited when it binds to its ligand and activated upon removal of stress stimuli (ligand) (Fukushima et al., 2008). *S. aureus* uses these TCSs to sense and respond to various hostile conditions including growing in presence of antibiotics within or outside of its host which leads to development of antibiotic resistance and poses challenge to treat *S. aureus* infections using antibiotic drugs. GraRS and SaeRS are cationic antimicrobial peptides (CAMPs) sensing TCSs which enables *S. aureus* to sense and survive in presence of CAMPs and vancomycin (Li et al., 2007; Yang et al., 2013). The ArlRS (autolysis-related locus) HK in *S. aureus*, upon activation, leads to expression of global response regulators; MgrA and SpxS. SpxS response regulator controls the *S. aureus* response to β-lactam antibiotics and help bacteria to evolve into methicillin-resistant *S. aureus* (MRSA). It has been shown that mutation in arlRS gene makes these MRSA multiple fold sensitive to oxacillin (β-lactam) antibiotics. On the other hand, MgrA controls the virulence of *S. aureus* by regulating the expression of over hundreds of effector genes (Bai et al., 2019). Recently, it has been reported that ArlS is necessary for the activation of ArlR (response regulator) when the bacteria face manganese starvation due to calprotectin and glucose scarcity (Parraga Solorzano et al., 2021).

Kinases and phosphatases in biotic conditions

Upon infection, bacterial pathogens grow and multiply in their hosts. Conditions withing the host create biotic stressful environment through anaerobic, oxidative stress and immune response. Host immune system creates very challenging and hostile conditions for invading pathogenic bacteria by employing its immune cells called phagocytes (macrophages, neutrophils, dendritic cells, monocytes, and osteoclasts). Phagocytes are responsible for fighting against infectious microorganisms including bacteria through a process called as phagocytosis. These phagocytes internalize and engulf the bacteria into phagosomes or phagosomal vacuoles through endocytosis (a process in which host cell membrane invaginate to internalize outside object within the vesicles inside the host cell), and merges to lysosomes to form phagolysosomes. This leads to maturation of phagolysosomes by converting prelytic enzymes into lytic enzymes and lyse the engulfed bacteria. Maturation of phagolysosomes leads to presentation of bacterial antigens to lymphocytes and activate adaptive immune response. However, some of the bacteria have evolved their genetic machinery to encode functional enzymes to neutralize host defense response initiated by host immune cells. Bacterial **Protein Kinase G (PknG)** is an appropriate example of such evading tactic used by bacteria to evade host immune system. PknG has been shown to play a key role in *Mycobacterium tuberculosis* (Mtb) and other similar bacteria to survive and fight against hostile conditions including biotic (metabolic adaptation, lysosomal escape, inhibition of phagocytosis, promoting antibiotic resistance, and intracellular survival) and abiotic (hypoxic, oxidative, and acidic stress) hostile conditions. Inhibition of PknG reduces the survival of drug resistant and drug sensitive Mtb. Moreover, in adjunct therapy by deleting and inhibiting PknG has been reported to reduce drug tolerance withing the murine macrophages (Khan and Nandicoori, 2021; Scherr et al., 2007). Apart from this, PknG helps Mtb in lysosomal escape by preventing maturation of phagolysosomes (Fig. 6.2).

Polo-like kinases (PLKs) type of kinases play an important role in cell cycle in mammals, where PLK2 and PLK3 are two central close knit acidophilic kinases. The major enzyme involved in pharmaceutical and physiological aspects is PLK1, while further investigations on PLK2 and PLK3 have led to their importance in cellular functionality (Cozza and Salvi, 2018).

In the process of host defense, phagocytes (macrophages, neutrophils, dendritic cells, monocytes, and osteoclasts) are responsible for fighting against infectious microorganisms including bacteria by a process called as phagocytosis. Once phagocytes engulf the wild type bacteria and form phagosomal vacuole, bacteria secrete PknG

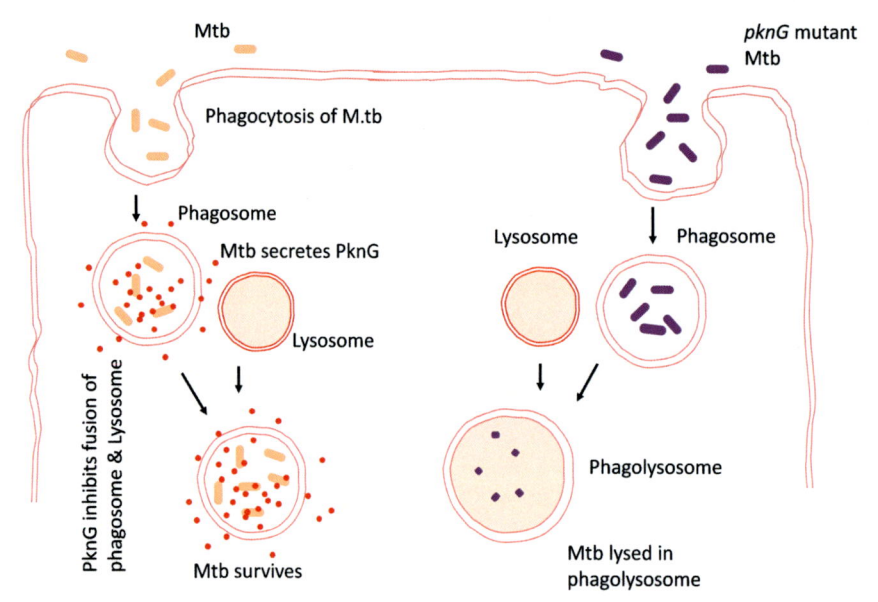

Figure 6.2 Macrophage phagocytose Mtb. Mtb secretes functional PknG that inhibits fusion of phagosome and lysosome leading to survival of the bacteria in the phagosome and immune evasion (left side). In PknG mutant Mtb, PknG could not inhibit fusion of phagosome and lysosome and forms phagolysosome leading to lysis of Mtb (right side) (Nguyen and Pieters, 2005). *Adapted with permission from Nguyen, L., Pieters, J., 2005. The Trojan horse: survival tactics of pathogenic mycobacteria in macrophages. Trends Cell Biol. 15 (5), 269–276. © 2005 Elsevier Ltd.*

virulence factor in the macrophages to prevent the formation of phagolysosomes and evading bacteria are released inside the phagocytes, thus promotes survival of nonreplicating latent Mtb (Fig. 6.2) (Warner and Mizrahi, 2007). Recently, it has been reported, PknG acts as an ubiquitinating enzyme and inhibits host innate response by inhibiting activation of NF-kB signaling.

Two-component system in host immune evasion

Upon infection with *S. aureus*, phagocytes at infection site and in the blood remove the bacterium by opsonin-based phagocytosis and present its antigens to lymphocytes leading to clearing of the infection. *S. aureus* has developed AirRS TCS to face anaerobic conditions in the tissue and the blood within the host and make this bacterium to survive. Under anaerobic conditions, sensor kinase AirR is activated leading to AirS response regulator mediated expression and secretion of virulence factors which inhibits opsonin-mediated phagocytosis of genes (Hall et al., 2015).

References

Bai, J., Zhu, X., Zhao, K., Yan, Y., Xu, T., Wang, J., et al., 2019. The role of ArlRS in regulating oxa-cillin susceptibility in methicillin-resistant *Staphylococcus aureus* indicates it is a potential target for anti-microbial resistance breakers. Emerg. Microbes Infect. 8, 503–515.

Bleul, L., Francois, P., Wolz, C., 2021. Two-component systems of *S. aureus*: signaling and sensing mechanisms. Genes. (Basel) 13.

Brigulla, M., Hoffmann, T., Krisp, A., Völker, A., Bremer, E., Völker, U., 2003. Chill induction of the SigB-dependent general stress response in *Bacillus subtilis* and its contribution to low-temperature adaptation. J. Bacteriol. 185, 15.

Cozza, G., Salvi, M., 2018. The acidophilic kinases PLK2 and PLK3: structure, substrate targeting and inhibition. Curr. Protein Pept. Sci. 19 (8), 728–745.

Engin, E.D., 2021. Bacterial protein kinases. Adv. Exp. Med. Biol. 1275, 323–338.

Fukushima, T., Szurmant, H., Kim, E.J., Perego, M., Hoch, J.A., 2008. A sensor histidine kinase co-ordinates cell wall architecture with cell division in *Bacillus subtilis*. Mol. Microbiol. 69, 621–632.

Hall, J.W., Yang, J., Guo, H., Ji, Y., 2015. The AirSR two-component system contributes to *Staphylococcus aureus* survival in human blood and transcriptionally regulates sspABC operon. Front. Microbiol. 6, 682.

Hecker, M., Völker, U., 2001. General stress response of *Bacillus subtilis* and other bacteria. Adv. Microb. Physiol. 44, 35–91.

Inouye, S., Nariya, H., 2008. Dual regulation with Ser/Thr kinase cascade and a His/Asp TCS in *Myxococcus xanthus*. Adv. Exp. Med. Biol. 631, 111–121.

Kennelly, P.J., 2002. Protein kinases and protein phosphatases in prokaryotes: a genomic perspective. Fems Microbiol. Lett. 206, 1–8.

Kennelly, P.J., 2003. Archaeal protein kinases and protein phosphatases: insights from genomics and bio-chemistry. Biochem. J. 370, 373–389.

Khan, M.Z., Nandicoori, V.K., 2021. Deletion of pknG abates reactivation of latent *Mycobacterium tubercu-losis* in mice. Antimicrob. Agents Chemother. 65.

Landry, B.P., Palanki, R., Dyulgyarov, N., Hartsough, L.A., Tabor, J.J., 2018. Phosphatase activity tunes two-component system sensor detection threshold. Nat. Commun. 9, 1433.

Li, M., Cha, D.J., Lai, Y., Villaruz, A.E., Sturdevant, D.E., Otto, M., 2007. The antimicrobial peptide-sensing system aps of *Staphylococcus aureus*. Mol. Microbiol. 66, 1136–1147.

Mijakovic, I., Grangeasse, C., Turgay, K., 2016. Exploring the diversity of protein modifications: special bacterial phosphorylation systems. FEMS Microbiol. Rev. 40, 398–417.

Nguyen, H.A., El Khoury, T., Guiral, S., Laaberki, M.H., Candusso, M.P., Galisson, F., et al., 2017. Expanding the Kinome world: a new protein kinase family widely conserved in bacteria. J. Mol. Biol. 429, 3056–3074.

Nguyen, L., Pieters, J., 2005. The Trojan horse: survival tactics of pathogenic mycobacteria in macro-phages. Trends Cell Biol. 15 (5), 269–276.

Parraga Solorzano, P.K., Shupe, A.C., Kehl-Fie, T.E., 2021. The sensor histidine kinase ArlS is necessary for *Staphylococcus aureus* to activate ArlR in response to nutrient availability. J. Bacteriol. 203, e0042221.

Pereira, S.F., Goss, L., Dworkin, J., 2011. Eukaryote-like serine/threonine kinases and phosphatases in bacteria. Microbiol. Mol. Biol. Rev. 75, 192–212V.

Sajid, A., Arora, G., Singhal, A., Kalia, V.C., Singh, Y., 2015. Protein phosphatases of pathogenic bacte-ria: role in physiology and virulence. Annu. Rev. Microbiol. 69, 527–547.

Scherr, N., Honnappa, S., Kunz, G., Mueller, P., Jayachandran, R., Winkler, F., et al., 2007. Structural basis for the specific inhibition of protein kinase G, a virulence factor of *Mycobacterium tuberculosis*. Proc. Natl. Acad. Sci. USA 104, 12151–12156.

Shah, I.M., Laaberki, M.H., Popham, D.L., Dworkin, J., 2008. A eukaryotic-like Ser/Thr kinase signals bacteria to exit dormancy in response to peptidoglycan fragments. Cell 135, 486–496.

Shi, Y., 2009. Serine/threonine phosphatases: mechanism through structure. Cell 139, 468–484.

van Hoek, M.L., Hoang, K.V., Gunn, J.S., 2019. Two-component systems in *Francisella* species. Front. Cell Infect. Microbiol. 9, 198.

Warner, D.F., Mizrahi, V., 2007. The survival kit of *Mycobacterium tuberculosis*. Nat. Med. 13, 282–284.

Yang, S.J., Xiong, Y.Q., Yeaman, M.R., Bayles, K.W., Abdelhady, W., Bayer, A.S., 2013. Role of the LytSR two-component regulatory system in adaptation to cationic antimicrobial peptides in *Staphylococcus aureus*. Antimicrob. Agents Chemother. 57, 3875–3882.

CHAPTER 7

Antimicrobial resistance—a serious global threat

Keerthi Rayasam[1], Palkar Omkar Prakash[1], Rajani Chowdary Akkina[1] and Vidyullatha Peddireddy[2]

[1]Department of Microbiology and Food Science and Technology, GITAM Institute of Science, GITAM (Deemed to be University), Visakhapatnam, Andhra Pradesh, India
[2]Department of Nutrition Biology, School of Interdisciplinary and Applied Sciences, Central University of Haryana, Mahendragarh, Haryana, India

Introduction

Cells which can be observed microscopically are known as microorganisms. They may be single celled (unicellular) or multicelled (multicellular). Life initially originated and derived from microbes. They succeeded to adapt and survive in hostile conditions forced by the environment like low nutrients (oligotrophy), pressure, temperature, pH, and radiation. Immense diversity within the same category of microorganisms has been studied by metagenomics (genome sequenced collectively for microbial sample directly acquired from environment). Despite vast diversity, an American plant ecologist, Robert Harding Whittaker in 1969 simply categorized all organisms into five kingdoms: Protista, Fungi, Monera, Plantae, and Animalia (Verma, 2016). Most of the microorganisms in nature are beneficial to life, which include bacteria and yeast involved in fermentation, microorganisms involved in vitamins and enzymes' production, antibiotics, and probiotics. Nevertheless, some are pathogenic and cause serious life-threatening diseases and infections to human, plants, and domestic animals. Once microorganisms invade into the human body either by the means of ruptured skin, mouth, nose or during open surgery, they can reach various parts of body causing diseases. During preantibiotic age, millions of people died due to bacterial diseases. Antibiotics or antibacterial agents like streptomycin, penicillin, and chloramphenicol were discovered to cure these bacterial diseases. The antibiotic golden age started with discovery of modern-day penicillin by Alexander Fleming (Bbosa et al., 2014). Antimicrobial agents are the chemicals or the drugs that are used to treat or prevent diseases due to microbial infections. Antimicrobial agents eradicate life-threatening infections by disrupting various processes of microorganisms like protein synthesis or cell wall formation or nucleic acid synthesis (Liwa and Jaka, 2015). However, microorganisms acquire resistance to antimicrobial agents due to various environmental or

Bacterial Survival in the Hostile Environment
DOI: https://doi.org/10.1016/B978-0-323-91806-0.00016-3

patient-related factors and no longer remain susceptible to antimicrobial agents (Prestinaci et al., 2015). This ability of microorganism to resist activity of antimicrobial agents is called antimicrobial resistance (AMR). The inception of antibiotic resistance in pathogenic bacteria made prevention of diseases burdensome. Under these circumstances, an in-depth knowledge of antibiotics classification, their mode of action, and mechanism of resistant bacteria against antibiotics is required for further action (Begum et al., 2021).

Mechanism of action of probiotics, challenges faced, and their evolution

Food and Agriculture Organization (FAO) and World Health Organization (WHO) define probiotics as live microorganisms that when consumed in adequate amount deliver health benefit to the consumer (Hotel and Cordoba, 2001). Dominantly, *Lactobacillus* and *Bifidobacterium* species are grouped as probiotics which are also part of human gut microbiota. These microorganisms have been used in many dietary supplements and functional foods (Gourbeyre et al., 2011). Probiotics have various mechanisms by which humans are benefitted, including increased adhesion of probiotics to intestinal mucosa, inhibition of pathogen attachment, enhancement of epithelial barrier, synthesis of antimicrobial substances, modulation of immune system, and exclusion of pathogens (Gourbeyre et al., 2011).

* Enhancement of the epithelial barrier:
    * The human intestinal epithelium is in constant touch with intestinal fluids and inconsistent microbiota. The integrity of epithelium is maintained by the intestinal barrier, a major defense mechanism which includes secretion of antimicrobial peptides, mucous, epithelial junction adhesion complex and secretory IgA (Gourbeyre et al., 2011). Numerous reports suggest that enhancement of gene expression associated with tight junction signaling can strengthen the integrity of intestinal barrier. Probiotics like *Lactobacillus* regulate several genes encoded for E-cadherin and β-catenin, some of the adherence junction proteins in trials done on T84 cells. Likewise, *Lactobacillus* influences phosphorylation of adherence junction proteins and production of protein kinase C (PKC) isoforms like PKCγ in abundance which modulates intestinal barrier positively (Hummel et al., 2012).
* Increased adhesion to intestinal mucosa:
    * Attachment on the surface of intestinal mucosa is a prerequisite step for colonization or biofilm formation. Probiotics lactic acid bacteria exhibit several surface reactive components that interact with mucus and intestinal epithelial cells (IECs). IECs are known for secretion of mucin, a complex glycoprotein that inhibits pathogens from attaching the intestinal cell surface, but probiotics alter

the intestinal mucin and get attached to the surface. Probiotics also induce epithelial cells to release defensins (Castagliuolo et al., 2005).

- Competitive expulsion of pathogenic microorganisms:
  - When one species of bacteria vigorously competes with other species of bacteria resulting in expulsion of other species is known as "competitive expulsion." Bacteria can create hostile microenvironment, production of antimicrobial components, competitive utilization of nutrients, utilize available bacterial receptors, and eliminating adherence of pathogen (Blankenship, 2012). *Bifidobacterium* and *Lactobacillus* can inhibit pathogens like *Salmonella*, *Helicobacter pylori*, *Escherichia coli*, *Rotavirus*, and *Listeria monocytogenes* (Chenoll et al., 2011).
- Production of antimicrobial substances:
  - Probiotics produce organic acids like lactic acid and acetic acid that act as strong inhibitory compounds against Gram-negative bacteria by entering the bacterial cell and dissociating itself in the cytoplasm leading to accumulation of ionized form of organic acid and subsequent death of the pathogen (Alakomi et al., 2000).

Antimicrobial agents

The battle between human and pathogenic microbes has been long lasting. Humans have studied the mechanism of virulence of pathogenic microbes. To combat these microbes, antimicrobial agents were discovered. Antimicrobial agents are any substances in the form of drugs or chemicals which are able to either kill or slow down the growth of microbes, that is microbiocidal or microbiostatic, respectively. They are classified on the basis of target organism against which they act. They can be antibacterial, antiprotozoan, antifungal, antiviral agents, or antibiotics like penicillin, penicillin G, penicillin V, procaine, and benzathine (Liwa and Jaka, 2015).

Antibacterial agent can be defined as an agent capable of disinfecting the surfaces by elimination of harmful bacteria. Antibacterials act only on bacteria. They involve chemical disinfectants, antiseptic drugs, and skin care products (Yao and Moellering, 2011). Antiprotozoan agents are also called antiparasitic agents. They are drugs capable of acting against human parasites like roundworms, protozoa, flatworms, and ectoparasites like fleas, mites, lice, and ticks. They are included in the treatment of parasitic diseases like malaria. Antifungal agents or antimycotic agents are medications with fungicidal or fungistatic activity. They are involved in the drugs for yeast infection like ringworm, candidiasis, athlete's foot, and cryptococcal meningitis. Antiviral drugs are a class of medication used for treating viral infections and diseases such as herpes, HIV, and corona virus. Unlike antibiotics, antiviral drugs do not aim to destroy target pathogen; however, they hinder its development (Safrin, 2021). Antibiotics are produced by microorganisms naturally to halt the growth or eliminate other microorganisms, mainly bacteria. These can be used as drugs consumed orally or by injection to treat bacterial infections (Walsh, 2003).

Antiseptics and disinfectants

Antiseptics are the antimicrobial substances applied on or in living tissue to prevent sepsis (hazardous condition in which the body responses to an infection and damages own tissues). They are used in hospitals as they reduce the possibility of infections during surgery. Disinfectants are the compounds that are applied on inanimate or inert surfaces to destroy and inactivate harmful microorganisms. They are not effective against bacterial spores. Sterilization refers to any process that can kill or remove or inactivate microbial life completely. It is effective against spores (Patterson, 1932). The term biocide describes any substance that is capable of destroying living things. They include fungicides, pesticides, and herbicides. Antiseptics and disinfectants are biocides that kill or hinder the growth of germs. The only basic difference is antiseptics are applied to living tissue and disinfectants are used on inorganic matter (Chapman, 2003).

Glutaraldehyde associates binding with unprotonated amines of bacterial outer cell wall strongly, especially those of *Staphylococcus aureus* and *E. coli*. Formaldehyde has sporicidal effect due to its ability to penetrate into bacterial spores and its interaction with deoxyribonucleic acid (DNA) (Fraenkel-Conrat, 1961), ribonucleic acid (RNA), and proteins (Fraenkel-Conrat et al., 1945; Fraenkel-Conrat and Olcott, 1946). The ability of alcohol to denature proteins leads to inhibition of bacterial sporulation and spore germination. Antifungal and antiviral effects of phenols are due to their ability to damage plasma membrane leading to the leakage of intracellular constituents (Russell and Furr, 1996). Hydrogen peroxide is been used as disinfectant and antiseptic due to its ability to produce hydroxyl-free radicals that react with DNA, proteins, and lipids which are essential constituents of the cell (Block, 2001). Quaternary ammonium compounds (QACs) or cationic agents can be used as antiseptics and disinfectants that induce damaging effect on structural integrity of bacterial cytoplasmic membrane (Denyer, 1995).

Antibiotics/antibacterial agents

The word antibiotics is a Greek word, with meaning of "anti" being "in opposition of" and of "biotikos" being "memoir." According to S.A. Waksman, definition of antibiotic is "any drug produced by microbes with antimicrobial activity" (Waksman, 1956). As mentioned earlier, antibiotics are natural substances produced by microorganisms to dominate by killing or hindering the growth of surrounding microorganisms. The crucial sources of antibacterial agents include *Penicillium*, *Actinomyces*, and *Streptomyces*. The antibiotic golden age started with the discovery of modern-day penicillin by Alexander Fleming in 1928. It is the era where entire bacterial infections were treated with antibiotics. Penicillin is the foremost β-lactam antibiotic discovered, followed by sulfonamide, aminoglycoside streptomycin plus streptothricin. These antibiotic drugs are worthwhile to combat bacterial infections like erythema, peptic ulcer, pneumonia, meningitis, and urethritis (Bbosa et al., 2014).

Classification of antibiotics

1. Based on chemical structures:
   a. Sulfonamides: Sulfadiazine, sulfamethoxazole
   b. Penicillin: Ampicillin, penicillin G (Benzylpenicillin), penicillin V, amoxicillin, nafcillin
   c. Cephalosporin: Cefazolin, cefamandole, ceftriaxone, cefuroxime, cefotaxime
   d. Aminoglycosides: Gentamycin, streptomycin, neomycin, kanamycin, tobramycin
   e. Chloramphenicol: Chloramphenicol, thiamphenicol
   f. Tetracyclines: Minocycline HCl, chlortetracycline, doxycycline, oxytetracycline
   g. Macrolides: Azithromycin, roxithromycin, erythromycin, spiramycin
   h. Polyenes: Nystatin, amphotericin B
   i. Lincomycins: Lincomycin, clindamycin
   j. Polymyxins: Polymyxin B, polymyxin E
2. Based on the sources:
   a. Antibiotic from microbes: Penicillin from *Penicillium notatum*
   b. Antibiotics from lichens: Usnic acid from *Usnea* spp.
   c. Antibiotics from higher plants: Allicin from *Allium sativum*
3. Based on mechanism of action:
   a. Disruption of cell membrane function: Polymyxin (polymyxin B)
   b. Inhibition of cell wall synthesis (bactericidal effect): Glycopeptides—teicoplanin, vancomycin *and* β-lactams—penicillins, cephalosporins
   c. Inhibition of protein synthesis:
      Inhibit ribosome on the 30S subunit—streptomycin, tetracycline, gentamycin, kanamycin
      Inhibit ribosome on the 50S subunit—azithromycin, chloramphenicol, erythromycin
   d. Inhibits the transcription process of microbes—rifampin, fluoroquinolones
   e. Inhibits specific metabolic reaction: sulfonamides, inhibits folic acid biosynthesis in bacteria.
4. Based on spectrum of action:
   a. Broad spectrum: Effective to Gram + ve, Gram −ve bacteria, mycoplasmas, chlamydia, rickettsia, sometimes protozoa—chloramphenicol, tetracycline
   b. Narrow spectrum: Effective to Gram + ve/Gram −ve bacteria only—penicillin, cephalosporin, erythromycins, polymyxin

Action of antimicrobial agents

1. Antimicrobial agents are successful in killing or inhibiting growth of pathogenic microorganisms by targeting the essential cellular process required for their growth.

Following are the mechanisms of how antibiotics work to inhibit bacterial growth—cell wall synthesis

2. Protein synthesis
3. RNA synthesis
4. DNA synthesis
5. Intermediary metabolism

Intruding with cell wall synthesis

Cell wall is a semipermeable layer protective structure that provides strength and avert cell lysis to endure high cytoplasmic osmotic pressure. Cell wall is composed of peptidoglycan layer with N-acetylmuramic acid (NAM) and N-acetylglucosamine (NAG) residues alongside an attached peptide bridge with β-(1,4)-glycosidic linkage. Despite of strength, cell wall has been a targeted area for antimicrobial agents. Due to selective cell wall components, which are lacking in human have been selected as potential targets by antimicrobial agents (Liwa and Jaka, 2015). During the synthesis of cell wall, transglycosylase enzymes create the glycosidic bonds linearly between NAM and NAG residues. Linear linked glycosidic bonds between NAM and NAG establish an immature peptidoglycan layer. Transpeptidase enzymes cleaves the 2-D alanines of peptide bridge creating the mature peptidoglycan layer (Liwa and Jaka, 2015).

β-Lactam antibiotics like penicillin and cephalosporin impede peptidation reaction obstructing formation of mature peptidoglycan layer. Linearly linked immature peptidoglycan layer is weak and is vulnerable to cell lysis and die. β-Lactam, especially penicillin, forms penicilloyl-enzyme intermediate, hindering the transpeptidase activity. Due to this interaction, transpeptidase is named as penicillin-binding protein (PBP). Vancomycin hinders cell wall cross-linking by coupling with acyl-D-ala-D-ala fragment of the cell wall (Waxman and Strominger, 1983).

Hindrance of bacterial protein synthesis

Protein synthesis of bacterial cell is the process which takes place on 70S ribosomes with 50S and 30S as subunits. Translation is the step to produce protein by decoding mRNA formed in transcription. It occurs in cytoplasm where ribosomes are located. Antimicrobial agents like macrolides, aminoglycosides, tetracyclines, ketolides, streptogramins, chloramphenicol, and oxazolidinones target ribosomal subunits to prevent protein synthesis. Peptidyl transferase is enzyme involved in formation of peptide bond between two amino acids. Macrolides are antimicrobial agents with lactone ring that cohere with 50S subunit of ribosomes and disable peptidyl transferase activity (Beringer and Rodnina, 2007).

Translocation is the step-in elongation where two tRNA molecules along with mRNA move on ribosome during translation. Aminoglycosides are antimicrobial agents that kill bacteria by targeting 30S ribosomal subunit leading to misread of genetic code and terminating translocation. Aminoacyl-tRNA binds to acceptor site of the mRNA—ribosome complex for proper functioning and replication of RNA. Tetracyclines target 30S ribosomal subunit, inhibiting protein synthesis. Chloramphenicol ties up with ribosomal 50S subunit and impedes attachment of aminoacyl-transfer RNA with peptidyl transferase enzyme, hence ward off peptide formation (Liwa and Jaka, 2015).

Termination of nucleic acid synthesis

1. Retardation of DNA topoisomerase 2

   Topoisomerase 2 supervise topology of DNA by unraveling and intertwining both strands of DNA which is crucial during the process of DNA synthesis and mRNA transcription. DNA gyrase is topoisomerase 2 enzyme that convert circular and closed DNA into linear form using ATP, thus facilitating replication and transcription (Reece and Maxwell, 1991). Quinolones like gemifloxacin, ciprofloxacin, moxifloxacin, norfloxacin, levofloxacin, oxolinic acid, and nalidixic acid are dynamic antimicrobial agents that block DNA gyrase from unwinding and rewinding of DNA strands by interacting with subunit A (GyrA) (Fàbrega et al., 2009).

2. Retardation of microbial RNA synthesis

   Rifamycins are the antibiotics synthesized naturally by *Amycolatopsis rifamycinica*. They bind to β-subunit (encoded by rpoB) of RNA polymerase—DNA complex with great affinity to terminate synthesis of RNA strand (Floss and Yu, 2005).

Inhibition of microbial metabolic pathways

Unlike mammals, bacteria synthesize folic acid for final DNA, RNA, and cell wall formation. Trimethoprim and sulfonamides have greater affinity to competitively block pteridine synthetase enzyme, which synthesizes tetrahydrofolic acid from pteridine and p-aminobenzoic acid. Tetrahydrofolic acid carries one carbon fragment for terminal synthesis of bacterial cell wall, RNA, and DNA (Liwa and Jaka, 2015).

Disruption and increased permeability of cytoplasmic membrane

Cytoplasmic membrane is a lipid matrix along with globular proteins scattered all over the lipid bilayer. It is semipermeable and selectively transports nutrients and ions across the membrane. Despite cytoplasmic membrane being a selectively permeable barrier, antimicrobial agents (cationic/anionic/neutral) permeate into the cell through plasma membrane. Polymyxin B and cationic antibacterial agent facilitate and enhance permeability of lysozyme, thus suppress the metabolic processes (Liwa and Jaka, 2015).

Polymixin B acts by: (1) impairing lipid composition and membrane design; (2) destruction of ionic gradient of the outer membrane; and (3) depolarization of the outer membrane. Polymyxin B, by acting on cell membrane, shoots up the permeability of the cytoplasmic membrane to "self-promoted" uptake. Daptomycin aggregation results in alteration of the curvature in the membrane leading to the formation of holes leading to leaking of ions (Liwa and Jaka, 2015).

Antimicrobial resistance

Antibiotic resistance is the resistance of microorganisms to antimicrobial agents or antibiotic drugs. Microorganisms on exposure to stress by antibiotics either get killed or become resistant to antibiotics. Bacteria which are susceptible get killed. However, those acquired resistance naturally/intrinsically (resistant bacteria) survive (Harbottle et al., 2006). Intrinsic resistance is the ability of bacterial species to resist antibiotic action against them. It is a natural phenomenon exhibited by all members of species due to bacterial structural and functional makeup. Many antibiotic resistomes intrinsic resistance to β-lactams, fluoroquinolones, and aminoglycosides antibiotics were identified. These resistant bacteria exhibit one of the following mechanism of resistance (Blair et al., 2015).

Antibiotics resistance in bacteria can be acquired because of chromosomal modifications (regulatory or structural) in genes or through foreign resistant genes in the form of either plasmids or transposons from intrinsically resistant bacteria via horizontal gene transfer (HGT) (Lerminiaux and Cameron, 2019). Bacteria when exposed to low dosage of minimum bacterial concentration, that is dose needed to kill all the bacteria in the given population, might mutate and become resistance to antibiotics through natural mutation or by acquisition of resistant gene. Besides overuse of antibiotics and low dosage, there are many other factors contributing to antibiotic resistance such as (1) environmental factors: increased national and international travel; (2) drug-related factors: fake drugs; and (3) patient-related factors: self-medication and misconceptions (Duong and Jaelin, 2015).

Resistant bacteria never react to antibiotics and the disease or infection that it causes can never be treated or cured. This is because it hinders or terminates efficacy of antibiotic activity. The inception of antibiotic resistance in pathogenic bacteria made prevention of diseases burdensome. In these circumstances, an in-depth knowledge of antibiotics classification, their mode of action, and mechanism of resistant bacteria against antibiotics are required for further conclusions.

Causes of antimicrobial resistance

1. Inappropriate use and prescribing of antibiotics:

    The inadvisable prescription and use of antibiotics has become a foremost issue in the rise of AMR among bacterial pathogens. The subinhibitory concentration of antibiotics will boost bacterial resistance against it (Luyt et al., 2014). The transmission of resistance gene cassettes among strains or species or genera by HGT facilitates increase in AMR. Absence of strict laws and regulations in some countries, over the counter supply of improper medical prescriptions is leading to misuse of antimicrobials (Ayukekbong et al., 2017).

2. Inadequacy of supervision on susceptibility of antimicrobials:

   Surveillance on susceptibility testing of antimicrobials need to be carried out periodically for better understanding of local epidemiology of AMR and to plan out an appropriate response in case of any critical situation. Nevertheless, the insufficient human manpower, cost and infrastructure, and unawareness about susceptibility testing in rural areas make it hard to control the supervision (Moo et al., 2020).

3. Large scale use of antimicrobials in nonhuman subjects:

   Consumption of food from livestock medicated with antimicrobials or contact with medicated animals assists transmission of resistant microbes to the humans. Nearly about 90% of antibiotics prescribed for livestock are excreted through urine and stool which affects the microbial ecosystem increasing chances of evolution of antimicrobial resistant microbiota. This excreted material enters the ground-water ecosystem exposing the environmental microbes to a low dosage of antibiotics (Moo et al., 2020).

Mechanism of antimicrobial resistance

In bacterial cells, resistance toward antibiotics can be acquired through inherent structural integrities or by genetic mutations or through HGT (Blair et al., 2015). The innate difference in the plasma membrane composition (structural integrities) of diderm (Gram-negative bacteria) and monoderm (Gram-positive bacteria) makes antibiotic-daptomycin potent against monoderms but ineffective against diderms. The cell membrane in diderm cells comprises of less anionic phospholipids in contrast to monoderm bacteria which affects the Ca^{2+}-dependent insertion efficacy of daptomycin into the cytoplasmic membrane required for the antibacterial activity (Blair et al., 2015).

High throughput screening and high-density genome mutant libraries studies indicate role of several genes in innate resistance against classes of antibiotics like β-lactams, aminoglycosides, and fluoroquinolones. Some of the resistant genes include D-Ala-D-Ala carboxypeptidase, SapC, thioredoxin A (TrxA), FabI, thioredoxin reductase (trxB), and DacA. For generation of libraries, mutagenesis via targets' insertions or random transposon insertion were carried out in *Pseudomonas aeruginosa*, *S. aureus*, and *E. coli* (Blake and O'Neill, 2013). Creating library of such kinds will effectively aid in discovery of possible drug combinations to inhibit the inherent resistant mechanism (Blair et al., 2015). The three fundamental mechanisms of antibiotic resistance are:

1. Blockage of access to target
   a. Decreased permeability
   b. Increased efflux
2. Alteration and safeguarding the targets
3. Enzymatic degradation of antibiotics
   a. By hydrolysis
   b. Relocation of chemical group

Blockage of access to target

Decreased permeability

Antibiotics aim at DNA, RNA, and metabolic process of cytoplasm of bacteria by penetrating through the outer membrane and cell wall. Porins are the proteins that serve as pores/channels through which specific molecules enter the cell by passive diffusion. Resistant bacteria combat antibiotic activity by modulation of porins with better selective channels (Begum et al., 2021). Resistant *E. coli, Acinetobacter* spp., *Enterobacter* spp., and *Pseudomonas* spp. acquired genetic mutation that regulates expression of porin, defeating antibiotic carbapenem. The resistance of the family *Enterobacteriaceae* toward carbapenems due to its less permeable membrane is one of the examples that hinders the assess of antibiotics to the target. Porin proteins assist diffusion of hydrophilic antibiotics into the bacterial cell, OmpF, and OmpC (from *E. coli*) are examples that represent porins in the *Enterobacteriaceae* family. blaNDM-1 gene is the mutated gene of *Acinetobacter and Pseudomonas* responsible for decreased permeability of porins (Begum et al., 2021).

Increased efflux

Efflux pumps are transport proteins that permit bacteria to synchronize their internal environment by extrusion of toxic substances like antibiotics, metabolites, antimicrobial agents, and quorum sensing signal molecules. These proteins exist in prokaryotic monoderm and diderm bacteria along with eukaryotic organisms. Quorum sensing signal molecules are sensing system that checks bacterial internal population density. AMR in biofilm structures is due to N-acyl-homoserine lactones, a quorum sensing signaling molecule. The assess to the antibiotics target can be prevented by increasing efflux of the antibiotics. Five types of efflux pumps have been recognized until now which include the small multidrug resistance family, the major facilitator superfamily, the multidrug and toxic compound extrusion family, the resistance-nodulation-cell division family (RND) and ATP-binding cassette family (ABC) along with multidrug resistant (MDR) efflux pumps that can channelize array of structurally different antibiotics (Munita and Arias, 2016). Several MDR efflux pumps encoded genes are positioned on the bacterial chromosomes and some of these genes can get mobilized onto the plasmids (Blair et al., 2015). IncH1 plasmid, a plasmid isolated from *Citrobacter freundii* has found to encode gene cassette for RND pump along with genes encoded for New Delhi metallo-β-lactamase 1 (NDM1) which indicates transmission of specific resistance between bacterial cells by plasmids, facilitating spread of resistance among clinically important bacterial pathogens (Dolejska et al., 2013). *S. aureus, P. aeruginosa*, and *Enterobacteriaceae* express RND efflux pumps on greater extent, which are controlled by regulators such as QacR from *S. aureus*, EmrR from *E. coli* while MtrA boosts mtrCDE transcription in *Neisseria gonorrhoeae* (Blair et al., 2015). MdeA, FuaABC, KexD, LmrS, AcrB, and MexB are efflux pumps in *Streptococcus mutans, Stenotrophomonas maltophilia, Klebsiella pneumoniae, S. aureus, E. coli*, and *P. aeruginosa*, respectively (Blair et al., 2015).

Alteration and safeguarding the targets

Antibiotics attach to the bacterial targets with great affinity, halting target's activity but changes in the target structure can prevent antibiotic binding. Spontaneous mutation in chromosomal gene of bacterial target site, to which antibiotics bind results in resistance. Mutation in the genes encoded for antibiotic target can lead to resistance toward the antibiotic. Resistance toward rifampin develops through single point mutation resulting in substitution of amino acid in the rpoB gene that decreases rifampin's affinity for the target and transcription continues (Blair et al., 2015; Munita and Arias, 2016).

Aminoglucosides target 30S ribosomal subunit to inhibit protein synthesis. Bacterial enzymes acetyltransferases, nucleotidyltransferases, and phosphotransferases modify hydroxyl or amino groups of aminoglycoside or modify ribosome through methylation (Luthra et al., 2018). Macrolides and lincosamides target 50S ribosomal subunit to halt protein synthesis. Bacteria resist macrolides and lincosamides by methylation or mutation of target site. Erm (A), erythromycin resistance methylase gene responsible for methylation of 16S rRNA, modifies drug-binding spot in ribosome, reducing the macrolides binding affinity. By these mechanisms of resistance, *Mycobacterium abscessus*, tuberculosis, and leprosy causing bacterium, acquired resistance to aminoglycosides, macrolides, and lincosamides (Leclercq, 2002). Quinolones target GyrA part of DNA gyrase, which is responsible for DNA strands. Metamorphosis of GyrA brought alterations in target site. Metamorphosis in DNA gyrase resulted in resistance to quinolones and metamorphosis in RNA polymerase resulted in resistance to rifamycins (Blair et al., 2015).

Resistance in *S. pneumoniae* and *S. aureus* toward antibiotic linezolid has been conferred due to multiple copies of 23S rRNA, which have a high recombination rate, producing resistant bacterial population. Additionally, resistance can be conferred by possessing a gene that is homologous to the target (original) of the antibiotic.

Methicillin-resistant S. aureus (MRSA) is conferred with resistance due to the presence of staphylococcal cassette chromosome mec, a genetic element that contains mecA gene responsible for the expression of penicillin-binding protein (PBP2a), a β-lactam insensitive protein which is a mutated PBP that continues cell wall synthesis even when original PBP is targeted by the antibiotics (Lerminiaux and Cameron, 2019).

Erythromycin ribosome methylase (erm) family is responsible for 16S rRNA methylation which results in modification in the antibiotic attachment site preventing binding of macrolides, streptogramin, and lincosamides (Woodford and Ellington, 2007). Another instance is of the chloramphenicol–florfenicol resistance gene (cfr) that is responsible for methylation of A2503 site (23S rRNA), causing the target to develop resistance toward array of antibiotics like oxazolidinones, pleuromutilins, streptogramins, lincosamides, and phenicols. Both erm and cfr genes can be located on plasmids which aid in spreading AMR among bacterial community (Zhang et al., 2014). Two genes armA and rmt have been located from clinical isolates of family *Enterobacteriaceae*, which encode methyltransferase. Additionally, qnr gene, a resistance gene for quinolone has also been present on plasmids which encodes for pentapeptide repeat proteins that bind to topoisomerase IV and DNA gyrase protecting them from quinolone activity (Lerminiaux and Cameron, 2019) as shown in Fig. 7.1.

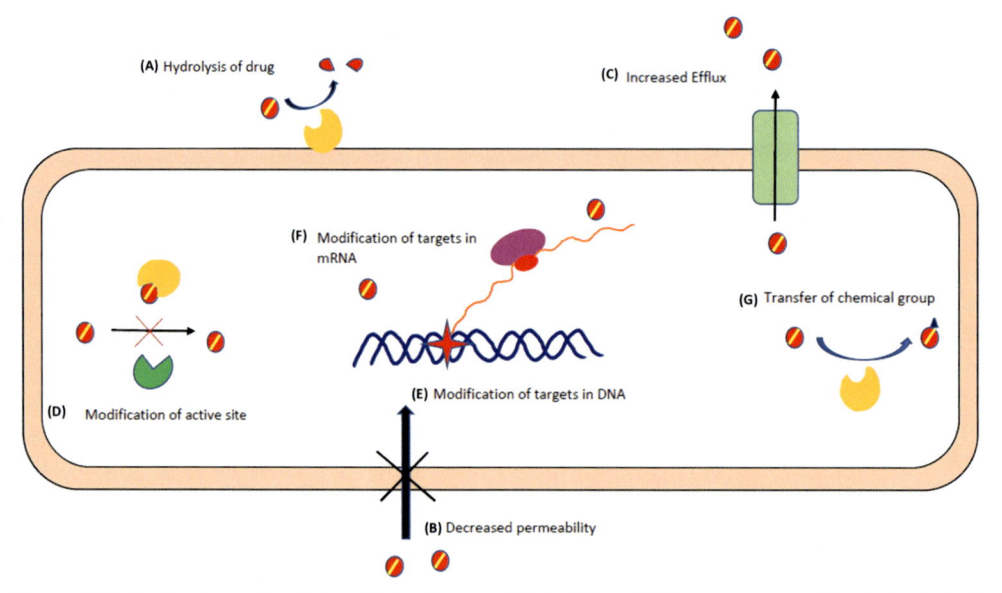

Figure 7.1 Mechanism of antimicrobial resistance. (A) Activity of β-lactamase resulting in hydrolysis of β-lactam ring; (B) decreased permeability to prevent access to targets; (C) increased efflux to prevent access to target; (D) modifying active site of enzyme; (E) modification of target in DNA; (F) modification of target in mRNA; and (G) enzymatic degradation of antibiotic by transfer of chemical group.

Enzymatic degradation of antibiotics

By hydrolysis

β-Lactam antibiotics like penicillin, carbapenem, cephalosporin, cephamycin, and oxapenam possess β-lactam ring which inactivates bacterial transpeptidases. Transpeptidases catalyze various reactions for cross-linking of peptidoglycan layer in bacteria. Emergence of bacterial resistance against β-lactam antibiotics arise due to the activity of β-lactamase that cleaves β-lactam ring. β-Lactamases are diverse class of antibiotic degrading enzymes with various hydrolytic spectrum activity against penicillin (penicillinase) and cephalosporin (cephalosporinase) (Begum et al., 2021).

Relocation of the chemical group

Bacterial community has developed tactics to modify the structure of antibiotics that hinders their activity. Enzymes such as chloramphenicol acetyltransferases and carbapenemases can alter or degrade different varieties of antibiotics such as aminoglycosides, β-lactams, phenicols, and macrolides. Bacterial cells basically inactivate antibiotics by adding a chemical group to the antibiotic molecule that creates a steric hindrance, preventing it from binding target site. Acyl, nucleotidyl, ribitol, and phosphate groups are few chemical groups that are transferable onto antibiotics. Broad range of enzymes is also responsible in transfers of chemical groups such as nucleotidyltransferases, phosphotransferases, and acetyltransferases which modify the aminoglycoside family from antibiotics—streptomycin, gentamicin, and amikacin (Wright, 2005). Acetylation is the mechanism used against chloramphenicol and fluoroquinolones (Blair et al., 2015).

Resistance versus persistence

Bacteria gain the AMR due to AMR genes present in plasmid. Plasmid is an extrachromosomal DNA, which holds genetic material. This plasmid DNA is expressed when cell encounters stress conditions like oxygen and nutrients' deprivation, exposure to antimicrobial agents, oxidative stress, acidification, and host immune responses in its vicinity (Schrader et al., 2020).

The lowest concentration of antibiotics at which microbial growth is prevented is known as minimum inhibitory concentration (MIC). Bacteria with antibiotic resistance grow even at higher levels of MIC. Resistant bacteria acquire resistance to antibiotics either by chromosomal gene mutations or by receiving foreign resistant genes via plasmids or transposons. This mechanism of bacterial resistance is termed as genetic antibiotic resistance (Schrader et al., 2020).

Not all bacteria in the bacterial population exhibit genetic antibiotic resistance. Few bacterial subsets exhibit phenotypic resistance, a phenomenon through which bacterial population escape stress not by resistance but by tolerance or persistence. Unlike antibiotic resistance where bacteria reduce the drug efficiency, persistence increases the tolerance to drugs and the duration of treatment by entering stationary growth phase (dormant). This bacterial subpopulation is termed as persisters. Persisters are neither susceptible to antibiotic drug nor contain resistant genes (Cohen et al., 2013).

Susceptible bacterial cells (nonresistant and nonpersistent) on exposure to treatment with antibiotics gets killed and nonsusceptible cells survive either by resistance or persistence. By isolation of resistant cells, regrowth and exposure to treatment, all the new grown cells in culture media exhibit resistance. However, persister cells (dormant) upon regrowing and reexposure to treatment remain dormant and new grown cells will still be susceptible to antibiotics. Persistent subpopulation, enter the growth phase only with removal of antibiotic drug treatment (Vogwill et al., 2016).

Transmission of resistance

As mentioned earlier, antibiotics resistance in bacteria can be acquired because of chromosomal modifications (regulatory or structural) in genes or through exchange of resistant genes in the form of either plasmids or transposons between recipient and donor via HGT (Holmes et al., 2016). Spread of resistance throughout the strains and species is taking place at rapid pace due to HGT, which allows easy transfer of genetic material throughout the strains and species (Lerminiaux and Cameron, 2019). Potential impact of HGT can be illustrated by *Shigella* outbreak in the United Kingdom emerging out of transfer of antibiotic resistance gene (ARG)-encoded plasmid. Routine antibiotics were no longer effective even on feeble pathogens due to dispersal of azithromycin resistance encoded plasmid which allowed spread of infectious diseases. The accession of the same plasmid by individual pathogens led to multiple outbreaks in the community (Baker et al., 2018).

1. Transformation by extracellular DNA
2. Conjugation by plasmids
3. Transduction by bacteriophages

Transduction

Another method of genetic transfer is "transduction." Transduction is controlled by bacterial viruses that can replicate independently and are known as "bacteriophages" or "phages." Transduction is well known as a major contributor in spread of ARGs, mostly in-between same species. Transduction takes place when a bacteriophage infects a bacterial host and while viral assembly takes up the bacterial gene encoding antibiotic resistance in the capsid. The virus then infects another microbial cell releasing the ARGs which the newly infected bacterial cell takes up making it antibiotic resistant (Carattoli, 2013) (Fig. 7.2).

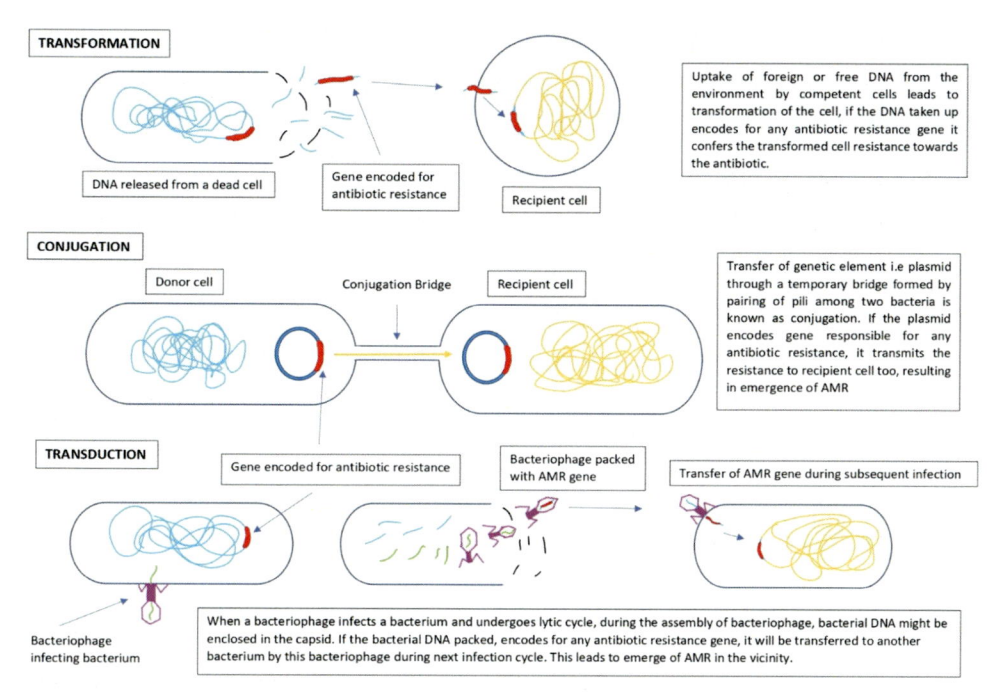

Figure 7.2 Horizontal transmission of resistant gene by conjugation, transformation, and transduction.

Stanczak-Mrozek et al. isolated bacteriophages from hospital-acquired MRSA infections that were able to transduce ARGs for penicillin and tetracycline through transfer of chromosomal elements from *S. aureus* to other strains in laboratory. Bacteriophage-mediated transduction significantly contributes to the initiation of AMR in MRSA (Stanczak-Mrozek et al., 2015). Studies indicate transfer of β-lactamase genes through transduction among *Acinetobacter* strains. Transfer of 5667 bp plasmid encoded for aminoglycoside and tetracycline via bacteriophage transduction has been studied among *S. aureus* strains. Studies even report transduction of 5620 bp plasmid encoded for kanamycin among *Serratia* and *Kluyvera* strains (Lerminiaux and Cameron, 2019).

Conjugation

Gram- negative bacteria have been largely endorsed for the distribution of AMR due to interspecific and intraspecific DNA exchanges. Plasmids are extrachromosomal genes, encoding antibiotic resistance against aminoglycosides, β-lactams, sulfonamides, chloramphenicol, quinolones, trimethoprim, tetracyclines, and macrolides. Plasmids can possess mobile genetic elements like transposons and insertion sequences that aid in mobilization of resistance gene. Plasmids control their own replication cycle and can transfer across bacteria of different species, genera, and kingdom. Horizontal transfer of ARG encoded on plasmid has increased chances of antibiotic resistance among wide range of microorganism creating a threat of nosocomial cases. Conjugation takes place in-between two cells in same vicinity through physical contact by the means of bridge formed by joining of pili. The bridge, an interconnecting module helps in plasmid translocation (one copy) from a donor to a recipient (Carattoli, 2013) (Fig. 7.2).

How HGT amplifies AR-related challenges in hospitals is exemplified by resistance toward β-lactam antibiotics, which is commonly carried through plasmids. *Acinetobacter*, *Enterobacteriaceae*, and *Pseudomonas* acquired resistance to β-lactam antibiotics, including carbapenems, penicillin, and cephalosporins through interspecies or intraspecies conjugation. Plasmid pCT has been a carrier of $bla_{CTX-M-14}$ gene responsible for AMR spreading among Enterobacteriaceae across globe (Lerminiaux and Cameron, 2019).

Transformation

Transformation is a natural process of genetic transfer, where a competent cell takes up extracellular DNA present in its vicinity. The foreign DNA having gene encoding antibiotic resistance is taken up by new competent cell it utilizes it to make itself resistant to the encoded antibiotic (Lerminiaux and Cameron, 2019) (Fig. 7.2). Naturally proficient bacteria especially *Haemophilus*, *Acinetobacter*, *Neisseria*, *Staphylococcus*, *Streptococcus*, and *Pseudomonas* can take up recombined extracellular DNA with ARG into their genome through transformation. Strains of nonvirulent *S. pneumoniae* acquire genes via transformation and transform into virulent pathogen in an infected mice or human models as proved by famous "Griffith's experiment" (Griffith, 1928). Likewise, acquisition of ARGs have been observed in *H. pylori* in human and mouse models (Lerminiaux and Cameron, 2019).

Spread of antimicrobial resistance

Majority of AMR genes are isolated from soil and hence soil is regarded as a reservoir of AMR genes. Furthermore, water polluted by fecal microorganism when used on food crops can extend the resistance in the soil microflora. In many of the wastewater samples huge amount of antibiotic resistant bacteria and genes can be detected due to disposal of unused antibiotics into wastewater (Prestinaci et al., 2015).

Excessive and inappropriate use of antibiotics has led to proliferation of antimicrobial resistant genes imparting serious threat to public health of human and veterinary medicine. Exposure of antibiotics to the environment may change or loose the diversity of microbial community in the soil (Han et al., 2018). AMR genes enter and persist in ecosystem through various pathways, they can spread from soil, crops, gut microbiota of wild and domesticated animals, and humans too (Su et al., 2015).

Transfer of mobile genetic elements such as transposons, plasmids, integron gene cassettes, and phages through HGT facilitates AMR gene spread. Bacteria might acquire AMR through spontaneous mutations or through the HGT from other bacteria. Evolution of AMR genes is facilitated by mutations producing hundreds of variants which are barely recognized but increase a danger to the environment (Woodford and Ellington, 2007).

In the context of an ecosystem, the total amount of resistance genes also known as "resistome" is mediated by plasmid transfers. Resistome grants microorganism resistance toward heavy metals and antibiotics, enhancing their chance in survival under extreme conditions (Su et al., 2015). Plasmid IncP-1 is a common plasmid group that is known for the transferability among range of Gram-negative bacteria. Likewise, Plasmids IncQ, IncF, and IncI have less transferability in narrower range of hosts. All these plasmids have an important role in dissemination of AMR gene transfers in *E. coli* and other *Enterobacteriaceae* (Lerminiaux and Cameron, 2019; Heuer et al., 2012).

Integron, another genetic element plays a vital role in spread of AMR genes. These elements encode genes for an integrase and an integration site which has the ability to acquire, convert, and express genes (Mazel, 2006). Integrons get themselves integrated into plasmids or transposons and are effectively transferred from one host to other. Integrases are basically divided into Inti1 (class1), Inti2 (class2), Inti3 (class3), and Inti4 (class4) among which Inti1 is well known for spreading AMR among bacteria (Heuer et al., 2012).

Conclusion and future prospects

Concerning the increase in AMR in medical and food sectors, a need of defining new strategies against resistance mechanism of microorganisms is crucial. The alternative ways to combat AMR emerging in research include extraction and isolation of antimicrobial compounds from natural products, that is antimicrobial peptides, phytochemicals, efflux pump inhibitors, lipopolysaccharide inhibitors, metallo-antibiotics, etc. These antimicrobial compounds isolated from natural sources and further synthesized chemically with few modifications can be effective antibiotics in future. Compounds targeting bacterial virulence can be developed, for example agents that target genes controlling multidrug resistance. Structural modification of existing antibiotics or synthesis of dynamic structures should be developed to solve the current situation. Probiotics which are live and beneficial microorganisms for the host can be selected to limit the transmission of antibiotic resistance and not carry transferable antibiotic resistance.

References

Alakomi, H.-L., Skytta, E., Saarela, M., Mattila-Sandholm, T., Latva-Kala, K., Helander, I.M., 2000. Lactic acid permeabilizes gram-negative bacteria by disrupting the outer membrane. Appl. Environ. Microbiol. 66 (5), 2001−2005.

Ayukekbong, J.A., Ntemgwa, M., Atabe, A.N., 2017. The threat of antimicrobial resistance in developing countries: causes and control strategies. Antimicrob. Resist. Infect. Control. 6 (1), 1–8.

Baker, K.S., Dallman, T.J., Field, N., Childs, T., Mitchell, H., Day, M., et al., 2018. Horizontal antimicrobial resistance transfer drives epidemics of multiple Shigella species. Nat. Commun. 9 (1), 1–10.

Bbosa, G.S., Mwebaza, N., Odda, J., Kyegombe, D.B., Ntale, M., 2014. Antibiotics/antibacterial drug use, their marketing and promotion during the post-antibiotic golden age and their role in emergence of bacterial resistance. Health 6.

Begum, S., Begum, T., Rahman, N., Khan, R.A., 2021. A review on antibiotic resistance and way of combating antimicrobial resistance. GSC Biol. Pharm. Sci. 14 (2), 087–097.

Beringer, M., Rodnina, M.V., 2007. The ribosomal peptidyl transferase. Mol. Cell. 26 (3), 311–321.

Blair, J.M., Webber, M.A., Baylay, A.J., Ogbolu, D.O., Piddock, L.J., 2015. Molecular mechanisms of antibiotic resistance. Nat. Rev. Microbiol. 13 (1), 42–51.

Blake, K.L., O'Neill, A.J., 2013. Transposon library screening for identification of genetic loci participating in intrinsic susceptibility and acquired resistance to antistaphylococcal agents. J. Antimicrob. Chemother. 68 (1), 12–16.

Blankenship, L., 2012. Colonization Control of Human Bacterial Enteropathologens in Poultry. Academic Press.

Block, S.S., 2001. Disinfection, Sterilization, and Preservation. Lippincott Williams & Wilkins.

Carattoli, A., 2013. Plasmids and the spread of resistance. Int. J. Med. Microbiol. 303 (6–7), 298–304.

Castagliuolo, I., Galeazzi, F., Ferrari, S., Elli, M., Brun, P., Cavaggioni, A., et al., 2005. Beneficial effect of auto-aggregating Lactobacillus crispatus on experimentally induced colitis in mice. FEMS Immunol. Med. Microbiol. 43 (2), 197–204.

Chapman, J.S., 2003. Biocide resistance mechanisms. Int. Biodeterior. Biodegrad. 51 (2), 133–138.

Chenoll, E., Casinos, B., Bataller, E., Astals, P., Echevarría, J., Iglesias, J.R., et al., 2011. Novel probiotic Bifidobacterium bifidum CECT 7366 strain active against the pathogenic bacterium Helicobacter pylori. Appl. Environ. Microbiol. 77 (4), 1335–1343.

Cohen, N.R., Lobritz, M.A., Collins, J.J., 2013. Microbial persistence and the road to drug resistance. Cell Host Microbe 13 (6), 632–642.

Denyer, S.P., 1995. Mechanisms of action of antibacterial biocides. Int. Biodeterior. Biodegrad. 36 (3–4), 227–245.

Dolejska, M., Villa, L., Poirel, L., Nordmann, P., Carattoli, A., 2013. Complete sequencing of an IncHI1 plasmid encoding the carbapenemase NDM-1, the ArmA 16S RNA methylase and a resistance–nodulation–cell division/multidrug efflux pump. J. Antimicrob. Chemother. 68 (1), 34–39.

Duong, A., Jaelin, M., 2015. 6 Factors That Have Caused Antibiotic Resistance. InfectionControl.tips.

Fàbrega, A., Madurga, S., Giralt, E., Vila, J., 2009. Mechanism of action of and resistance to quinolones. Microb. Biotechnol. 2 (1), 40–61.

Floss, H.G., Yu, T.-W., 2005. Rifamycin mode of action, resistance, and biosynthesis. Chem. Rev. 105 (2), 621–632.

Fraenkel-Conrat, H., 1961. Chemical modification of viral ribonucleic acid: I. Alkylating agents. Biochimic. Biophys. Acta 49 (1), 169–180.

Fraenkel-Conrat, H., Olcott, H.S., 1946. Reaction of formaldehyde with proteins. II. Participation of the guanidyl groups and evidence of crosslinking. J. Am. Chem. Soc. 68 (1), 34–37.

Fraenkel-Conrat, H., Cooper, M., Olcott, H.S., 1945. The reaction of formaldehyde with proteins II. Participation of the guanidyl groups and evidence of crosslinking. J. Am. Chem. Soc. 67 (6), 950–954.

Gourbeyre, P., Denery, S., Bodinier, M., 2011. Probiotics, prebiotics, and synbiotics: impact on the gut immune system and allergic reactions. J. Leukoc. Biol. 89 (5), 685–695.

Griffith, F., 1928. The significance of pneumococcal types. Epidemiol. Infect. 27 (2), 113–159.

Han, X.-M., Hu, H.-W., Chen, Q.-L., Yang, L.-Y., Li, H.-L., Zhu, Y.-B., et al., 2018. Antibiotic resistance genes and associated bacterial communities in agricultural soils amended with different sources of animal manures. Soil. Biol. Biochem. 126, 91–102.

Harbottle, H., Thakur, S., Zhao, S., White, D.G., 2006. Genetics of antimicrobial resistance. Anim. Biotechnol. 17 (2), 111–124.

Heuer, H., Binh, C.T., Jechalke, S., Kopmann, C., Zimmerling, U., Krögerrecklenfort, E., et al., 2012. IncP-1ε plasmids are important vectors of antibiotic resistance genes in agricultural systems: diversification driven by class 1 integron gene cassettes. Front. Microbiol. 3, 2.

Holmes, A.H., Moore, L.S., Sundsfjord, A., Steinbakk, M., Regmi, S., Karkey, A., et al., 2016. Understanding the mechanisms and drivers of antimicrobial resistance. Lancet 387 (10014), 176−187.

Hotel, A.C.P., Cordoba, A., 2001. Health and nutritional properties of probiotics in food including powder milk with live lactic acid bacteria. Prevention 5 (1), 1−10.

Hummel, S., Veltman, K., Cichon, C., Sonnenborn, U., Schmidt, M.A., 2012. Differential targeting of the E-cadherin/β-catenin complex by Gram-positive probiotic lactobacilli improves epithelial barrier function. Appl. Environ. Microbiol. 78 (4), 1140−1147.

Leclercq, R., 2002. Mechanisms of resistance to macrolides and lincosamides: nature of the resistance elements and their clinical implications. Clin. Infect. Dis. 34 (4), 482−492.

Lerminiaux, N.A., Cameron, A.D., 2019. Horizontal transfer of antibiotic resistance genes in clinical environments. Can. J. Microbiol. 65 (1), 34−44.

Liwa, A.C., Jaka, H., 2015. Antimicrobial resistance: mechanisms of action of antimicrobial agents. Battle Microb. Pathogens: Basic. Science, Technol. Adv. Educ. Prog. 5, 876−885.

Luthra, S., Rominski, A., Sander, P., 2018. The role of antibiotic-target-modifying and antibiotic-modifying enzymes in Mycobacterium abscessus drug resistance. Front. Microbiol. 9, 2179.

Luyt, C.-E., Bréchot, N., Trouillet, J.-L., Chastre, J., 2014. Antibiotic stewardship in the intensive care unit. Crit. Care 18 (5), 1−12.

Mazel, D., 2006. Integrons: agents of bacterial evolution. Nat. Rev. Microbiol. 4 (8), 608−620.

Moo, C.-L., Yang, S.-K., Yusoff, K., Ajat, M., Thomas, W., Abushelaibi, A., et al., 2020. Mechanisms of antimicrobial resistance (AMR) and alternative approaches to overcome AMR. Curr. Drug. Discov. Technol. 17 (4), 430−447.

Munita, J.M., Arias, C.A., 2016. Mechanisms of antibiotic resistance. Microbiol. Spectr. 4 (2), 4.2. 15.

Patterson, A.M., 1932. Meaning of "Antiseptic," "Disinfectant" and related words. Am. J. Public. Health Nations Health 22 (5), 465−472.

Prestinaci, F., Pezzotti, P., Pantosti, A., 2015. Antimicrobial resistance: a global multifaceted phenomenon. Pathog. Glob. Health 109 (7), 309−318.

Reece, R.J., Maxwell, A., 1991. DNA gyrase: structure and function. Crit. Rev. Biochem. Mol. Biol. 26 (3−4), 335−375.

Russell, A., Furr, J., 1996. Biocides: mechanisms of antifungal action and fungal resistance. Sci. Prog. 79, 27−48.

Safrin, S., 2021. Antiviral agents 15th ed In: Katzung, B.G., Vanderah, T.W. (Eds.), Basic & Clinical Pharmacology. McGraw-Hill, New York, NY.

Schrader, S.M., Vaubourgeix, J., Nathan, C., 2020. Biology of antimicrobial resistance and approaches to combat it. Sci. Transl. Med. 12 (549).

Stanczak-Mrozek, K.I., Manne, A., Knight, G.M., Gould, K., Witney, A.A., Lindsay, J.A., 2015. Within-host diversity of MRSA antimicrobial resistances. J. Antimicrob. Chemother. 70 (8), 2191−2198.

Su, J.-Q., Wei, B., Ou-Yang, W.-Y., Huang, F.-Y., Zhao, Y., Xu, H.-J., et al., 2015. Antibiotic resistome and its association with bacterial communities during sewage sludge composting. Environ. Sci. Technol. 49 (12), 7356−7363.

Verma, A.K., 2016. Evolution, merits and demerits of five kingdom system. Flora Fauna 22 (1), 76−78.

Vogwill, T., Comfort, A.C., Furió, V., MacLean, R.C., 2016. Persistence and resistance as complementary bacterial adaptations to antibiotics. J. Evolut. Biol. 29 (6), 1223−1233.

Waksman, S.A., 1956. Definition of antibiotics. Antibiot. Chemother. 6 (2), 90−94.

Walsh, C., 2003. Antibiotics: Actions, Origins, Resistance. American Society for Microbiology (ASM) 13 (11), 3059−3060.

Waxman, D.J., Strominger, J.L., 1983. Penicillin-binding proteins and the mechanism of action of beta-lactam antibiotics. Annu. Rev. Biochem. 52 (1), 825−869.

Woodford, N., Ellington, M.J., 2007. The emergence of antibiotic resistance by mutation. Clin. Microbiol. Infect. 13 (1), 5−18.

Wright, G.D., 2005. Bacterial resistance to antibiotics: enzymatic degradation and modification. Adv. Drug Deliv. Rev. 57 (10), 1451–1470.

Yao, J.D., Moellering, R.C., 2011. Antibacterial agents, Manual of Clinical Microbiology, *10th ed* American Society of Microbiology, pp. 1043–1081.

Zhang, W.-J., Xu, X.-R., Schwarz, S., Wang, X.-M., Dai, L., Zheng, H.-J., et al., 2014. Characterization of the IncA/C plasmid pSCEC2 from Escherichia coli of swine origin that harbours the multiresistance gene cfr. J. Antimicrob. Chemother. 69 (2), 385–389.

CHAPTER 8

Combination of virulence and antibiotic resistance: a successful bacterial strategy to survive under hostile environments

Arif Hussain[1], Razib Mazumder[1], Md. Asadulghani[2], Taane G. Clark[3] and Dinesh Mondal[1]

[1]Laboratory Sciences and Services Division, International Centre for Diarrhoeal Disease Research (icddr,b), Dhaka, Bangladesh
[2]Biosafety and BSL3 Laboratory, Biosafety Office, International Centre for Diarrhoeal Disease Research (icddr,b), Dhaka, Bangladesh
[3]London School of Hygiene & Tropical Medicine, London, United Kingdom

Introduction

Bacterial infection outcomes are determined by two important properties associated with the bacterial strain; the pathogenicity of the strain and its antibiotic-resistant phenotype. Pathogenicity is the property of a bacterial pathogen that enables it to infect and cause disease in a suitable host (Beceiro et al., 2013). Most pathogens harbor virulence factors that enhance their degree of pathogenicity. Virulence factors possessed by the pathogenic bacteria are mainly related to two broad categories: (i) toxins, these are the molecules produced by pathogenic bacteria that determine the degree to which an etiological agent can cause disease, and (ii) invasins, which are the protein molecules that enable a pathogen to infect and penetrate the host/cells (Beceiro et al., 2013). Whereas antibiotic resistance is the ability of a bacteria to survive and no longer respond to antibacterial agents/antibiotics to which it was previously susceptible. The final outcome of the infection process in a host is dependent on the level of pathogenicity the microbe carries and the susceptibility demonstrated by the host to a given infection. The susceptibility is driven by factors, such as immunological strength, comorbidities, age, gender, and stress levels (Beceiro et al., 2013).

Bacteria and hosts are coevolving over the course of millions of years. During the evolution, hosts have modified their defense system and consequently, bacteria have developed ways to escape the host defense mechanisms (Beceiro et al., 2013). The development of antibiotic resistance in bacteria, particularly the pathogenic strains, is a relatively recent phenomenon (Fair and Tor, 2014). The medical practice has restricted the severe outcomes of infectious diseases and prevented the spread of pathogens by the successful application of antibiotics as therapeutic agents. However, the overuse

Bacterial Survival in the Hostile Environment
DOI: https://doi.org/10.1016/B978-0-323-91806-0.00004-7

and misuse of antibiotics are in a significant way responsible for the predominance of antimicrobial resistance (AMR) among bacterial pathogens. The development and transmission of antibiotic resistance started relatively recently, after introducing the first antibiotic in the medical practice. Therefore the time scales of the evolution of virulence factors and antibiotic resistance determinants in bacteria are different (Beceiro et al., 2013).

Despite the varying evolutionary time scales between virulence and antibiotic resistance mechanisms, bacterial pathogens share these two features in several aspects. First, both of these mechanisms are required to overcome adverse (hostile) conditions from the pathogen survival point of view. Virulence is necessary to thwart the defense responses of the host. Similarly, antibiotic resistance enables bacteria to counter the adverse effects of antibiotics so that they can survive and spread to new niches. Host immune mechanisms and the use of antibiotics present stress to the bacteria by restricting their growth and survivability (Feldman and Laland, 1996; Martínez and Baquero, 2002). Second, the molecular determinants encoding virulence and resistance can be exchanged between bacterial strains of the same species and/or different species by vertical and horizontal gene transfer (HGT) mechanisms. HGT is the most common mechanism for transmission of molecular determinants of antibiotic resistance and virulence genes/pathogenicity islands; therefore HGT is primarily responsible for coselecting antibiotic resistance and virulence in bacterial strains. Besides, compensatory and adaptive mutations can also help coselect resistance and virulence in bacteria (Burrus and Waldor, 2004; Handel et al., 2006). Third, both antibiotic resistance and virulence genes are associated with clinically relevant bacteria and their interplay is a cause of concern for medical practitioners. For example, the interplay of virulence and resistance proves a lethal combination by establishing a successful biofilm formation (Patel, 2005; Seral et al., 2003). Fourth, some components of bacterial cells share their function with both resistance and virulence mechanisms; such components comprise porins (Tsai et al., 2011) and efflux pumps (Barbosa and Levy, 2000). Alteration in the cell wall components and two-component systems is also involved in inducing and repressing the gene expression that encodes resistance and virulence factors.

The use of antibiotics as therapeutics for treating infectious diseases and as prophylactics in poultry and cattle farms has affected the natural evolution of bacteria by inducing selective pressure and selectively increasing the resistant bacterial population over the susceptible bacterial population. This has led to a significant threat to public health (Mahmud et al., 2020; Hussain et al., 2017; Ranjan et al., 2015). However, the development of antibiotic resistance in bacteria results in reduced bacterial fitness as the antibiotic mutations disrupt essential biological functions of the bacterial cell. Thus the genetic burden imposed by resistance in a bacterial cell would be detrimental to bacteria in antibiotic-free environments (Andersson and Levin, 1999). The most common strategy proposed to reduce the increasing AMR trend is to restrict antibiotic use because these mutations are believed to impose huge fitness costs on bacteria in antibiotic-free environments, which helps to eradicate or at least contain the spread of antibiotic-resistant bacteria once it has been evolved (Andersson and Hughes, 2011).

Several factors can alleviate the cost of resistance in bacteria, and some conditions may even selectively encourage the most resistant and virulent strains. Mechanisms such as hypermutation, compensatory mutations, cross coselection, antibiotic tolerance, and persistence help maintain antibiotic resistance by compensating the fitness costs. In addition, the role of these factors in offsetting the fitness cost depends on its interaction with four major factors: the bacterial species, the ecological niche, the type of virulence and resistance factors involved, and the host species (Beceiro et al., 2013). This chapter focuses on exploration of mechanisms involved in compensating the fitness cost induced due to antibiotic resistance. A thorough understanding of different factors/mechanisms that alleviate the costs of antibiotic resistance is essential in order to mitigate the emergence of AMR and manage the rising trend of antibiotic resistance.

Cross coselection

In general, it was observed that an increase in antibiotic resistance in bacterial pathogens is linked either directly or indirectly with a decrease in virulence, which means there exists a compromise/trade-off between virulence and resistance leading to a reduction in overall bacterial fitness (Andersson and Levin, 1999; Patel, 2005). Recent evidence suggests that this scenario might not be true in all cases (Patel, 2005; Seral et al., 2003; Hussain et al., 2014; Shaik et al., 2017; Hussain et al., 2012). In fact, in some cases, the resistance and virulence mechanisms are so adjusted that the association proves to be advantageous over pathogens that have either increased resistance or increased virulence. Such a phenomenon can be explained in the light of the Darwinian model, in which bacterial pathogens with high resistance and high virulence gene repertoire with a distinct edge over other pathogens, in a particular niche, is selected and fixed (Patel, 2005; Seral et al., 2003). Further, genetic events allow pathogens to recoup for any functional deficiencies, resulting in the emergence of multiresistant, multivirulent clones of public health significance. A few examples of such clones are provided in Table 8.1.

Coselection mediated by horizontal gene transfer

HGT is the second most widespread mechanism for mediating antibiotic resistance between bacteria, as good as spontaneous mutations and vertical transmission. It enables DNA sequence transfers laterally, not vertically; these transferable DNA elements are termed mobile genetic elements (MGEs). HGT can occur between bacterial pathogens of the same or different species. Plasmids, integrative and conjugative elements (ICEs), integrons and transposons constitute MGEs. Plasmids and ICEs are primarily involved in mediating AMR spread (Aminov, 2011). Among the MGEs, the

Table 8.1 High-risk clones that successfully carry antibiotic resistance genes plus virulence factors in single clinical lineages.

SI no.	Pathogenic strains/lineages	References
1	Epidemic methicillin-resistant *Staphylococcus aureus*-15 (EMRSA-15)	Aanensen et al. (2016)
2	Carbapenem-resistant hypervirulent *Klebsiella pneumonia* (CR-HvKP) ST11	Gu et al. (2018)
3	*Pseudomonas aeruginosa* ST235	Treepong et al. (2018)
4	Clonal complex 17 (CC17) *Enterococcus faecium*	Raven et al. (2016)
5	*Escherichia coli* ST131	Nicolas-Chanoine et al. (2014)

ICEs and plasmids play a crucial role in coselecting resistance and virulence factors that spread via conjugation, transformation, or transduction (Botelho and Schulenburg, 2021). Additional elements, such as sublethal concentration of antibiotic residues and bacterial biofilms, also favor the HGT (Aminov, 2011).

Role of plasmids

Plasmids are self-replicating extra chromosomal double-stranded DNA molecules. Plasmids do not make the essential components of bacterial genetic structure; however, they confer bacteria with characteristics that help them adapt to different environmental conditions (Johnson and Nolan, 2009). Certain plasmids can encode both virulence and resistance factors and in niches like infective environments, coselection of such plasmids may occur in bacterial pathogens. Specific conditions, such as bacterial biofilm and the presence of subinhibitory concentrations of antibiotics, may encourage plasmid dissemination through conjugation and transformation (Cepas and Soto, 2020). Depending on the type of bacterial population and favorable conditions, there is a possibility of generating plasmid chimeras from different plasmids (Bennett, 2008). Different lineages of bacterial species have varying affinities for plasmid acquisition and once these plasmids are selected within a particular lineage, their dissemination is driven by clonal expansion (Beceiro et al., 2013) (Fig. 8.1). Many such plasmids have contributed to the development of high-risk clones (Mazumder et al., 2021). For instance, the IncFII group plasmids have contributed to selecting CTX-M-15 β-lactamase within *E. coli* ST131, responsible for most bloodstream and urinary tract infections worldwide (Nicolas-Chanoine et al., 2014; Mazumder et al., 2020; Shaik et al., 2017; Hussain et al., 2019).

Role of integrative and conjugative elements

ICEs are a category of MGEs detected in diverse gram-negative and gram-positive bacteria; these are horizontally transferred between cells by conjugation (Wozniak and Waldor, 2010). ICEs are integrated into host chromosomes or plasmids by site-specific recombination. When the ICE gene expression is induced, a series of steps are

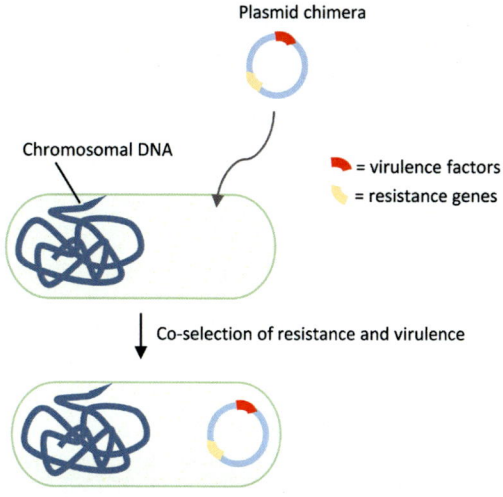

Figure 8.1 *Coselection mediated by plasmid chimeras.* The bacteria, mainly through transformation and conjugation, can take up plasmids that harbor both antibiotic resistance and virulence genes. Once bacteria acquire such plasmids under conducive conditions, these plasmids are coselected within the strain and often, this leads to the emergence of virulent multiresistant lineages with robust fitness.

executed that include excision of ICE, followed by synthesis of conjugation machinery, mainly mediated via a type IV secretion system and finally DNA is transferred to respective recipient bacteria (Johnson and Grossman, 2015) (Fig. 8.2). In addition to the genes for conjugative machinery, the ICEs carry genes that do not have any role in the ICE life cycle but confer advantageous phenotypes such as antibiotic resistance, virulence, or metabolic traits to recipient bacteria. Once they are integrated into the chromosome or plasmid, they propagate passively by replicating chromosomes or plasmids during cell division or propagate actively between or within the cell via induction of ICE gene expression (Wozniak and Waldor, 2010). Both virulence and resistance factors can sometimes be captured on ICE leading to coselection of resistance and virulence mechanisms in a bacteria–containing such ICE (Beceiro et al., 2013). For instance, the ICE-SXT and R391 are documented widely in *Vibrio cholerae* clinical isolates. These have been involved in the transmission of both virulence and resistance genes (Marrero and Waldor, 2007; Botelho and Schulenburg, 2021; Baddam et al., 2020).

Role of genomic islands

Genomic islands (GIs) are vast segments of DNA (>10 kb in size), which are transferred horizontally (Langille et al., 2010). GIs are integrated into the host chromosome and contribute to the genome plasticity. GIs get excised from the chromosome and

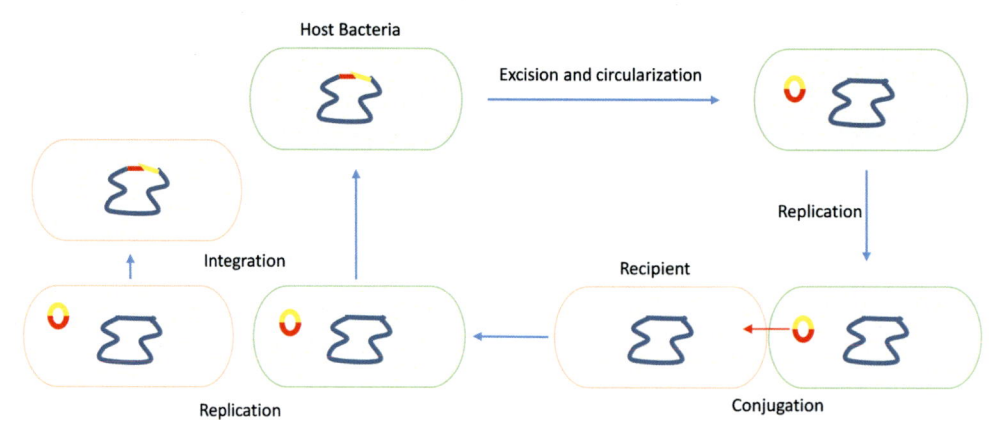

Figure 8.2 *Illustration of an ICE life cycle.* The cargo genes present in the ICE elements can comprise resistance or virulence genes, which together with ICE genes can undergo excision and circularization. It then directs the synthesis of conjugation machinery and transfers the DNA to the recipient bacteria via rolling circle replication. The host and recipient cells then replicate the DNA into double-stranded circular DNA. Finally, these elements can integrate into host chromosomes by a recombination event. ICEs can confer a range of phenotypes to their host, including antibiotic and heavy metal resistance, metabolic traits, virulence, and biofilm formation. Their clinical relevance is substantial due to their involvement in the widespread dissemination of AMR genes (Wozniak and Waldor, 2010).

are horizontally transferred to a new recipient bacterium by either conjugation, transformation, or transduction (Langille et al., 2010). They are usually flanked by repeat sequences and contain elements such as bacteriophages, plasmids, and insertion sequences (IS). GIs confer fitness to the bacteria; they are known to frequently provide a gain-of-function to the recipient bacterium (Juhas et al., 2009). As GIs are usually very large segments of DNA, they carry multiple genes that are able to change the entire phenotype of the bacteria in a single successful event of HGT. For example, the SGI1 GI in Salmonella confers resistance to several antibiotics. Isolates harboring this GI are at risk for rapid dissemination (Vo et al., 2007). The presence of pks GI in specific lineages of extraintestinal pathogenic *E. coli* (ExPEC) is reported to confer bacterial pathogenicity as it encodes a genotoxin called colibactin (Suresh et al., 2018).

Bacteriophage-mediated transduction

Bacteriophage-mediated transduction is an important means of dissemination of antibiotic resistance genes (ARGs) between bacterial pathogens. In particular, bacteriophages play a crucial role in spreading ARGs from environmental bacteria to animal and human bacterial pathogens (Colavecchio et al., 2017). Bacteriophages are present in diverse environments and are one of the most abundant organisms. Bacteriophages take up fragments of host bacterial DNA during transduction, which are included as part of viral DNA. The temperate bacteriophages then infect new bacterial cells and integrate their DNA into the bacterial chromosome, which is replicated passively during cell division (Fig. 8.3) (Colavecchio et al., 2017). The prophages can subsequently get excised and cause lysis of host cells upon induction by some stress signals, thereby contributing to the cycle of transmission of bacterial DNA segments together with the viral particles (Colavecchio et al., 2017). The prophages can also disseminate virulence factors to bacteria already possessing resistance mechanisms that favor the coselection of resistance and virulence phenotypes (Beceiro et al., 2013). For example, an outbreak involving 3842 people in Germany with the hemolytic uremic syndrome was caused by an *E. coli* strain. This *E. coli* strain was later found to be evolved from enteroaggregative *E. coli* strain horizontally acquiring the genes for Shiga toxin 2a (Stx2a)— from a bacteriophage (Buchholz et al., 2011).

Hypermutations

Mutator or hypermutable bacteria are those that demonstrate a high frequency of spontaneous mutations because of the defects or deficiencies in their DNA repair mechanisms. More specifically, the mutator phenotype is developed due to flaws in their mismatch repair (MMR) system. The MMR system genes are referred to as antimutator genes (Oliver and Mena, 2010; Blázquez, 2003). This system functions to correct errors, such as mismatches, insertions or deletions. The essential genes that constitute the MMR system are *mutS*, *mutL and mutH*, or *uvrD*. Inactivation of any of

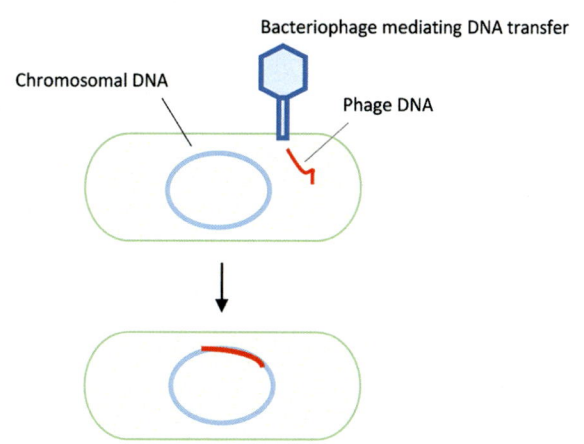

Phage DNA integrated into the chromosome mediates co-selection
of resistance and virulence, based on the type of genes it carries
and the host species it has infected

Figure 8.3 Temperate phage life cycle: temperate phage integrates their DNA into the bacterial chromosome, a prophage that can confer advantageous phenotypes to host bacteria such as resistance or virulence. They may remain fixed in the chromosome if they have defects in their mechanism or may get excised and cause subsequent lysis upon induction by some stress signals.

these genes increases the mutation frequency from 100- to 1000-fold and may induce DNA errors, such as transversions, deletions, or insertions. In addition to the role of MMR in correcting the replication errors, it plays an essential role in preventing recombination of weakly or moderately homologous DNA sequences between genomes (Modrich and Lahue, 1996). Therefore any defects in the MMR system increase the mutation rate and the rate of recombination. Mutation in proofreading subunit of DNA polymerase III (*mutD*) can also lead to mutator phenotype in a bacteria with mutation rates increased up to 1000-fold. Unlike stable mutations produced by defects in mutator genes, a transient hypermutable phenotype can also be induced by the DNA damage. The vulnerable DNA polymerases (polymerase IV and V) are induced due to SOS response. Some antibiotics can also induce transient mutator phenotype favoring the development of antibiotic resistance (Miller, 1996).

Laboratory and theoretical models have shown that bacteria undergo passive evolution and, through various mechanisms, accumulate mutations at frequent intervals that can render them resistant to most of the antibiotics they are exposed to (Blázquez, 2003). Most of these mutations are deleterious or neutral, but some may help in the directional selection of strains under stressful conditions such as antibiotic pressure or an infective environment. Mutator strains are found in the bacterial population in the natural habitats at a much higher frequency than expected. If such mutator pathogenic bacteria are present in an infective environment, they will likely become antibiotic resistant. Moreover, a few antibiotics do not just exert selective pressure for antibiotic resistance but can promote hypermutable phenotype in bacterial clones (Beceiro et al., 2013).

Hypermutation in bacteria that are responsible for chronic infections is a significant concern due to antibiotic therapy failure. It is frequently observed in infections caused by *P. aeruginosa* (Oliver and Mena, 2010), *Haemophilus influenzae* (Román et al., 2004), and *S. aureus* (Prunier et al., 2003). The first and the abundant evidence for the role of hypermutable strains in therapy failures came from studies on *P. aeruginosa* in cystic fibrosis (CF) patients (Oliver et al., 2000). *P. aeruginosa* exhibits an enhanced ability to accumulate chromosomal mutations conferring resistance against antibiotics. In CF, it has been observed that hypermutable strains were enriched in the lungs that led to resistance against antipseudomonal antibiotics with higher frequency than the nonmutator strains. It has been reported that *P. aeruginosa* adapts in the lungs of CF patients in such a way that the mutator strains showed enhanced resistant phenotype with altered virulence. This favored the strains to reduce acute injury and alternatively induce chronic inflammation. Such strains were also shown to have changes in their metabolic pathways in order to adapt to the specific conditions of the lungs in CF patients (Oliver et al., 2000; Beceiro et al., 2013).

Hypermutable strains were also shown to be involved in the pathogenesis of meningitis. A study observed a high prevalence of mutator strains in a collection of 95 *Neisseria meningitidis* isolates (Richardson et al., 2002). The hypermutable phenotype incidence was also observed to be frequent among *E. coli* strains involved in chronic urinary tract infections (Baquero et al., 2005). An in vivo infection study using strains of *Salmonella typhimurium* in a mouse model detected that minimum inhibitory concentration (MIC) for two antibiotics were higher in mutator strains than in the wild-type strains. However, when the metabolism of several carbon sources was compared in an *in vitro* assay between the mutator and wild-type strains, the mutator strains were found to have lost some metabolic functions to utilize all carbon sources, which led to the conclusion that the strains were explicitly adapted to infection niche in mice (Nilsson et al., 2004).

Therefore hypermutation attributes several advantages to bacteria by accumulating rapid chromosomal mutations that confer antibiotic resistance, attenuation, or enhancement of virulence or metabolism in order to adapt to a specific stressful niche in the host. Nevertheless, hypermutation often confers a biological cost to the bacteria outside its primary environment or niche.

Antibiotic tolerance and persistence

An organism with the genetic basis involved with the resistant phenotype growing at higher concentrations of a particular antibiotic than the MIC is called resistance (Brauner et al., 2016). Unlike resistance, antibiotic tolerance can be defined as the property of microbes, whether heritable or not, to withstand short-term exposure to higher antibiotic concentrations without having a change in MIC for that particular antibiotic (Fig. 8.4A). The tolerance to the antibiotic is achieved by turning down the pace of an essential/critical biological process that decreases bacterial replication (Brauner et al., 2016; Westblade et al., 2020). While the whole bacterial population achieves resistance and tolerance, persistence is the property demonstrated by a sub-population of bacteria in a given niche; this subpopulation can persist in environments with elevated antibiotic concentrations without having underwent any change in the MICs for that particluar antibiotic. Persistence is charcaterized by killing of the majority of a clonal bacterial population due to antibiotic exposure and a subpopulation of this bacteria survives for a longer time irrespective of whether the population is highly clonal in nature. The phenotype of persistence is nonheritable because experiments demonstrated that when these bacteria (persisters) were isolated and the fresh cultures were exposed to the same antibiotic, it again yielded a biphasic killing curve as that of the previous persister cells (Fig. 8.4B) (Brauner et al., 2016; Balaban et al., 2019).

Stress responses against starvation, host immune defense or antibiotic itself can enable bacteria to become antibiotic tolerant. Essentially the conditions that cause slow growth in bacteria can induce antibiotic tolerance (Westblade et al., 2020). Antibiotics such as fluoroquinolones, β-lactams, polymixins, and aminoglycosides induce various forms of stress in bacteria (Westblade et al., 2020); the rate of killing by these antibiotics is directly linked to the bacterial growth rate. Therefore if the bacteria somehow manages to slow down its

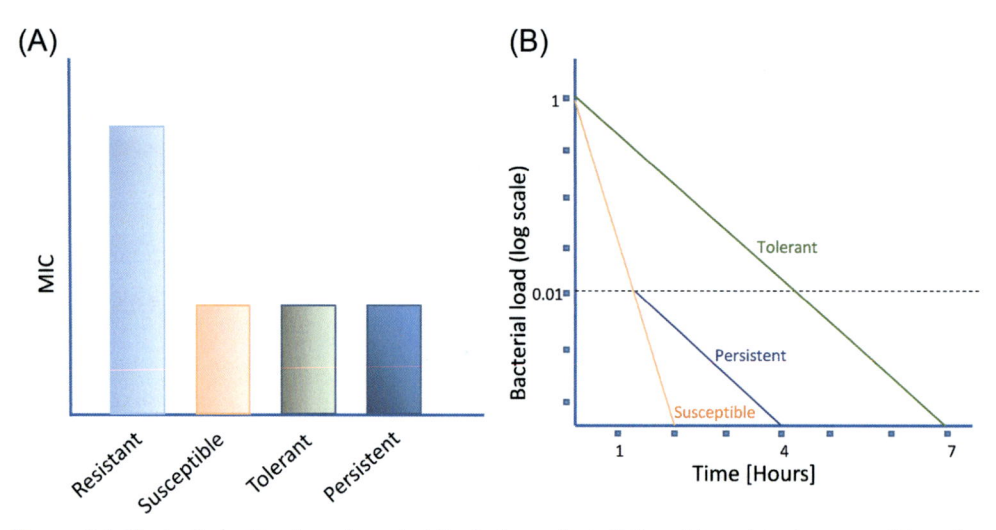

Figure 8.4 Illustrations showing characteristic features that distinguish resistant, susceptible, tolerant and persistent bacterial populations. (A) Antibiotic resistance is associated with a higher MIC of the antibiotic than for susceptible bacteria, whereas bacteria that demonstrate tolerance and persistence do not show any increase in MIC compared to susceptible strains. (B) Compared to antibiotic susceptibility, bacteria that demonstrate tolerance can withstand high antibiotic concentration and show slow killing by an antibiotic. Persistence initially demonstrates a killing curve similar to susceptible bacteria. However, after some time, a subpopulation of bacteria survives extended time than the susceptible population, producing a biphasic killing curve.

growth, the rate of killing by these antibiotics also comes down, resulting in a state of tolerance. Tolerance can be inherited if the bacterium is inherently slow-growing or noninherent if the bacteria grows slow just because of the induced stress or poor growth conditions in a particular niche like a biofilm or an environment having growth inhibitors (Brauner et al., 2016). Various gram-negative and gram-positive bacteria were reported to demonstrate tolerance by slow growth. Pathogens such as *Mycobacterium tuberculosis* have slow growth with a doubling time of 24 hours in nutrient-rich media (Manina et al., 2015). This organism demonstrates inherent drug tolerance within the granulomas of the lungs. Intracellular bacteria such as *Salmonella enterica* serovar Typhimurium exhibits noninherent drug tolerance in infected macrophages (Helaine et al., 2010). Tolerance is shown transiently during the lag phase, the growth-arrested phase wherein the bacteria adapt to a new environment before resuming exponential growth. Bacteria can adapt to extend their lag phase duration to enable tolerance to different antibiotics, leading to inherited tolerance by lag. Such tolerance is seen in *E. coli* isolates having intermittent exposure to β-lactam antibiotics (Fridman et al., 2014).

Persistence also occurs because of the same reason as does tolerance. However, persistence is the survival against higher antibiotic concentrations only by a subpopulation of bacteria in a given niche compared to the rest of the bacterial population. The tolerant subpopulation exhibits either an extended lag time known as tolerance by lag or a decreased growth rate, which is called tolerance due to slow growth. For example, the presence of the *hipA7* allele in *E. coli* produces persister strains, leading to the composition of two subpopulations with different durations of lag phase. The subpopulation with a more extended lag phase will persist longer and contributes to antibiotic tolerance (Brauner et al., 2016).

Therefore a high level of antibiotic tolerance is exhibited by clinical isolates of significant bacterial pathogens against many antibiotics without having any biological cost. Tolerance and persistence are likely to be important contributors to treatment failures only next to antibiotic resistance.

Compensatory mutations

Compensatory mutations are those mutations that recover a loss of fitness in bacteria resulting due to an earlier mutation (Levin et al., 2000). They can stabilize the resistant bacterial strains even in antibiotic-free environments and make them fit as the original susceptible strain. Numerous studies have reported that antibiotic-resistant bacteria can recover their lost fitness, including virulence, by secondary mutations that compensate for the lost fitness (Brown et al., 2010). The level of compensation achieved by compensatory mutations is linked to several factors such as the type of bacterial species, the antibiotic resistance or virulence mechanisms in question, and the bacteria's niche (Beceiro et al., 2013).

Bacterial strains utilize three ways to compensate the fitness of resistant mutations (Andersson and Levin, 1999): (i) direct restoration, (ii) replacement by similar function, and (iii) reduced dependence on the function lost. The same is discussed in detail below:

1. A recent study on linezolid resistant and susceptible *S. pneumoniae* transformants by whole-genome sequence analysis revealed mutations associated with resistance and compensation (Billal et al., 2011). Comparative genomics revealed that the strains acquired mutations that caused overexpression of ABC transporter (spr1021) to compensate for the chromosomal mutations of linezolid resistance (23S rRNA mutation). Therefore the excess accumulation of linezolid can be taken care of by overexpression of ABC transporters that can pump out the accumulated linezolid. They also found other mutations in 50S ribosomal proteins L3 and L16 that could compensate for the decreased fitness in the strains with linezolid resistance.

2. Another way to regain the fitness cost is to replace that function with another function having a similar effect. An excellent example of such a mechanism is seen in the case of resistance to actinonin, a peptide deformylase inhibitor (PDFI) in *Salmonella. enterica* (Nilsson et al., 2006). For initiation of bacterial translation, a formylated Met-tRNAi is required, which is cleaved by the peptide deformylase

(PDF) upon completion of translation. The PDFIs act against the peptide formylase activity. In order to gain resistance to PDFIs, the bacteria accumulate mutations in specific genes such as *fmt* and *folD* that are responsible for forming formylated Met-tRNAi; this decreases the chances of translation initiation and reduces the bacterial growth due to scarcity of formylated Met-tRNAi. To compensate for the decreased formylated Met-tRNAi, the bacteria carry out amplification of *metZ* and *metW* genes that encode tRNAi. This increases tRNAi levels, compensating for the scanty translation initiation by directly binding to the ribosomal P site and initiating the translation at higher rates (Nilsson et al., 2006).

3. The third approach that bacteria adapt to resist losing biological fitness is decreasing the dependency to alter the function. For instance, *Streptococcus pneumoniae gains resistance to* β-lactam antibiotics by undergoing mutations in three penicillin-binding proteins—PBP1a, PBP2x, and PBP2b. These proteins are critical for the viability of *S. pneumoniae* strains as they are essential for cell wall synthesis during bacterial fission (Albarracín Orio et al., 2011). However, strains of *S. pneumoniae* can overcome this biological cost of cell wall defects by gaining mutations in *pbp1a*, *pbp2x* genes sparing the pbp2b gene. The bacteria do this by increasing the stability and mislocalization of proteins. Therefore mutations in the two *pbp* mutant alleles compensate for the mutation in *pbp2b* and increase the level and spectrum of β-lactam resistance.

The above examples highlighted the importance of compensatory mutations in bacteria of clinical significance to adapt and compensate for the biological costs associated with antibiotic resistance. Laboratory and mathematical models also suggest that when the resistant mutant has a particular fitness cost, it can restore its fitness in an environment outside its primary niche and could even outcompete with the original susceptible strain if it can survive for more extended periods to accumulate more than one compensatory mutations.

Conclusion

The development of antibiotic resistance in pathogenic bacteria is most often biologically disadvantageous as it is associated with a fitness price because antibiotic mutations often disrupt essential biological functions of the bacterial cells. Despite this, bacteria have developed ways to combine virulence and resistance mechanisms by overcoming the evolutionary trade-offs and compensating for the fitness cost (Fig. 8.5). Several mechanisms can alleviate the cost of resistance in bacteria and some agents may even selectively encourage the most resistant and virulent strains. These mechanisms are hypermutation, compensatory mutations, cross coselection, antibiotic tolerance, and persistence. Hypermutation is a phenomenon where bacteria demonstrate a high frequency of spontaneous mutations and these occur as a result of deficiencies and defects in error avoidance and DNA repair

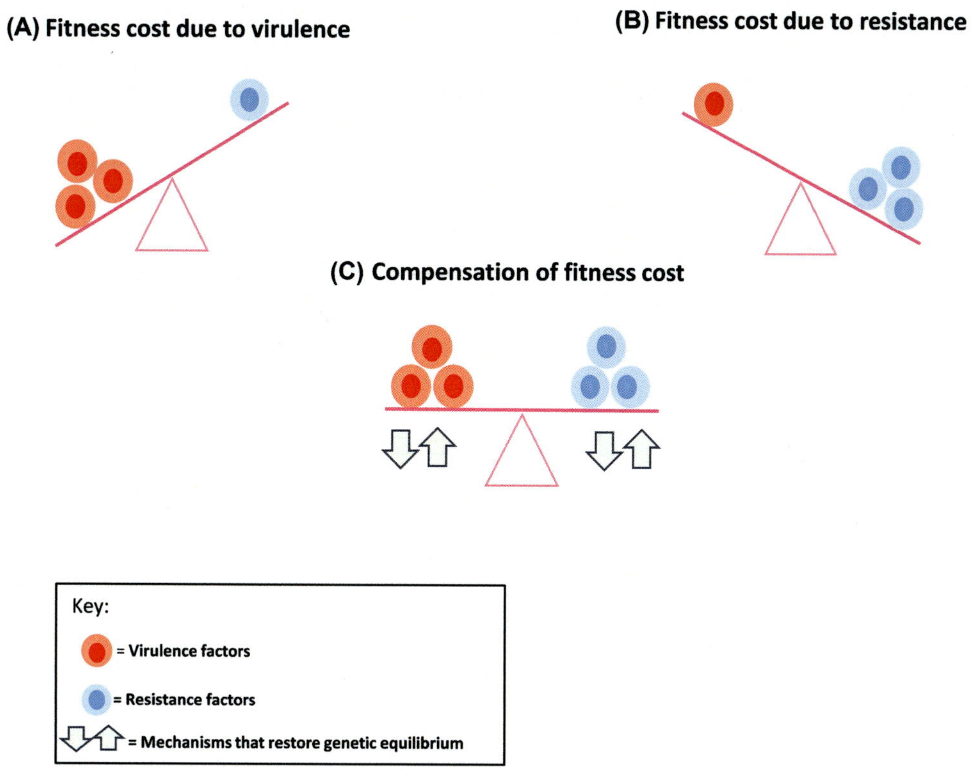

(A) Fitness cost due to virulence

(B) Fitness cost due to resistance

(C) Compensation of fitness cost

Key:

= Virulence factors

= Resistance factors

= Mechanisms that restore genetic equilibrium

Figure 8.5 Evolutionary dynamics of resistance and virulence during bacterial adaptation under ceratin hostile conditions. (A) The reduction of bacterial fitness due to enhanced virulence implies that there will be excessive cellular energy expenditures or limitations of some bacterial processes; this may lead to selection against virulent pathogens under certain conditions. (B) The reduction of bacterial fitness due to enhanced antibiotic resistance implies that bacteria acquiring either chromosomal mutations or horizontally acquired genes conferring phenotypes such as resistance to antibiotics will have defects in biological processes such as reduced growth and transmission rate and less invasiveness compared to its susceptible counterpart. (C) Nonetheless, several mechanisms can alleviate the fitness cost in a bacteria species induced either due to virulence or resistance. These mechanisms include hypermutation, compensatory mutations, cross coselection, antibiotic tolerance, and persistence. The interplay of these mechanisms in a pathogen through a context-dependent manner can lead to the emergence of specialized strains/ high-risk clones of high public health importance.

mechanisms. This mutator phenotype contributes several advantages to bacteria by accumulating rapid chromosomal mutations that confer antibiotic resistance, attenuation, or enhancement of virulence or metabolism in order to adapt to a specific stressful environment. Compensatory mutations in bacteria also help them adjust and make up for the fitness costs connected with antibiotic resistance. These mutations work by restoring fitness and virulence and such strains could even outcompete with the original susceptible strain in fitness. Another way to compensate for the fitness cost in bacteria is by combining (coselecting) AMR genes and virulence-associated genes by HGT of MGEs that mainly comprise plasmids and ICEs. HGT mediated by bacteriophage is also an important driver of coselection of resistance and virulence-associated genes. Tolerance and persistence by bacterial population is another mechanism to survive antibiotic exposure to varying periods. This property can be heritable in the case of strains that demonstrate antibiotic tolerance and nonheritable in the case of persisters. In this way, the virulent pathogens can withstand higher antibiotic exposure without experiencing any fitness loss. The mechanisms described in this chapter should not be seen in isolation of bacterial and host factors. They must not be construed as general mechanisms that apply to all pathogens and work under all circumstances. Instead, the level of success (fitness) achieved by bacteria employing these mechanisms depends on several considerations such as the type of species, the virulence/antibiotic resistance mechanism in question, the environmental factors, the ecological niche of the bacteria and the host. A comprehensive understanding of these mechanisms is essential to direct research and developmental efforts to prevent severe infection outcomes and combat AMR.

Acknowledgments

icddr,b is grateful to the governments of Bangladesh, Canada, Sweden, and the United Kingdom for providing core/unrestricted support for its operations and research. icddr,b also thanks the Royal Society, UK and The London School of Hygiene and Tropical Medicine (LSHTM), for supporting the capacity building on genomic surveillance of antimicrobial-resistant bacterial pathogens.

References

Aanensen, D.M., Feil, E.J., Holden, M.T.G., Dordel, J., Yeats, C.A., Fedosejev, A., et al., 2016. Whole-genome sequencing for routine pathogen surveillance in public health: a population snapshot of invasive *Staphylococcus aureus* in Europe. MBio. Available from: https://doi.org/10.1128/mBio.00444-16.

Albarracín Orio, A.G., Piñas, G.E., Cortes, P.R., Cian, M.B., Echenique, J., 2011. Compensatory evolution of Pbp mutations restores the fitness cost imposed by β-lactam resistance in *Streptococcus pneumoniae*. PLoS Pathog. Available from: https://doi.org/10.1371/journal.ppat.1002000.

Aminov, R.I., 2011. Horizontal gene exchange in environmental microbiota. Front. Microbiol. Available from: https://doi.org/10.3389/fmicb.2011.00158.

Andersson, D.I., Levin, B.R., 1999. The biological cost of antibiotic resistance. Curr. Opin. Microbiol. Available from: https://doi.org/10.1016/S1369-5274(99)00005-3.

Andersson, D.I., Hughes, D., 2011. Persistence of antibiotic resistance in bacterial populations. FEMS Microbiol. Rev. Available from: https://doi.org/10.1111/j.1574-6976.2011.00289.x.

Baddam, R., Sarker, N., Ahmed, D., Mazumder, R., Abdullah, A., Morshed, R., et al., 2020. Genome dynamics of *Vibrio cholerae* isolates linked to seasonal outbreaks of cholera in Dhaka, Bangladesh. MBio 11 (1). Available from: https://doi.org/10.1128/mBio.03339-19.

Balaban, N.Q., Helaine, S., Lewis, K., Ackermann, M., Aldridge, B., Andersson, D.I., et al., 2019. Definitions and guidelines for research on antibiotic persistence. Nat. Rev. Microbiol. Available from: https://doi.org/10.1038/s41579-019-0196-3.

Baquero, M.R., Galán, J.C., Del Carmen Turrientes, M., Cantón, R., Coque, T.M., Martínez, J.L., et al., 2005. Increased mutation frequencies in *Escherichia coli* isolates harboring extended-spectrum β-lactamases. Antimicrob. Agents Chemotherap. Available from: https://doi.org/10.1128/AAC.49.11.4754-4756.2005.

Barbosa, T.M., Levy, S.B., 2000. Differential expression of over 60 chromosomal genes in *Escherichia coli* by constitutive expression of MarA. J. Bacteriol. Available from: https://doi.org/10.1128/JB.182.12.3467-3474.2000.

Beceiro, A., Tomás, M., Bou, G., 2013. Antimicrobial resistance and virulence: a successful or deleterious association in the bacterial world? Clin. Microbiol. Rev. Available from: https://doi.org/10.1128/CMR.00059-12.

Bennett, P.M., 2008. Plasmid encoded antibiotic resistance: acquisition and transfer of antibiotic resistance genes in bacteria. Br. J. Pharmacol. Available from: https://doi.org/10.1038/sj.bjp.0707607.

Billal, D.S., Feng, J., Leprohon, P., Légaré, D., Ouellette, M., 2011. Whole genome analysis of linezolid resistance in *Streptococcus pneumoniae* reveals resistance and compensatory mutations. BMC Genom. Available from: https://doi.org/10.1186/1471-2164-12-512.

Blázquez, J., 2003. Hypermutation as a factor contributing to the acquisition of antimicrobial resistance. Clin. Infect. Dis. Available from: https://doi.org/10.1086/378810.

Botelho, J., Schulenburg, H., 2021. The role of integrative and conjugative elements in antibiotic resistance evolution. Trends Microbiol. Available from: https://doi.org/10.1016/j.tim.2020.05.011.

Brauner, A., Fridman, O., Gefen, O., Balaban, N.Q., 2016. Distinguishing between resistance, tolerance and persistence to antibiotic treatment. Nat. Rev. Microbiol. Available from: https://doi.org/10.1038/nrmicro.2016.34.

Brown, K.M., Costanzo, M.S., Xu, W., Roy, S., Lozovsky, E.R., Hartl, D.L., 2010. Compensatory mutations restore fitness during the evolution of dihydrofolate reductase. Mol. Biol. Evolution. Available from: https://doi.org/10.1093/molbev/msq160.

Buchholz, U., Bernard, H., Werber, D., Böhmer, M.M., Remschmidt, C., Wilking, H., et al., 2011. German outbreak of *Escherichia coli* O104:H4 associated with sprouts. N. Engl. J. Med. Available from: https://doi.org/10.1056/nejmoa1106482.

Burrus, V., Waldor, M.K., 2004. Shaping bacterial genomes with integrative and conjugative elements. Res. Microbiol. Available from: https://doi.org/10.1016/j.resmic.2004.01.012.

Cepas, V., Soto, S.M., 2020. Relationship between virulence and resistance among gram-negative bacteria. Antibiotics. Available from: https://doi.org/10.3390/antibiotics9100719.

Colavecchio, A., Cadieux, B., Lo, A., Goodridge, L.D., 2017. Bacteriophages contribute to the spread of antibiotic resistance genes among foodborne pathogens of the *Enterobacteriaceae* family - a review. Front. Microbiol. Available from: https://doi.org/10.3389/fmicb.2017.01108.

Fair, R.J., Tor, Y., 2014. Antibiotics and bacterial resistance in the 21st century. Perspect. Medicinal Chem. Available from: https://doi.org/10.4137/PMC.S14459.

Feldman, M.W., Laland, K.N., 1996. Gene-culture coevolutionary theory. Trends Ecol. Evolution. Available from: https://doi.org/10.1016/0169-5347(96)10052-5.

Fridman, O., Goldberg, A., Ronin, I., Shoresh, N., Balaban, N.Q., 2014. Optimization of lag time underlies antibiotic tolerance in evolved bacterial populations. Nature. Available from: https://doi.org/10.1038/nature13469.

Gu, D., Dong, N., Zheng, Z., Lin, D., Huang, M., Wang, L., et al., 2018. A fatal outbreak of ST11 carbapenem-resistant hypervirulent *Klebsiella pneumoniae* in a Chinese hospital: a molecular epidemiological study. Lancet Infect. Dis. Available from: https://doi.org/10.1016/S1473-3099(17)30489-9.

Handel, A., Regoes, R.R., Antia, R., 2006. The role of compensatory mutations in the emergence of drug resistance. PLoS Comput. Biol. Available from: https://doi.org/10.1371/journal.pcbi.0020137.

Helaine, S., Thompson, J.A., Watson, K.G., Liu, M., Boyle, C., Holden, D.W., 2010. Dynamics of intracellular bacterial replication at the single cell level. Proc. Natl. Acad. Sci. USA . Available from: https://doi.org/10.1073/pnas.1000041107.

Hussain, A., Ewers, C., Nandanwar, N., Guenther, S., Jadhav, S., Wieler, L.H., et al., 2012. Multiresistant uropathogenic *Escherichia coli* from a region in India where urinary tract infections are endemic: genotypic and phenotypic characteristics of sequence type 131 isolates of the CTX-M-15 extended-spectrum-β- lactamase-producing lineage. Antimicrob. Agents Chemother. 56 (12), 6358—6365. Available from: https://doi.org/10.1128/AAC.01099-12.

Hussain, A., Ranjan, A., Nandanwar, N., Babbar, A., Jadhav, S., Ahmed, N., 2014. Genotypic and phenotypic profiles of *Escherichia coli* isolates belonging to clinical sequence type 131 (ST131), clinical non-ST131, and fecal Non-ST131 lineages from India. Antimicrob. Agents Chemother. 58 (12), 7240—7249. Available from: https://doi.org/10.1128/AAC.03320-14.

Hussain, A., Shaik, S., Ranjan, A., Nandanwar, N., Tiwari, S.K., Majid, M., et al., 2017. Risk of transmission of antimicrobial resistant *Escherichia coli* from commercial broiler and free-range retail chicken in India. Front. Microbiol. 8 (November), 2120. Available from: https://doi.org/10.3389/fmicb.2017.02120.

Hussain, A., Shaik, S., Ranjan, A., Suresh, A., Sarker, N., Semmler, T., et al., 2019. Genomic and functional characterization of poultry *Escherichia coli* from India revealed diverse extended-spectrum β-lactamase-producing lineages with shared virulence profiles. Front. Microbiol. 10 (December), 2766. Available from: https://doi.org/10.3389/fmicb.2019.02766.

Johnson, C.M., Grossman, A.D., 2015. Integrative and conjugative elements (ICEs): what they do and how they work. Annu. Rev. Genet. Available from: https://doi.org/10.1146/annurev-genet-112414-055018.

Johnson, T.J., Nolan, L.K., 2009. Pathogenomics of the virulence plasmids of *Escherichia coli*. Microbiol. Mol. Biol. Rev. Available from: https://doi.org/10.1128/mmbr.00015-09.

Juhas, M., Van Der Meer, J.R., Gaillard, M., Harding, R.M., Hood, D.W., Crook, D.W., 2009. Genomic islands: tools of bacterial horizontal gene transfer and evolution. FEMS Microbiol. Rev. Available from: https://doi.org/10.1111/j.1574-6976.2008.00136.x.

Langille, M.G.I., Hsiao, W.W.L., Brinkman, F.S.L., 2010. Detecting genomic islands using bioinformatics approaches. Nat. Rev. Microbiol. Available from: https://doi.org/10.1038/nrmicro2350.

Levin, B.R., Perrot, V., Walker, N., 2000. Compensatory mutations, antibiotic resistance and the population genetics of adaptive evolution in bacteria. Genetics. Available from: https://doi.org/10.1093/genetics/154.3.985.

Mahmud, Z.H., Kabir, M.H., Ali, S., Moniruzzaman, M., Imran, K.M., Nafiz, T.N., et al., 2020. Extended-spectrum beta-lactamase-producing *Escherichia coli* in drinking water samples from a forcibly displaced, densely populated community setting in Bangladesh. Front. Public. Health 8 (June), 228. Available from: https://doi.org/10.3389/FPUBH.2020.00228/BIBTEX.

Manina, G., Dhar, N., McKinney, J.D., 2015. Stress and host immunity amplify *Mycobacterium tuberculosis* phenotypic heterogeneity and induce nongrowing metabolically active forms. Cell Host Microb. Available from: https://doi.org/10.1016/j.chom.2014.11.016.

Marrero, J., Waldor, M.K., 2007. The SXT/R391 family of integrative conjugative elements is composed of two exclusion groups. J. Bacteriol. Available from: https://doi.org/10.1128/JB.01902-06.

Martínez, J.L., Baquero, F., 2002. Interactions among strategies associated with bacterial infection: pathogenicity, epidemicity, and antibiotic resistance. Clin. Microbiol. Rev. Available from: https://doi.org/10.1128/CMR.15.4.647-679.2002.

Mazumder, R., Abdullah, A., Ahmed, D., Hussain, A., 2020. High prevalence of blactx-m-15 gene among extended-spectrum β-lactamase-producing *Escherichia coli* isolates causing extraintestinal infections in Bangladesh. Antibiotics. Available from: https://doi.org/10.3390/antibiotics9110796.

Mazumder, R., Hussain, A., Abdullah, A., Islam, M.N., Sadique, M.T., Muniruzzaman, S.M., Tabassum, A., et al., 2021. International high-risk clones among extended-spectrum β-lactamase—producing *Escherichia coli* in Dhaka, Bangladesh. Front. Microbiol. 12 (October), 2843. Available from: https://doi.org/10.3389/FMICB.2021.736464/BIBTEX.

Miller, J.H., 1996. Spontaneous mutators in bacteria: insights into pathways of mutagenesis and repair. Annu. Rev. Microbiol. Available from: https://doi.org/10.1146/annurev.micro.50.1.625.

Modrich, P., Lahue, R., 1996. Mismatch repair in replication fidelity, genetic recombination, and cancer biology. Annu. Rev. Biochem. Available from: https://doi.org/10.1146/annurev.bi.65.070196.000533.

Nicolas-Chanoine, M.-H., Bertrand, X., Madec, J.-Y., 2014. *Escherichia coli* ST131, an intriguing clonal group. Clin. Microbiol. Rev. 27 (3), 543—574. Available from: https://doi.org/10.1128/CMR.00125-13.

Nilsson, A.I., Kugelberg, E., Berg, O.G., Andersson, D.I., 2004. Experimental adaptation of *Salmonella typhimurium* to mice. Genetics. Available from: https://doi.org/10.1534/genetics.104.030304.

Nilsson, A.I., Zorzet, A., Kanth, A., Dahlström, S., Berg, O.G., Andersson, D.I., 2006. Reducing the fitness cost of antibiotic resistance by amplification of initiator TRNA genes. Proc. Natl. Acad. Sci. USA . Available from: https://doi.org/10.1073/pnas.0602171103.

Oliver, A., Mena, A., 2010. Bacterial hypermutation in cystic fibrosis, not only for antibiotic resistance. Clin. Microbiol. Infect. Available from: https://doi.org/10.1111/j.1469-0691.2010.03250.x.

Oliver, A., Cantón, R., Campo, P., Baquero, F., Blázquez, J., 2000. High frequency of hypermutable *Pseudomonas aeruginosa* in cystic fibrosis lung infection. Science. Available from: https://doi.org/10.1126/science.288.5469.1251.

Patel, R., 2005. Biofilms and antimicrobial resistance. Clin. Orthop. Relat. Res. Available from: https://doi.org/10.1097/01.blo.0000175714.68624.74.

Prunier, A.L., Malbruny, B., Laurans, M., Brouard, J., Duhamel, J.F., Leclercq, R., 2003. High rate of macrolide resistance in *Staphylococcus aureus* strains from patients with cystic fibrosis reveals high proportions of hypermutable strains. J. Infect. Dis. Available from: https://doi.org/10.1086/374937.

Ranjan, A., Shaik, S., Hussain, A., Nandanwar, N., Semmler, T., Jadhav, S., et al., 2015. Genomic and functional portrait of a highly virulent, CTX-M-15-producing H 30-Rx subclone of *Escherichia coli* sequence type 131.". Antimicrob. Agents Chemother. 59 (10), 6087−6095. Available from: https://doi.org/10.1128/AAC.01447-15.

Raven, K.E., Reuter, S., Gouliouris, T., Reynolds, R., Russell, J.E., Brown, N.M., et al., 2016. Genome-based characterization of hospital-adapted *Enterococcus faecalis* lineages. Nat. Microbiol. Available from: https://doi.org/10.1038/nmicrobiol.2015.33.

Richardson, A.R., Yu, Z., Popovic, T., Stojiljkovic, I., 2002. Mutator clones of *Neisseria meningitidis* in epidemic serogroup A disease. Proc. Natl. Acad. Sci. USA. Available from: https://doi.org/10.1073/pnas.092568699.

Román, F., Cantón, R., Pérez-Vázquez, M., Baquero, F., Campos, J., 2004. Dynamics of long-term colonization of respiratory tract by *Haemophilus influenzae* in cystic fibrosis patients shows a marked increase in hypermutable strains. J. Clin. Microbiol. Available from: https://doi.org/10.1128/JCM.42.4.1450-1459.2004.

Seral, C., Van Bambeke, F., Tulkens, P.M., 2003. Quantitative analysis of gentamicin, azithromycin, telithromycin, ciprofloxacin, moxifloxacin, and oritavancin (LY333328) activities against intracellular *Staphylococcus aureus* in mouse J774 macrophages. Antimicrob. Agents Chemother. Available from: https://doi.org/10.1128/AAC.47.7.2283-2292.2003.

Shaik, S., Ranjan, A., Tiwari, S.K., Hussain, A., Nandanwar, N., Kumar, N., et al., 2017. Comparative genomic analysis of globally dominant ST131 clone with other epidemiologically successful extraintestinal pathogenic *Escherichia coli* (ExPEC) lineages. MBio 8 (5), Available from: https://doi.org/10.1128/mBio.01596-17e01596-17.

Suresh, A., Ranjan, A., Patil, S., Hussain, A., Shaik, S., Alam, M., et al., 2018. Molecular genetic and functional analysis of Pks-harboring, extra-intestinal pathogenic *Escherichia coli* from India. Front. Microbiol. 9, 2631. Available from: https://doi.org/10.3389/FMICB.2018.02631.

Treepong, P., Kos, V.N., Guyeux, C., Blanc, D.S., Bertrand, X., Valot, B., et al., 2018. Global emergence of the widespread *Pseudomonas aeruginosa* ST235 Clone. Clin. Microbiol. Infect. Available from: https://doi.org/10.1016/j.cmi.2017.06.018.

Tsai, Y.K., Fung, C.P., Lin, J.C., Chen, J.H., Chang, F.Y., Chen, T.L., et al., 2011. *Klebsiella pneumoniae* outer membrane porins OmpK35 and OmpK36 play roles in both antimicrobial resistance and virulence. Antimicrob. Agents Chemother. Available from: https://doi.org/10.1128/AAC.01275-10.

Vo, A.T.T., van Duijkeren, E., Fluit, A.C., Gaastra, W., 2007. A novel *Salmonella* genomic island 1 and rare integron types in *Salmonella typhimurium* isolates from horses in The Netherlands. J. Antimicrob. Chemother. Available from: https://doi.org/10.1093/jac/dkl531.

Westblade, L.F., Errington, J., Dörr, T., 2020. Antibiotic tolerance. PLoS Pathog. Available from: https://doi.org/10.1371/journal.ppat.1008892.

Wozniak, R.A.F., Waldor, M.K., 2010. Integrative and conjugative elements: mosaic mobile genetic elements enabling dynamic lateral gene flow. Nat. Rev. Microbiol. Available from: https://doi.org/10.1038/nrmicro2382.

CHAPTER 9

Mechanisms of biofilm-based antibiotic resistance and tolerance in *Mycobacterium tuberculosis*

Amit Singh[1,2], Anil Kumar Gupta[3], Arti Singh Katiyar[4] and Divakar Sharma[5]
[1]Centralized Core Research Facility (CCRF), All India Institute of Medical Sciences, New Delhi, Delhi, India
[2]Department of Gastroenterology and HNU, All India Institute of Medical Sciences, New Delhi, Delhi, India
[3]IRCH, All India Institute of Medical Sciences, New Delhi, Delhi, India
[4]Department of Chemistry, MVGU, Jaipur, Rajasthan, India
[5]Department of Microbiology, Maulana Azad Medical College, New Delhi, Delhi, India

Introduction

Tuberculosis (TB) has been surviving in the human population since antiquity. In the word of Frederic Marais "the germ may be as old as the earth itself- surviving in the primeval mud at the very beginning of time." Its cause remained mysterious till 24th March 1984, when Robert Koch discovers the bacillus responsible and subsequently named *Mycobacterium tuberculosis (Mtb)*. TB is spread when the infected person expels TB bacteria in to air by coughing. TB is mainly affecting lungs (pulmonary TB) but can spread to other sits of body (extrapulmonary TB). Fragment of the *Mycobacterium* genus currently comprises more than 170 species (Esteban and García-Coca, 2018). Majority of species are noninfectious to humans and are environmental, moreover *Mtb* and other *Mtb* complex (MTBC) are oldest and most common cause of human infection. About one-fourth of population is infected with *Mtb*, equivalent to approx. 2 billion populations. Only small proportion (5%—10%) may develop TB disease in their lifetime. TB is a worldwide emergency with an estimated 9.9 million new cases and more than 1.5 million deaths were reported to the World Health Organization (WHO) globally in 2020 and India tops the list by having highest burden of infected people (WHO, 2021).

The situation has worsened with the emergence of multidrug resistant (MDR) forms of the causative agent *Mtb*, that is, resistance to both isoniazid (INH) and rifampicin (RIF) and now, extensively drug-resistant tuberculosis (XDR-TB) strains that is virtually untreatable. Globally, 71% (2.1/3.0 million) of people in year 2020 diagnosed with bacteriologically confirmed PTB were tested for resistance top rifampicin, up from 61% in 2019. It resulted mainly due to limited laboratory infrastructure, treatment on the basis of least sensitive diagnostics, irrational and nonadherence to the standard therapeutic regimens (Singh et al., 2017a,b; Singh et al., 2014). These drug-resistant

Bacterial Survival in the Hostile Environment
DOI: https://doi.org/10.1016/B978-0-323-91806-0.00001-1

strains are more infectious by benefit of their high transmissibility in the population (Singh et al., 2017a,b).

Biofilms in mycobacterium are serious health apprehension, worldwide due to their antibiotic tolerance abilities, host defense mechanisms and other external stresses; hence, it contributes to persistent chronic infections. Biofilms are accumulation of immobile microorganisms which colonize and grow on the surface with the help of extracellular polymeric substances (EPSs) produced by cells and cause chronic infections. The biofilm formation inside the host gives numerous advantages to the pathogens like protection from the host immune system, relapse infections, and treatment failure (Chakraborty et al., 2021). The infection of TB shows these features. During the infection TB remains hidden or escaped from immune mechanisms of host. The formation of biofilm during infection is still unknown but *Mtb* naturally form cords and adhere to the surface in liquid culture medium. The biofilm formation in mycobacterial cell is more resistant to environmental disinfectants and aggressions then the planktonic forms. The biofilm developed by *Mtb* plays an important role in the development of caseous necrosis and cavity formation in lungs which leads to the infection in hosts. It may also develop biofilms in clinical biomaterials such as catheters, dental implants, sutures, and other surgical equipment (Sharma et al., 2019a; Sharma et al., 2019b). The development of biofilm is an important factor for drug resistance, as it provides protection against drugs/antibiotics which are normally active against the same organisms in normal state. This drug resistance in microorganisms forming biofilm may result in failure of treatment and to resolve the infection biofilm should physically remove. Formation of biofilm therefore is considered as common persistent strategy for microbes in various growth conditions. In this chapter, we explored the role of biofilm in antibiotic resistance and tolerance in mycobacterium.

What is biofilm?

A biofilm is a cluster of cells that have formed an attachment and are bordered by a self-produced matrix. EPS is the matrix, which is made up of proteins, polysaccharides, and other components. It acts as a haven for microorganisms that are resistant to medications and disinfection.

Origins of biofilm hypothesis

Antonie van Leeuwenhoek described microbes were the first communities of individual cells in 1684 when he observes the human dental scuff (plaque) as "animalcules." In 1978 the first report of modern concept of biofilm was published (Esteban and García-Coca, 2018). After a decade researchers started study of biofilm on environmental mycobacterium (Schulze-Röbbecke et al., 1992; Wallace et al., 1983). Even though the occurrence of mycobacterial aggregates or pellicles were described in earlier reports. Bacteriology, blooming under the influence of Robert Koch's. He described the appearance of "cells which are pressed together and arranged in bundles" (Koch, 1882). Successive studies by Calmette entirely similar to the biofilm described in modern time. In 1970s and 1980s, Costerton et al. (1981) published literature through electron microscopy on wide range of samples from body fluids to medical implants, showing microcolonies of adherent microbes encapsulated by exopolysaccharides. Moreover, observed that these sessile populations are distinct in terms of their resistance to antimicrobial agents. Various studies were carried out with extensive microscopic observations of many surfaces for discoveries of biofilm. Biofilms are the predominant lifestyle of microbes in both clinical as well as environmental settings and stance important challenges in the cure of microbial infections (Richards and Ojha, 2014). Recent studies by Chakraborty et al. (2021) proved experimentally the formation of biofilm inside in vitro granulomas and also inside the infected mouse lungs. Biofilm also plays important role in protection from host and antimicrobials.

Characteristics of biofilm

EPS is a complex mixture of proteins, lipids, nucleic acids, polysaccharides, and water created by microbes. Table 9.1 shows the percentages of these components. The carbohydrate moiety is responsible for biofilm adhesion to the surface as well as nutrition conning. Microorganisms benefit from this particular trait because it gives structural and defensive support against antimicrobial drugs and disinfectants (Jamal et al., 2015).

The backbone of a biofilm typically consists of two components: a water channel that transports critical nutrients and an area of densely packed cells with no visible holes. Biofilm is typically too thin in thickness. The thickness of biofilm is influenced by a variety of environmental conditions, including warm or stagnant water, water movement, or water stress, which is an important element that can impede biofilm growth. Biofilm production can also be influenced by the type of organism. Some species create a lot of EPS and build a thicker biofilm as well (Sutherland, 2001).

Table 9.1 Components of biofilm.

S. No	Components	Percentage in EPS (%)
1	Microbial cells	2−5
2	Nucleic acid	<1−2
3	Polysaccharides	1−2
4	Proteins (including enzymes)	<1−2
5	Water	∼97

How biofilm is formed?

The formation of biofilm is a complex procedure, which is regulated uniquely. The biofilm-producing microbes such as *E. coli*, *P. aeruginosa*, *S. aureus*, *Staphylococcus epidermidis*, *E. cloacae*, and *Mtb* convert their growth from planktonic to sessile mode (Limoli et al., 2015). Bacterial biofilm communities differ from planktonic bacterial populations in a variety of ways, including growth rate, gene function, transcription, and translation, since they live in microhabitats with higher osmolarity, nutrient scarcity, and higher cell density of heterogeneous bacterial populations (Sharma et al., 2019a). These changes are solely depending on the differential expression of specific genes, whose product established the direction of the biofilm formation and keto mycolic acid (Sambandan et al., 2013). Biofilm formation is a multistep procedure following a series of events in which the microorganisms experience numerous changes after adhering to a surface (Kumar et al., 2017). Biofilm formation commonly occurs in three main stages (Fig. 9.1).

Adhesion of microbial cells

Cells' attachments start, when a single free-floating bacterium comes close to the surface/support and land on the surface by reducing its motion. The bacterium makes a revocable connection with the surface and /or already adhered another microbe to the surface. The solid−liquid interface (e.g., blood, water) provides an ideal environment for the attachment and growth of microorganisms. For most common attachment and biofilm formation rough, hydrophilic and coated surfaces will provide a healthier environment (Floyd et al., 2017). The attachment may also be enhanced because of flow velocity, water temperature, or nutrient availability. The presence of motion structures on microbial surfaces such as flagella, pili, fimbriae, proteins, or polysaccharides are also important and may provide an added value during biofilm formation (Morales and Kolter, 2014). Soon after

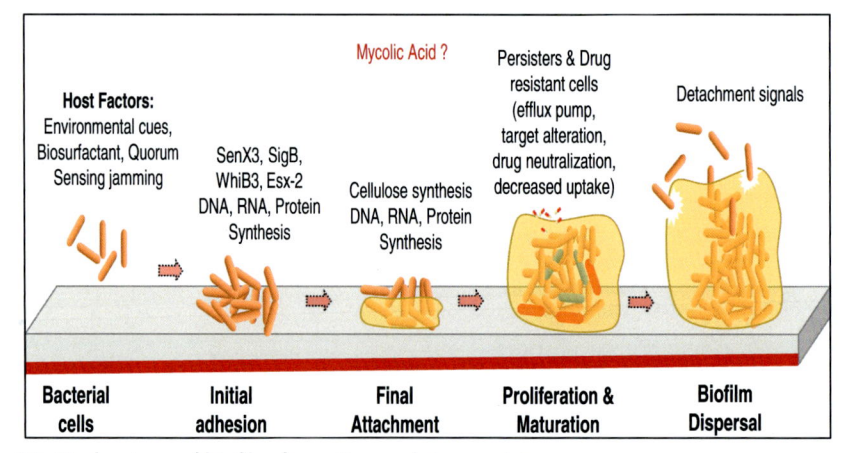

Figure 9.1 Mechanisms of biofilm formation and drug resistance.

attachment of the bacterium to the physical surface or biological tissue, the formation of microcolony is happening via secretion of specific chemicals by bacteria, which induce multiplication and formation of biofilm (Sutherland, 2001). The activation genetic mechanism behind the production of exopolysaccharides is initiated once the signal intensity crosses a certain threshold. Subsequently, bacterial cell divisions take place within the embedded exopolysaccharide matrix, which finally result in formation of microcolony (Guzmán-Soto et al., 2021).

Proliferation and maturation

Following the establishment of a microcolony, the expression of certain biofilm-related genes begins. The product of these genes is necessary for the synthesis of EPS (key component of biofilm). Self-attachment by bacteria has been shown to activate the synthesis of the extracellular matrix. Subsequently, water-filled channels formed, allowing nutrients to be transported inside the biofilm (Kuang et al., 2018). This dynamic process of various bacterial populations forming the three-dimensional architecture of biofilm is indeed a dynamic process. Bacteria residing in biofilms are guarded against a variety of environmental challenges, including as desiccation, antimicrobials, immune system attack, and protozoa ingestion, and as a result of this structure, biofilm communities progress faster than planktonic colonies (López et al., 2010).

Biofilm dispersal

It is often assumed that bacteria predictably leave biofilm. The bacteria can multiply and disperse quickly as a result of this. The detachment of planktonic microbial species from the biofilm is a controlled process that follows a natural pattern (Zhang et al., 2018). Bacteria are periodically separated from the colony and released into the environment due to mechanical stress. However, in the majority of cases, bacteria stop producing EPS and are released into the environment. Biofilm cells disperse due to the separation of newly produced cells from developing cells, dispersion of biofilm aggregates due to quorum sensing or flowing effects (Zhao et al., 2017). Cells are eliminated from biofilms due to an enzyme action that causes alginate breakdown. The manner of biofilm dispersion appears to alter the phenotypic characteristics of organisms (Kostakioti et al., 2013). Certain biofilm features, such as antibiotic sensitivity, can be retained by dispersed cells from the biofilm. The cells which are dispersed from biofilm as a result of growth may return swiftly to their normal planktonic phenotype (Kaplan, 2010). The different steps in the biofilm life cycle are shown in Fig. 9.1.

Quorum sensing

Many bacterial species can interact with one another during biofilm development through a technique known as quorum sensing. It is a type of cell-to-cell communication that involves the accumulation of signaling molecules in the extracellular environment, which leads to the regulation of specific gene expression of various virulence factors and biofilm differentiation (Kostakioti et al., 2013). It works as a stimulus-response-based mechanism that allows cells to coordinate gene expression with other cells and respond by the density of the local population. During quorum sensing, signaling molecules connect to new bacterial receptors and promote gene transcription both within a single or different species of bacteria (Fuqua et al., 2001). Quorum sensing plays important role in development of biofilm. The connection between biofilm and quorum sensing is known as sociomicrobiology. In terms of biofilm formation, environmental stress conditions and food shortages, such as antibiotics, disinfectants, bacterial colonization, identification of annoying species, the formation of typical intestinal flora, and the preclusion of harmful intestinal flora, the quorum sensing system enables the communication between intraspecies and interspecies as well (Frederix and Downie, 2011). Quorum sensing is used by many clinically associated bacteria to control the collective synthesis of virulence components (Rutherford and Bassler, 2012). Efflux system also involved in the regulation of quorum sensing.

Drug resistance mechanisms of biofilm

Biofilms formation is possibly an important commitment from microbial communities not only because a sessile form limits their spatial freedom, but also because architectural growth needs energy disbursement. In return, the microbes encapsulated in biofilms enjoy more protection from numerous types of environmental pressures such as antibiotics, pH, chemicals, and host immunity (Richards and Ojha, 2014). For many infectious bacterial pathogens, biofilms are participated in the process of caseous necrosis, cavitation formation in lung tissue (site in which *Mtb* form a biofilm) their survival against host defense mechanisms and antibiotics(Richards et al., 2019; Richards and Ojha, 2014). Biofilms of mycobacterial in vitro demonstrate more phenotypic resistance to antibiotics than liquid cultures. Bacteria in the biofilms are frequently 1000 times more tolerable against antimicrobial agents than their planktonic (free floating). Though the drug tolerance mechanisms remain unclear, intrinsic and extrinsic factors contribute to the phenomenon. The waxy extracellular matrix restricts the exposure to the antimicrobials, limits growths in biofilm microenvironments could lead to drug-resistant bacteria. Moreover, nutrients depletion and limited oxygen supply help in the development of persister cells of mycobacterium. In mycobacteria, phenotypic drug tolerance also linked with biofilms because asynchronous growth of bacterial cells due to nonuniform microenvironments which leads to asymmetric cell division, cell size, and gene expression. The mature stage biofilm structure shows maximum antibiotic resistance. Many antibiotic resistance mechanisms in biofilm structures are responsible like:

1. *Persister cells: the subpopulation of cells that resist killing by antimicrobials*
2. *Various physiological variations due to slow growth rate and malnourishment responses (environmental stress, hypoxia condition, or nutrient deficiency)*

3. *Reduced dispersion of antimicrobial agents through the biofilm, while some antibiotics are still able to enter inside biofilm*
4. *Changes in the phenotypic characteristics of bacterial cells forming the biofilm*
5. *Genetic changes on target cells or target hiding*
6. *Enzyme-mediated resistance*
7. *Quorum sensing*
8. *Efflux pumps*

Escape from host defense mechanisms

Bacterial biofilm can escape or bypass the host defense system in various ways. Reduced phagocytosis activity (frustrated phagocytosis) or neutrophil and macrophages could not able to engulf the biofilm bacterial cells, enzymes, and WBCs could not get enter inside the biofilms (Rodis et al., 2020).

Persistence and drug tolerance: role of mycobacterial biofilms

Mycobacterial biofilm contains persister cells which are extremely tolerant to isoniazid and also can survive in higher concentration of rifampicin (Boldrin et al., 2020). Study on *Mtb* mutants find out the involvement of various genes in the formation of biofilm, moreover also involved in stress and drug tolerance. The strict environment of biofilm selects a bacterial population naturally tolerant to drug exposure and exogenous stresses, thus growing the population of persister cell (Richards et al., 2019). Lee et al. (2019) reported the catalytic shift of trehalose to other intermediates in order to maintain ATP and antioxidant biosynthetic activities. The alteration of trehalose mono- and dimycolate to the synthesis of its intermediates to the central carbon metabolism for sustain the important functions, which lead to the formation of dormant bacteria. *Mtb* cells form big bunches extracellularly (in necrotic lesions) and intracellularly resembling to the biofilm which is clinically pertinent and plays important role in persistence or induction of tolerance to slow down the treatment effects (Boldrin et al., 2020).

Extracellular polymeric substances: matrix and capsule

The EPS is an vital component of biofilms, helps in the stability and attachment of the biofilm with the solid surface, resulting in aid in biofilm growth (Chakraborty and Kumar, 2019; Kapellos and Alexiou, 2013). The composition of EPS is variable and regulated by types of biofilm formation, which allows harmful bacteria to flourish in harsh host environments. Various environmental factors affect the ingredients of biofilm capsules, such as glycoproteins and polysaccharides (Bjarnsholt, 2013). The matrix enhances bacteria's resistance to antibiotics and other antimicrobial agents. Antibacterial molecules can accumulate up to 25% of the weight of the EPS. The matrix's adsorption sites limit the delivery of biocides and act as an adhesive for extracellular enzymes. These extracellular enzymes protect the motility/absorption of the antibacterial compounds and provide the main source of substrate for biocide metabolite breakdown, resulting in a reduction in the activity of sensitive medications (Dincer et al., 2020; Singh et al., 2017a,b). The entrapments or inactivation of antibiotics by biofilm arise drug tolerance. Though, the drugs or antibiotics that do not interact with EPS components can easily diffuse through biofilm, suggesting some alternate mechanisms also exist for drug resistance mechanism of biofilm (Rodis et al., 2020).

Horizontal gene transfer

One of the established mechanisms of antimicrobial resistance is horizontal gene transfer. Inside the biofilm, the accumulation of genetic elements due to lysis of heterogeneous bacterial species occurs which provide the suitable environmental condition for the uptake of resistant genes to develop drug resistance. Some upregulated ABC transporter may also plays important role to increase the level of antibiotic efflux pump (Goel et al., 2021).

Enzyme-mediated resistance

Enzymes that offer resistance to biofilm mediate the transformation of bactericide into a harmless form. A few bacterial species have been reported as the ability to degrade hazardous chemicals such as phenolic, aromatic, and other heavy metals (cadmium, nickel, mercury, copper, zinc, silver, cobalt, lead, antimony, and others). Detoxification usually takes place by enzymatic reduction of metals and ions resistance genes. The presence of heavy metals induced a larger spectrum of resistant phenotypes (Singh et al., 2017a,b).

Metabolic state of the organisms in the biofilm

The restriction of biofilm-specific proliferation inside the biofilm has been proposed as a reason for biofilm resistance. Bacterial receptivity to antibacterial drugs can vary dramatically depending on the physiological condition of cells and the nature of the environment (Vestby et al., 2020). The configuration of the barrier and the bacterial cell membrane is mainly affected by the limited supply of nutrients. Resistance to antibiotic causes a morphological change in the biofilms, exposed to inhibitory concentrations (Sharma et al., 2019a). The *E. coli* expresses UV light or H_2O_2 resistance in response to heat or hunger stress (Arrage et al., 1993). Another example of Enterococcal strains, that upregulate the expression of antioxidative enzymes after induction of the oxidative stress and downregulation of prooxidative enzymes. However, the percentage of resistant alleles becomes dropped significantly after the removal of bactericides. It has been proposed that nutritional deficiency in biofilm causes it to slow down and enter a hungry state (Maira-Litrán et al., 2000). Because of their activity near the biofilm—bulk water interface, antimicrobial drugs used to treat biofilm diminish their respiratory action. The nongrowing cells are less sensitive to antibacterial compounds when the cells are propagated in high growth rate-rich media (Singh et al., 2017a,b).

In summary, bacterial populations develop systematic strategies like biofilms to survive antibiotic treatment. The increases in antibiotic resistance in biofilm is due to combination of mechanisms such as reduced diffusion, enzyme-mediated, efflux pump, binding of drugs to EPS, persister, and drug tolerant cells and expression of various genes in response to stress condition. Understanding the biofilm-mediated drug resistance will be helpful to shorten the treatment duration and improving clinical outcome for patient suffering from chronic biofilm-based infections.

References

Arrage, A.A., Phelps, T.J., Benoit, R.E., White, D.C., 1993. Survival of subsurface microorganisms exposed to UV radiation and hydrogen peroxide. Appl. Environ. Microbiol. 59, 3545—3550.

Bjarnsholt, T., 2013. The role of bacterial biofilms in chronic infections. APMIS 121, 1—58. Available from: https://doi.org/10.1111/apm.12099.

Boldrin, F., Provvedi, R., CioettoMazzabò, L., Segafreddo, G., Manganelli, R., 2020. Tolerance and persistence to drugs: a main challenge in the fight against *Mycobacterium tuberculosis*. Front. Microbiol. 11, 1924. Available from: https://doi.org/10.3389/fmicb.2020.01924.

Chakraborty, P., Bajeli, S., Kaushal, D., Radotra, B.D., Kumar, A., 2021. Biofilm formation in the lung contributes to virulence and drug tolerance of *Mycobacterium tuberculosis*. Nat. Commun. 12, 1606. Available from: https://doi.org/10.1038/s41467-021-21748-6.

Chakraborty, P., Kumar, A., 2019. The extracellular matrix of mycobacterial biofilms: could we shorten the treatment of mycobacterial infections? Microb. Cell 6, 105—122. Available from: https://doi.org/10.15698/mic2019.02.667.

Costerton, J.W., Irvin, R.T., Cheng, K.J., 1981. The role of bacterial surface structures in pathogenesis. Crit. Rev. Microbiol. 8, 303—338. Available from: https://doi.org/10.3109/10408418109085082.

Dincer, S., Uslu, F.M., Delik, A., 2020. Antibiotic resistance in biofilm, bacterial biofilms. IntechOpen . Available from: https://doi.org/10.5772/intechopen.92388.

Esteban, J., García-Coca, M., 2018. Mycobacterium biofilms. Front. Microbiol. 8, 2651. Available from: https://doi.org/10.3389/fmicb.2017.02651.

Floyd, K.A., Eberly, A.R., Hadjifrangiskou, M., 2017. Adhesion of bacteria to surfaces and biofilm formation on medical devices. In: Deng, Y., Lv, W. (Eds.), Biofilms and Implantable Medical Devices. Woodhead Publishing, pp. 47—95. (Chapter 3). Available from: https://doi.org/10.1016/B978-0-08-100382-4.00003-4.

Frederix, M., Downie, J.A., 2011. Quorum sensing: regulating the regulators. In: Poole, R.K. (Ed.), Advances in Microbial Physiology. Academic Press, pp. 23—80. (Chapter 2). Available from: https://doi.org/10.1016/B978-0-12-381043-4.00002-7.

Fuqua, C., Parsek, M., Greenberg, E., 2001. Regulation of gene expression by cell-to-cell communication: acyl-homoserine lactone quorum sensing. Annu. Rev. Genet. 35, 439—468. Available from: https://doi.org/10.1146/annurev.genet.35.102401.090913.

Goel, N., Fatima, S.W., Kumar, S., Sinha, R., Khare, S.K., 2021. Antimicrobial resistance in biofilms: exploring marine actinobacteria as a potential source of antibiotics and biofilm inhibitors. Biotechnol. Rep. 30, e00613. Available from: https://doi.org/10.1016/j.btre.2021.e00613.

Guzmán-Soto, I., McTiernan, C., Gonzalez-Gomez, M., Ross, A., Gupta, K., Suuronen, E.J., et al., 2021. Mimicking biofilm formation and development: recent progress in in vitro and in vivo biofilm models. iScience 24, 102443. Available from: https://doi.org/10.1016/j.isci.2021.102443.

Jamal, M., Tasneem, U., Hussain, T., Andleeb, S., 2015. Bacterial biofilm: its composition, formation and role in human infections. Res. Rev. J. Microbiol. Biotechnol. 4.

Kapellos, G.E., Alexiou, T.S., 2013. Modeling momentum and mass transport in cellular biological media: from the molecular to the tissue scale. In: Becker, S.M., Kuznetsov, A.V. (Eds.), Transport in Biological Media. Elsevier, Boston, pp. 1–40. (Chapter 1). Available from: https://doi.org/10.1016/B978-0-12-415824-5.00001-1.

Kaplan, J.B., 2010. Biofilm dispersal. J. Dent. Res. 89, 205–218. Available from: https://doi.org/10.1177/0022034509359403.

Koch, R., 1882. Classics in infectious diseases. The etiology of tuberculosis: Robert Koch. Berlin, Germany 1882. Rev. Infect. Dis. 4, 1270–1274.

Kostakioti, M., Hadjifrangiskou, M., Hultgren, S.J., 2013. Bacterial biofilms: development, dispersal, and therapeutic strategies in the dawn of the postantibiotic era. Cold Spring Harb. Perspect. Med. 3, a010306. Available from: https://doi.org/10.1101/cshperspect.a010306.

Kuang, X., Chen, V., Xu, X., 2018. Novel approaches to the control of oral microbial biofilms. BioMed. Res. Int. 2018, 6498932. Available from: https://doi.org/10.1155/2018/6498932.

Kumar, A., Alam, A., Rani, M., Ehtesham, N.Z., Hasnain, S.E., 2017. Biofilms: survival and defense strategy for pathogens. Int. J. Med. Microbiol. 307, 481–489. Available from: https://doi.org/10.1016/j.ijmm.2017.09.016.

Lee, J.J., Lee, S.-K., Song, N., Nathan, T.O., Swarts, B.M., Eum, S.-Y., et al., 2019. Transient drug-tolerance and permanent drug-resistance rely on the trehalose-catalytic shift in *Mycobacterium tuberculosis*. Nat. Commun. 10, 2928. Available from: https://doi.org/10.1038/s41467-019-10975-7.

Limoli, D.H., Jones, C.J., Wozniak, D.J., 2015. Bacterial extracellular polysaccharides in biofilm formation and function. Microbiol. Spectr. 3. Available from: https://doi.org/10.1128/microbiolspec.MB-0011-2014.

López, D., Vlamakis, H., Kolter, R., 2010. Biofilms. Cold Spring Harb. Perspect. Biol. 2, a000398. Available from: https://doi.org/10.1101/cshperspect.a000398.

Maira-Litrán, T., Allison, D.G., Gilbert, P., 2000. An evaluation of the potential of the multiple antibiotic resistance operon (mar) and the multidrug efflux pump acrAB to moderate resistance towards ciprofloxacin in Escherichia coli biofilms. J. Antimicrob. Chemother. 45, 789–795. Available from: https://doi.org/10.1093/jac/45.6.789.

Morales, D.K., Kolter, R., 2014. Microbial biofilms in human disease. Reference Module in Biomedical Sciences. Elsevier. Available from: https://doi.org/10.1016/B978-0-12-801238-3.00132-X.

Richards, J.P., Cai, W., Zill, N.A., Zhang, W., Ojha, A.K., 2019. Adaptation of *Mycobacterium tuberculosis* to biofilm growth is genetically linked to drug tolerance. Antimicrob. Agents Chemother. 63, e01213–e01219. Available from: https://doi.org/10.1128/AAC.01213-19.

Richards, J.P., Ojha, A.K., 2014. Mycobacterial biofilms. Microbiol. Spectr. 2. Available from: https://doi.org/10.1128/microbiolspec.MGM2-0004-2013.

Rodis, N., Kalouda, V.T., Potsios, C., Xaplanteri, P., 2020. Resistance mechanisms in bacterial biofilm formations: a review. J. Emerg. Intern. Med. 4, 1–8. Available from: http://doi.org/10.36648/2576-3938.100030.

Rutherford, S.T., Bassler, B.L., 2012. Bacterial quorum sensing: its role in virulence and possibilities for its control. Cold Spring Harb. Perspect. Med. 2, a012427. Available from: https://doi.org/10.1101/cshperspect.a012427.

Sambandan, D., Dao, D.N., Weinrick, B.C., Vilchèze, C., Gurcha, S.S., Ojha, A., et al., 2013. Keto-mycolic acid-dependent pellicle formation confers tolerance to drug-sensitive Mycobacterium tuberculosis. mBio 4. Available from: https://doi.org/10.1128/mBio.00222-13e00222-00213.

Schulze-Röbbecke, R., Janning, B., Fischeder, R., 1992. Occurrence of mycobacteria in biofilm samples. Tuber. Lung Dis. Off. J. Int. Union. Tuberc. Lung Dis 73, 141–144. Available from: https://doi.org/10.1016/0962-8479(92)90147-C.

Sharma, D., Garg, A., Kumar, M., Rashid, F., Khan, A.U., 2019b. Down-regulation of flagellar, fimbriae, and pili proteins in carbapenem-resistant *Klebsiella pneumoniae* (NDM-4) clinical isolates: a novel linkage to drug resistance. Front. microbiology 10, 2865.

Sharma, D., Misba, L., Khan, A.U., 2019a. Antibiotics vs biofilm: an emerging battleground in microbial communities. Antimicrob. Resist. Infect. Control. 8, 76. Available from: https://doi.org/10.1186/s13756-019-0533-3.

Singh, A., Gopinath, K., Singh, N., Singh, S., 2014. Deciphering the sequential events during in vivo acquisition of drug resistance in *Mycobacterium tuberculosis*. Int. J. Mycobacteriol. 3, 36–40. Available from: https://doi.org/10.1016/j.ijmyco.2013.10.006.

Singh, A., Kumar Gupta, A., Gopinath, K., Sharma, P., Singh, S., 2017a. Evaluation of 5 Novel protein biomarkers for the rapid diagnosis of pulmonary and extra-pulmonary tuberculosis: preliminary results. Sci. Rep. 7, 44121. Available from: https://doi.org/10.1038/srep44121.

Singh, S., Singh, S.K., Chowdhury, I., Singh, R., 2017b. Understanding the mechanism of bacterial biofilms resistance to antimicrobial agents. Open. Microbiol. J. 11. Available from: https://doi.org/10.2174/1874285801711010053.

Sutherland, I., 2001. Biofilm exopolysaccharides: a strong and sticky framework. Microbiol. Read. Engl. 147, 3–9. Available from: https://doi.org/10.1099/00221287-147-1-3.

Vestby, L.K., Grønseth, T., Simm, R., Nesse, L.L., 2020. Bacterial biofilm and its role in the pathogenesis of disease. Antibiotics 9, 59. Available from: https://doi.org/10.3390/antibiotics9020059.

Wallace, R.J., Swenson, J.M., Silcox, V.A., Good, R.C., Tschen, J.A., Stone, M.S., 1983. Spectrum of disease due to rapidly growing mycobacteria. Rev. Infect. Dis. 5, 657–679. Available from: https://doi.org/10.1093/clinids/5.4.657.

WHO, 2021. Global Tuberculosis Report 2021.

Zhang, J., Li, W., Chen, J., Qi, W., Wang, F., Zhou, Y., 2018. Impact of biofilm formation and detachment on the transmission of bacterial antibiotic resistance in drinking water distribution systems. Chemosphere 203, 368–380. Available from: https://doi.org/10.1016/j.chemosphere.2018.03.143.

Zhao, X., Zhao, F., Wang, J., Zhong, N., 2017. Biofilm formation and control strategies of foodborne pathogens: food safety perspectives. RSC Adv. 7, 36670–36683. Available from: https://doi.org/10.1039/C7RA02497E.

CHAPTER 10

Biofilms: cities of microorganisms

Palkar Omkar Prakash[1], Keerthi Rayasam[1], Kolluru Viswanatha Chaitanya[1] and Vidyullatha Peddireddy[2]

[1]Department of Microbiology and Food Science and Technology, GITAM Institute of Science, GITAM (Deemed to be University), Visakhapatnam, Andhra Pradesh, India
[2]Department of Nutrition Biology, School of Interdisciplinary and Applied Sciences, Central University of Haryana, Mahendragarh, Haryana, India

Introduction

Microorganisms have to survive harsh ecological conditions like extreme cold, heat, pH and oxygen alterations, salts, pressures, etc. which make their existence quiet challenging. Very few microorganisms have the genetic make-up or survival strategies that aid them in surviving such extremities. Among the prominent strategies that help the microorganisms to sustain these conditions, biofilm formation is regarded as an extreme adaptation (Yin et al., 2019; Yoshida and Kuramitsu, 2002). Biofilm is a cluster or aggregate of microbial cells that are trapped inside self-created grid of extracellular (exocytosed) polymeric substances that helps in adherence to a tissue or an inert surface superficially. During biofilm formation, the planktonic cells, which are freely moving cells, migrate above tissue or solid surface, recognize the stratum and activates the adherence machinery. Activation of certain genes leads to attachment of planktonic cells to the surface reversibly, which initiates formation of microcolony. The cell mass produce exopolysaccharides that keep the cells intact and adhered to each other making a biofilm. The newly formed biofilm has the capacity to spread new microbial cells which can propagate to new niche and initiate formation of new biofilm (Yin et al., 2019; Yoshida and Kuramitsu, 2002; Chao and Zhang, 2012; Webb et al., 2003). As the microbial cells switch to biofilm mode, phenotypic alterations take place leading to regulations of large number of genes (Shakibaie, 2018). For example, in *Escherichia coli*, it was found that around 230 genes are regulated and expressed including those responsible for encoding adhesion protein, outer membrane proteins, auto-aggregation (OmpF, OmpC, OmpT, and Slp), and lipid A biosynthesis associated proteins (Schembri et al., 2003).

In natural conditions, biofilms are the typical model of perseverance for bacteria. It aids in increasing resistance toward antibiotics and antimicrobial agents, being responsible for protection of microbial community from the external environmental stresses. Biofilms pose a severe challenge to the medical practitioners to treat chronic infections which are

Bacterial Survival in the Hostile Environment
DOI: https://doi.org/10.1016/B978-0-323-91806-0.00017-5

not eradicated by the resistance of biofilms (Hernández-Jiménez et al., 2013). Biofilm can spread throughout the human circulatory system infecting or covering almost all organs such as lung, epithelium, heart, etc. Presence of biofilm has been noted in cystic fibrosis (CF) patients (lung tissue), in clinical wounds, in patients with otitis (middle ear), in chronic rhinosinusitis cases, on medical aids such as prosthetic heart valves and urinary catheters (Hernández-Jiménez et al., 2013; Bjarnsholt et al., 2009).

Biofilm helps in delivering planktonic bacteria into the host which can regulate certain components of host immune system, altering the activity of apoptosis, innate immune receptors, and by making host immunosuppressant (Thakur et al., 2019). In this chapter, we have discussed mechanisms used by few pathogenic and probiotic microorganisms to develop biofilms.

History of biofilms

Antonie van Leuwenhoek, in 1670s used primitive version of microscope to describe "animalcules," which was isolated from tooth surfaces of human. Claude Zobell, in 1934 observed marine cell cultures under direct microscope and concluded that there are bacterial cells that can get themselves attached to surfaces and become sessile. At Forsyth Dental Center during 1935, microbiologists Van Houte and Ron Gibbons studied bacterial biofilms that create oral plaque subsequently forming macroscopical deposits over teeth. Kevin Marshall and Ralph Mitchell in 1964 studied initial stage of biofilm formation in pure cultures, and their study revealed reversible adsorption by bacterial cells initially, followed by irreversible attachment which makes up the first stage of biofilm foundation (Costerton, 1999). Sputum samples from patient of CF, caused by *Pseudomonas aeruginosa*, showed that polysaccharides are the main components of biofilm when treated with ruthenium red dye and subsequently fixed by osmium tetroxide (Alam et al., 2019). New strategies for studying biofilms were developed in the subsequent years, which demonstrated the pervasive biofilm contaminations in medicine.

How are biofilms formed?

Bacterial cells can develop biofilm anywhere, and these mechanisms of biofilm development are extremely preserved in different species. For bacteria surviving in extreme environmental conditions, biofilm formation is a frequently observed phenomenon, in which adhesion and protection have highest priority (Alam et al., 2019). Quorum sensing (QS) signaling helps in triggering expression of gene encoded with biofilm formation. QS signaling molecules are secreted constantly at low levels and are sensed through respective receptors. The trigger for behavioral alteration is not issued until the adequate number of bacteria are concentrated. Once the threshold is reached, bacteria generally adept to biofilm formation (Yoshida and Kuramitsu, 2002).

Biofilm formation involves four stages: (1) attachment to surface, (2) microcolony formation, (3) development of three dimensional mature matrix, and (4) dispersal (Alam et al., 2019) (Fig. 10.1).

Initial attachment during biofilm construction is a reversible connection. The mobile planktonic cells can adhere to abiotic or biotic surfaces depending upon the genetic make-up of microorganism. The initial binding force can be Van der Waals force, hydrophobic interactions, etc., which means the planktonic cells can disengage themselves from the surface due to lack of proper adhesion site or lack of nutrition in the vicinity, unfavorable temperature, or pressure. If cells sense substratum as suitable, cell surface entities such as fimbriae, flagella, pili are utilized to initiate permanent (irreversible) attachment. Studies indicate that surface properties of substratum are extensively responsible to microbial adhesion. Aggregation of bacterial cells to form microcolony succeeds irreversible adhesion. In this stage, the cells undergo mass reproduction and colonize the substratum. To be connected with every cell and exchange

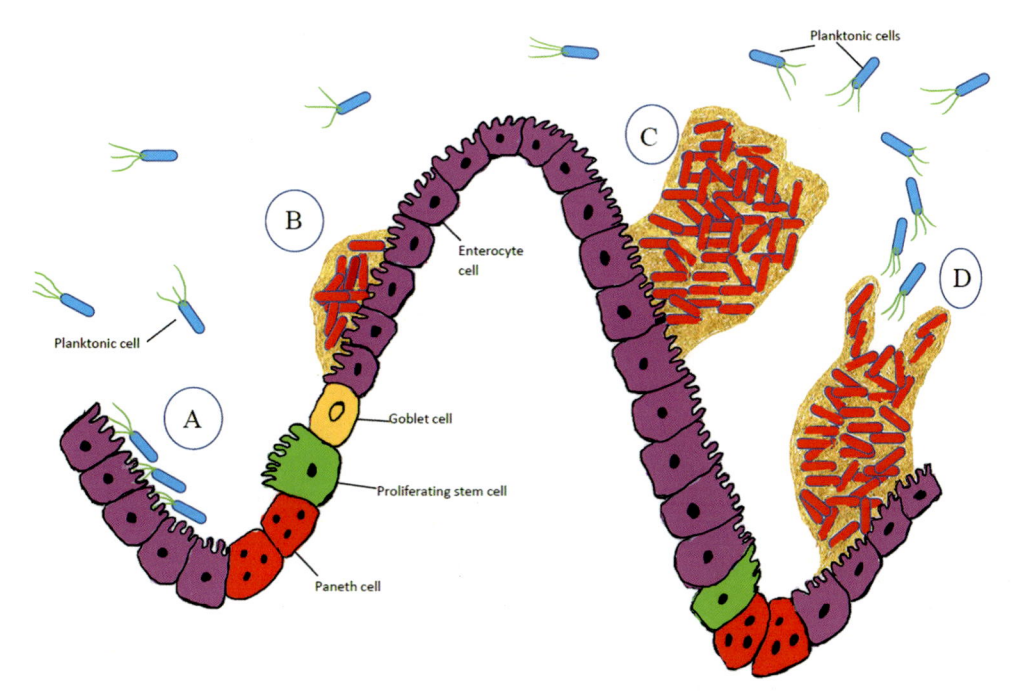

Figure 10.1 The process of biofilm establishment. (A) Attachment of planktonic cells to the epithelial cells; (B) attached planktonic cells form microcolony by secreting exopolysaccharides (EPS), cells lose their ability to swim; (C) biofilm formation by increased cell mass, EPS production, maintenance of anaerobic condition; (D) dispersion of planktonic cells due to nutrient deficiency or unfavorable condition, planktonic cells move over other surfaces to form a new biofilm.

signal, chemical signaling molecules are produced. Exchange of information triggers cells to produce signaling molecules, exopolysaccharide, peptides to improve structural integrity of the biofilm. Bacterial growth coordination is done by QS, a signaling system in which various autoinducers regulate cell density by genetic expression. Expression of genes related to exopolysaccharides (EPSs) production triggers formation of tri-dimensional structure. The intricated tri-dimensional matrix of a biofilm is facilitated with channels that help in nutrient absorption and delivery of waste material. Different bacterial biofilm contains different components in their EPSs giving a great phenotypic diversity to biofilms. EPSs may contain DNA, protein, lipids, amyloids, etc. Ultimate phase in biofilm is recognized by dispersion, disintegration of cells from the mature biofilm. The shift of lifestyle from sessile to motile may be due to fluid pressure around the biofilm or by dispersal due to environmental factors, limited nutrients, or other signals. Signals responsible for expression of enzyme for EPS degradation and motility is activated to detach cells. The detached cells get adapted to planktonic life or get spread out to colonize new niches (Alam et al., 2019).

Impact of biofilms on human

Detrimental biofilms bound to mucosal tissue have been linked to various human diseases such as gum diseases, surgical implant infections, lung infections, catheter related urinary, and some intestinal infections. A chronic stage in any disease condition emerges when an acute infection undergoes dormancy for a long time with repeated reoccurrence. A compromised immune system, destruction of mucosal barriers, postsurgery infections, and altered microbiota can lead to chronic infection (Harrell et al., 2021). Alike latent infections, chronic infections frequently do not cause instant threat to host as these infections are mostly asymptomatic. In some cases, reactivation of chronic/latent infections lead to medically significant diseases or life-threatening illness. Pathogen's survival and spreading are important factors for initiating reactivation; biofilms facilitate both to establish bacterial infections. Both the survival and propagation of bacteria are facilitated through biofilm formation. During chronic disease conditions, biofilm mediated increased bacterial growth helps bacteria to increase tolerance toward hostile environment posed by host responses such as antibodies, antimicrobial peptides, phagocytic response of macrophages and neutrophils, complementary systems, and antibiotics (Harrell et al., 2021).

Richard Chole stated that bacteria possess the ability of biofilm formation and establish in folds of tonsils which may act as a safehouse for bacterial survival and reinfection powerhouse (Shakibaie, 2018). Biofilms *of Staphylococcus epidermidis*, *P.aeruginosa*, and *Staphylococcus aureus* are most commonly identified biofilms. *S. epidermidis* is regarded as a model organism in dental implant associated biofilms (Harrell et al., 2021). *Proteus mirabilis* is responsible for a number of infections linked to catheter-associated urinary infections due to the capability of biofilm formation. The factors that are responsible for biofilm establishment in *P.mirabilis* include transcription factors, adhesion factors, proteins from lipopolysaccharide (LPS) synthesis, two component systems, and communication factors (Shakibaie, 2018).

The human gastrointestinal tract (GIT) is a suitable place for microorganisms to colonize and establish a biofilm. Various entero-pathogens are responsible for disease conditions, ranging from symptomless, mild signs and self-limiting to life-threating chronic infections. Numerous studies have concluded that these entero-pathogens (*Listeria monocytogenes*, *Salmonella enterica*, *Shigella* spp., *E.coli*, *Campylobacter* spp., *Yersinia enterocolitica*, etc.) can effectively form biofilm in vitro. Microscopy studies indicate presence of dense mucosa-linked biofilms covering tissues (Motta et al., 2021).

Biofilms and food sector

Biofilm formation in a food industrial ecosystem can impact not only on the health of consumer but also the market value of the product. Often the biofilms of concern in food industries are formed by human pathogens on various substratum like polyethylene, stainless steel, glass, rubber, wood, etc. (Abdallah et al., 2014; Nikolaev and Plakunov, 2007). Biofilm-coupled detrimental effects like variation in organoleptic properties and corrosion of metallic surfaces deteriorate the quality of food and cause significant economic losses. In food industries like dairies, numerous equipment or structures like raw milk holding tanks, pasteurizer, homogenizer, packaging facilities act as a perfect substratum for biofilm formation accompanied by different processing temperatures aid in colonization of variety of species (Mizan et al., 2015). Pathogens like psychrotrophic *Pseudomonas* spp. and thermophilic *Geobacillus stearothermophilus* are some of the examples of biofilm formers in dairies. *L.monocytogenes*, *Aeromonas hydrophila*, *Vibrio* spp., and *S.enterica* are responsible for biofilm development in fresh fish products. Bacterial biofilms, associated foodborne diseases, harboring on food matrixes and equipment may cause infections and intoxications. Food intoxication can lead to an outbreak if the biofilm capable of secreting toxin contaminates food product during food processing (Mizan et al., 2015).

Bacteria forming the biofilms can acquire genetic variations in biofilm associated genes, which in turn, might be responsible for formation of entirely diverse biofilm and according to surrounding environmental conditions. These genetic variations make it difficult to eliminate biofilms from food-processing units in food industry. Location of biofilms in food industry depends on the nature of food-processing units but mainly includes milk, water, membranes/filters, liquid pipeline, gloves, storage silos, surfaces in frequent contact, packaging material, etc. Existence of biofilm developers inside or around food industries can jeopardize human health and incur huge financial losses (Camargo et al., 2017).

Mechanisms used by various microorganisms to form biofilm

The symbiotic interaction of epithelial cells of GIT with the human intestinal microbiota is a complex process. The intestinal microbiota is crucial in forming the first line of defense toward attack of pathogenic microorganisms, providing nourishment, immune responses, and regulating epithelial development. Consecutively, the host offers steady conditions such as pH, temperature, food supply, and osmolarity for the microbiota. Microbiota utilize these favorable conditions to form microcolonies and subsequently biofilms (Vélez et al., 2010).

Different microorganisms have different sets of proteins or polysaccharides that help in adherence to the enterocytes and are involved in biofilm formation such as Bhp (Bap homologous protein) and Bap (biofilm-associated protein) in *S.epidermidis*, Mus20 (mutants unattached to seeds) in *Pseudomonas putida*, VPA1445 (calcium-binding protein) in *Vibrio parahaemolyticus*, Esp (enterococcal surface protein) in *Enterococcus faecalis*, Bap in *Burkholderia cepacian*, LapA (large adhesin protein) in *Pseudomonas fluorescens*, YeeJ (inverse autotransporter) in *E.coli* (Latasa et al., 2005).

Biofilm by pathogenic microorganisms

Studies prove that chronic infections are persistent owing to the biofilm forming ability of pathogenic microorganisms. Biofilms are formed when the host is in an immunocompromised state may be due to drug treatment, trauma, immune deficiency, or tissue damage. Pathogenic microorganisms get protection from the antimicrobial compounds when they develop biofilm, leading to increased pathogenesis and increased antimicrobial resistance (Shakibaie, 2018). Mechanisms or factors responsible for biofilm formation in few pathogenic microorganisms have been discussed below.

Pseudomonas

Pseudomonas aeruginosa is Gram-negative microorganism present in soil and water. *P.aeruginosa* is commonly known for its opportunistic pathogenesis in immunocompromised patients. *P.aeruginosa* genesis acute and chronic lung infection in CF cases resulting in substantial sickness and deaths. It is estimated to be responsible for 10%—20% hospital acquired infections. *P.aeruginosa* encodes several characters that helps it to occupy and sustain in acute and chronic condition which includes, capacity to form biofilms, enhanced virulence compounds or complexes, and high resistance to antimicrobials. Further, this microbe utilizes QS, a cellular interaction between two cells, a procedure which helps in regulating many of these factors leading to the pathogenesis of *P.aeruginosa* (Wagner and Iglewski, 2008).

P.aeruginosa utilizes QS to synchronize cellular behavior by producing autoinducer, a diffusible signaling chemical. It encodes two QS systems, that is, rhl and las systems. In rhl, RhlR protein and its autoinducer, C4-HSL (N-butyryl homoserine lactone) are present. Similarly in las system, LasR (transcriptional regulatory protein) and its related autoinducer signal molecule, 3O-C12-HSL (N-(3-oxododecanoyl) homoserine lactone) are present. The las system influences rhl system and are intertwined through third signal particle produced by *P.aeruginosa*, the Pseudomonas quinolone signal (PQS) or 2-heptyl-3-hydroxy-4-quinolone (Wagner and Iglewski, 2008). Davies et al. concluded that lasI encoding synthetase control production of 3O-C12-HSL [N-(3-oxododecanoyl) homoserine lactone] is crucial for formation of biofilm in *P.aeruginosa* PAO1 (Davies et al., 1998).

Alginate, an exopolysaccharide, when overproduced marks the transformation of *P.aeruginosa* to mucoidal form. The mucoidal *P.aeruginosa* under CF condition is connected to chronic infection (Wagner and Iglewski, 2008). Alginate, a major element of complex biofilm of mucoidal *P.aeruginosa*, confers high resistance to antimicrobial compounds (Hentzer et al., 2001). Alginate is not a prime constituent of biofilm matrix in nonmucoidal strains like *P.aeruginosa PAO1* and *P.aeruginosa PA14*, instead they have two loci, pel gene in *P.aeruginosa PA14* and psl gene in *P.aeruginosa PAO1*. Gene encoding for pel produces dextrose rich polysaccharides that is important for pellicle establishment and biofilm progress. Extracellular DNA has a major role in initial stages of biofilm development. Autoinducers 3O-C12-HSL and C4-HSL along with PQS help in modulating extracellular DNA throughout planktonic and biofilm growth. In early stages of biofilm, large amount of extracellular DNA is produced compared to matured biofilm in which it is located in the stalk structure of mushroom-shaped biofilm. It is proposed that the extracellular DNA might be released due to prophage attack or release of membrane vesicles. Biofilms that are short on extracellular DNA concentration are sensitive to SDS (Wagner and Iglewski, 2008; Davies et al., 1998; Hentzer et al., 2001; Yang et al., 2007).

Rhamnolipid is an important moiety for initiation, maturation, and dispersion of *P. aeruginosa* biofilm and it has been reported that rhlA is responsible for its synthesis (Wagner and Iglewski, 2008). Tolker-Nielsen and Pamp concluded that rhamnolipid is vital in development of cap in mushroom-structured biofilm. They also stated that rhlA has vital role in microcolony establishment as rhlA mutants exhibited flat, thin biofilms (Pamp and Tolker-Nielsen, 2007). Rhamnolipid is essential for scattering of bacteria from the biofilm. Dispersing bacteria are phenotypically altered and are known as colony morphotypic variants. Small colony variants (SCVs), isolated from *P. aeruginosa PAO1*, showed increase in antimicrobial resistance and increase in biofilm developing ability. Maintenance of anaerobic condition has a significant role in survival of *P.aeruginosa* in the lungs and inside biofilm too. Studies indicate, oxygen depletion can be observed around 30 μm from the base of the biofilm (Wagner and Iglewski, 2008). During formation of biofilm by *P.aeruginosa* in CF cases, a shift from aerobic to anaerobic metabolism is observed (Hassett et al., 2002). Oxygen gradients have been observed at the base and center of *P.aeruginosa* biofilm (Wagner and Iglewski, 2008).

Enterotoxigenic *Escherichia coli*

World Health Organization reported that about 2 million humans die each year from *Escherichia coli* related diarrheal condition. Most of the gastrointestinal diseases occur due to infection by *Enterotoxigenic E.coli* (ETEC). Two toxins—heat stable enterotoxin and heat labile enterotoxin are the primary roots of diarrhea in ETEC infections. Recognition and adherence of the bacterial cells to the host intestinal surfaces are the essential steps toward pathogenesis. In ETEC, TibA (potent bacterial adhesin) is regarded as potential bacterial adhesin, which facilitates attachment of bacterial cells to human intestinal cells. TibA is also responsible for invasiveness of ETEC. This adhesin protein (autotransporter) belongs exclusively to the bacterial glycoprotein group, which is presented on the surface. The glycosylated protein has the ability to bind and confer invasiveness toward human intestinal cells. The TibA protein was initially detected in standard *ETEC serotype O78:H11 strain H10407*. The TibA loci regulates TibA synthesis, a 104-kDa protein that is transported to outer membrane (Sherlock et al., 2005).

Translation product of TibA has 989 amino acid residues which undergoes post-translational alterations. Firstly, a 54 amino acid peptide is removed during transportation to periplasm. TibA contains repetitive amino acid sequence motifs as it is grouped under autotransporter subfamily which also includes AIDA, filamentous hemagglutinin, and antigen43 (Ag43). Alike this protein, TibA also contains N-terminal passenger region and C-terminal translocator region. Translocator portion undergoes modification to form β-barrel porin, a protein moiety in the outer membrane, which helps attachment portion to reach out to the surface. Subsequently, few more alterations take place as TibA is a carbohydrate-protein moiety. A second gene tibC is located to the upstream of tibA gene that encodes (406 amino acid residues) enzyme glycosyltransferase, apparently heptosyltransferase which transfers the glycosyl group from the TibA protein without which TibA will not bind to human intestinal cells. Studies indicate TibA is a multifaceted protein as it performs roles of adhesin, invasion, biofilm enhancer, and auto-aggregator (Sherlock et al., 2005).

Vibrio cholerae

Vibrio cholerae, a comma-shaped Gram-negative bacterium, is estimated to affect millions worldwide and causes thousands of mortalities each year. *V.cholerae* inhabits various water bodies such as lakes, rivers, oceans. *V.cholerae* exists in two forms, a free-swimming—planktonic state or sessile state inside a biofilm. The alteration from planktonic to biofilm state is controlled by c-di-GMP (bis-$(3' - 5')$ cyclic dimeric guanosine monophosphate), a subordinate signaling molecule. The expression of genes responsible for biosynthesis of vibrio polysaccharide and material essential for biofilm development is controlled by c-di-GMP. Two GTP molecules are utilized to synthesize c-di-GMP by using diguanylate cyclases comprising a GGDEF region. Phosphodiesterase (HD-GYP or EAL region) is utilized to disintegrate c-di-GMP into either two molecules of GMP or 5′-phosphoguanylyl-3,5′-guanosine (5′pGpG). Domains GGDEF, EAL/HD-GYP are labeled because of their conserved amino acid sequences. Biofilm formation is increased when high levels of c-di-GMP is attained while low level is required for motility or planktonic life (Young et al., 2021).

Diguanylate cyclases and phosphodiesterase perceive environmental signals directly with the sensory domain or through communication with other signaling proteins. In case of *V.cholerae*, the signaling molecules can be polyamines that are linear aliphatic carbon chains with multiple amine groups which are positively charged (Young et al., 2021). The polyamines spermidine and spermine inhibit *V.cholerae* biofilm development, while norspermidine increases biofilm development capacity (Joshua et al., 2006; Melo et al., 2017).

Salmonella

Along with *Salmonella typhi*, several additional serotypes belonging to *S.enterica* are capable of colonizing and subsequently forming biofilms. The biofilm development in the serovars is extremely conserved specifically in those serovars which colonize multiple hosts. The ability of forming biofilm is advantageous throughout the transmission and cycle of infection. Alike other biofilm forming microorganism, salmonella biofilm contains carbohydrate, protein, extracellular DNA (eDNA), primarily network of proteinaceous curli fimbriae. Additionally, EPS may contain cellular protuberance such as adhesive fimbriae or flagella or cell surface large biofilm-related protein, BapA (biofilm associated protein) and colanic acid—an exopolysaccharides or O-antigen capsule (Harrell et al., 2021).

The change of state from planktonic to biofilm is controlled by various environmental factors, which include nutrient availability, temperature fluctuations, encountering with harmful substances, etc. Several global regulators, small RNAs, two-component system activate expression of main components for biofilm. CsgD encodes site for curli biosynthesis operon. CsgD is master biofilm regulator. CsgD activates production of curli fimbriae along with cellulose by attaching the CsgB promoter in-between CsgBAC curli biogenesis operon and via triggering the activity of diguanylate cyclase AdrA, which later controls biogenesis of cellulose by activating (3′-5′)-cyclicdiguanosine monophosphate (c–di–GMP), a secondary signaling molecule (Harrell et al., 2021). CsgD might be regulating the activity of additional genes required for producing O-antigen capsule and BapA (Latasa et al., 2005).

Majority of *Salmonella* chronic infections occur within gallbladder and progresses when gallstone or other abnormalities are present. This makes the surface appropriate to attach and establish biofilm. *S. typhi* attaches gallstone cholesterol surfaces by the help of FliC on flagellar appendages (Harrell et al., 2021).

Campylobacter jejuni

Campylobacter jejuni is a Gram–negative microorganism often detected in foodborne bacterial infection cases. *C.jejuni* is known to cause neuropathy conditions such as Miller-Fisher and Guillain-Barre syndromes in which it is hypothesized that *C.jejuni* mimics human gangliosides and lipooligosaccharides which initiates autoimmune response. Studies indicate that *C.jejuni* survives in the environment by making biofilm and is usually present in biofilms of other bacterial species (Joshua et al., 2006). According to certain reports, flagellar expressions of flaA and flaB are essential for initiation of biofilm formation by *C.jejuni*. Yet in absence of flagellar expressions, the bacterium is not completely prevented from its sessile form. Previous studies have reported the involvement of various genes in adhesion of *C.jejuni*. Among these genes, cadF gene encoding binding protein CadF fibronectin is mainly important (Melo et al., 2017).

Few studies have detected presence of QS systems in *Campylobacter*. *Campylobacter* produces acyl-homoserine autoinducer (AI-1) which gets accumulated in the extracellular matrix (ECM) and can freely diffuse through bacterial cytoplasm. Upon high bacterial density, it attaches to luxS (cellular transcription enhancer), encoding for luciferase, a crucial enzyme for SAM recycling pathway (S-adenosylmethionine). The gene encoding for luxS is also linked with formation of autoinducer-2 (AI-2). Growing bacterial cells increases AI-2 levels in the surrounding. *C.jejuni* have functional luxS enzyme and has capacity to produce AI-2. But existence of nutrients is must for AI-2 production (Melo et al., 2017).

Streptococcus mutans

Streptococcus mutans has been mainly linked with human dental caries. The capability of forming biofilm (oral plaque) on dental surfaces is an important property of *S.mutans* (Yoshida and Kuramitsu, 2002). *S.mutans* has the ability to utilize sucrose and produce extracellular polysaccharides. *S.mutans* utilize sucrose to synthesize α-1,3 and α-1,6 linked glucan polymers by using 3 glucosyltransferases (gtfB, gtfC, and gtfD). The gtfB and gtfC encode enzyme which induce production of water-insoluble D–glucose polysaccharide with α–1,3 linkages. This water insoluble products have an important role in adhesion to teeth and effective initiation of dental caries. The gtfD gene encodes a product that catalyzes glucan production with α–1,6 linkages. This end product has greater solubility than gtfB and gtfC end product.

The glucan produced by *S.mutans* is essential for adhesion and establishment of an extracellular polysaccharide matrix that can confer resistance toward normal mechanical forces created during brushing teeth and from host immune system (Burne et al., 1997). Presence of carbon source in the surroundings triggers a signal of biofilm development (Loo et al., 2000). comB gene encoding additional protein for ComA-ABC transporter is essential for production of competence-stimulating peptide (CSP). Vast group of closely associated streptococci regulate CSP-mediated QS system for biofilm formation (Li et al., 2001).

Veillonella parvula

Veillonella parvula is a Gram-negative commensal microorganism and a causing agent for dental caries and periodontitis. It acts as opportunistic pathogen upon suitable growth conditions for the bacterium. *V.parvula*, an important colonizer of dental plaque, promotes multispecies growth and plays prime role in lactic acid fermentation. FNLLGLLA_00516 encoding T5SS type Vc trimeric autotransporter has been identified in *V.parvula* as an outer membrane protein precisely found in Gram-negative bacteria and are known for their binding capability to diverse surfaces including other bacteria. In *V.parvula SKV38*, a trimeric autotransporter, Veillonella trimeric autotransporter A (VtaA) was identified which is crucial for biofilm formation. VtaA also plays a role in autoaggregation of *V.parvula SKV38*. In *V.atypica*, along with Hag1, YadA-like transporters (present in *Yersinia* species) have been identified. VtaA acts as an adhesion protein on biotic and abiotic surfaces. In *V.parvula*, VtaA and eight gene group are essential in initiation of attachment to a glass surface independently (Béchon et al., 2020).

V.parvula resides over different parts of the body and is expected to have large number of adhesion mechanisms to get attached to different surfaces such as tooth enamel or epithelial cells owing to the differences in cell anatomy. Additionally, *Veillonella* has been known for its coaggregation with *Streptococci* which aids in producing lactate. It has been demonstrated that *Veillonella* coaggregates with *actinomyces* and *streptococcus* from the same microbiota as the coaggregation could permit them to have a strong hold on the niche and help in colonization. FNLLGLLA_01127, a homologous of *B.subtilis* YqeK regulating phosphatase activity, which has role in pellicle formation and biofilm development has been identified (Béchon et al., 2020).

Biofilm by beneficial microorganisms (probiotics)

A probiotic microorganism is a live microbe which when consumed in suitable quantities can confer health benefits to host. Probiotic is characterized by capacity to adhere to human intestinal cells by which they can colonize the host, act against pathogenic microorganisms, interact with host cells to improve the immunity of host (Vélez et al., 2010). Many probiotic microorganisms have been used up-to-date such as *Lactobacillus* species, *Bifidobacterium* species, etc. Mechanism of how probiotic microorganisms (*Lactobacillus rhamnosus* GG and *Bifidobacterium* species) form biofilms has been discussed below.

Lactobacillus rhamnosus GG

Lactobacillus rhamnosus GG possess good adherence capacity toward immobilized mucus and epithelial cells in human gut. *L.rhamnosus* GG facilitates biofilm development on abiotic substratum more efficiently than any other *Lactobacillus* species. Mostly two proteins have been analyzed to construct the mechanism of biofilm establishment in *L.rhamnosus* GG, out of which LGG_01865 is a protein chain made up of 2419 amino acids consisting of a C-terminal LPxTG motif linked with sortase enzyme activity, possessing important role in attaching of the protein to the bacterial surface. LGG_01865 has a role of adhesin during biofilm establishment and attachment to CaCo-2 cells (Vélez et al., 2010).

LGG_01866 is a presumed protein chain (496 amino acid residues). It has 50% similarity to transcriptional antiterminators (BglG) of other *Lactobacillus* species beside 24% similarity to M-protein trans-acting positive transcriptional regulators (Mga) in *Streptococcus* species, which have been proven to regulate several genes required for establishment on host tissues. The locus of LGG_01866 is upstream of LGG_01865 in a different direction (Vélez et al., 2010).

LGG_01866 and LGG_01865 have similar functioning like Mga and Embp of *group A streptococci* (GAS), respectively. Mga of GAS is regarded as global transcriptional activator which expresses the genes controlling transcription of ECM binding proteins. Additionally, Mga mutants of *group A streptococci* displayed reduced attachment toward ECM components and human skin tissue. Mga binding putative gene was situated upstream of LGG_01865 indicating that LGG_01866 may be a catalyst of LGG_01865. Phenotypic studies of LGG_01865 eliminated mutant showed decreased capability of biofilm formation toward abiotic substratum, colonic epithelial cells, and tissues of mouse GIT, hence proving that LGG_01865 has important function in *L.rhamnosus* GG and host interactions. Vélez et al. named LGG_01865 as modulator of adhesion and biofilm (MabA) (Vélez et al., 2010).

Bifidobacterium

Bifidobacterium species have been used as probiotics for human ingestion because of their valuable role in intestinal nutrition and health. Environmental stresses like nutrient starvation, pH, bile, and oxidative stress can trigger the biofilm formation. Oxidative stress can occur due to metabolic processes too, which generates reactive oxygen species (ROS) indicating a SOS (save our soul) trigger. SOS response is activated when a cell undergoes DNA damage, and LexA and RecA proteins have important role in regulation of this response. Oxygen treatment around 3% v/v can induce oxidative stress in *Bifidobacterium longum BBMN68* to initiate biofilm formation. *B.longum FGSZY16M3* encodes guaB, groS, sufD, recF, atpD, atpH, and atpF genes along with LexA-RecA proteins as a SOS response tactic, whereas dnaK and recA act as response to oxidative stress (Liu et al., 2021).

Almost all *Bifidobacterium strains* possess luxS genes which regulate production of standard QS signaling autoinducer-2 (AI-2). In earlier stages of biofilm development TadIV pili, the surface appendages help in sensing the surface for attachment. Once the cell concentration reaches an appropriate threshold, genes responsible for QS are expressed which includes autoinducer peptide (AIP), AI-2 (luxS) along with some signaling molecules like amyE and livK. Gram-positive bacteria use AI-2 and AIP signaling molecules. Homologous genes for RbsB-like receptors have been recognized within *B. longum FGSZY16M3* and their expression have shown that AI-2 acts as QS signaling factor. ABC transport system or twin arginine translocation (TAT) pathway is used to modify and deliver AI-2 peptides. Once the required threshold of AI peptide is reached, kinase protein is triggered to phosphorylate response regulator. Peptide transporting machinery can distinguish various substrates along with resources essential for biofilm development. For example, dppBCDF is behind uptake of dipeptides and tripeptides in ABC transport system (Liu et al., 2021).

Studies on *B.longum FGSZY16M3* indicate presence of three yidC genes, which are considered to play vital part in biofilm development. Basically, yidC proteins act as membrane integrated chaperones or insertases in association with SecYEG translocon. Knocking of yidC genes in streptococcus have shown changes in glucan production and the cell envelope, leading to disordered EPS composition and biofilm formation (Liu et al., 2021).

Strategies or future trends against biofilms

At present, clinicians are facing major challenge for biofilm treatment. Since discovery of antibiotics, it has been an ultimate option for any kind of infection, but microorganisms have their own survival tactics. Antibiotic treatment has become inadequate as a defense mechanism against biofilm associated infections. Counters against biofilm can be planned once we understand the nature of biofilms. Treatment against biofilm can include use of probiotics and their derivatives which have better penetration than regular antibiotics, compounds targeting QS or anti-QS drugs, and administration of high dose of antibiotics or in combination (Barzegari et al., 2020).

Striking the QS mechanism can infiltrate the biofilm development process, making bacterial virulence vulnerable and easy to taken down by host immune system. Compounds that are used to inhibit QS are regarded as new generation antimicrobial agents but this have not been successful. Numerous strategies have been presented to interrupt QS; one such strategy includes utilization of an antagonist for deactivation of LuxR utilizing homologs of N-acyl homoserine lactone (AHL), which conflicts with the natural AHL to attach with LuxR type receptor. The attachment of competitor inhibits AHL attachment causing inactivation of LuxR which inhibits the expression of virulence element (Barzegari et al., 2020; Hentzer et al., 2002).

Probiotic microorganisms have the ability to hamper the functionality of pathogenic bacteria by targeting the QS signaling molecules leading to inhibition and eradication of biofilm. Probiotics synthesis certain antagonistic substances such as organic acids, hydrogen peroxide, surfactants, some enzymes like amylase, lipase, and generate adverse conditions like pH alterations, competition for nutrients, and surface for colonization. Various probiotic strains which have been reported to possess biofilm inhibition potential include *L.brevis*, *B.lactis*, *L.rhamnosus*, *L.salivarius*, *L.fermentum*, *L. delbrueckii*, *Streptococcus salivarius*, *L.pentosus*, *B.longum*, *L.sporogenes*, *L.acidophilus*, *L.casei*, *L.plantarum*, and *S.oralis* (Barzegari et al., 2020).

Lactobacillus strains are known for producing various metabolites like bacteriocins, biosurfactants, EPSs, and oxygen reactive species which aid in antibiofilm activity. The EPS produced by *Lactobacillus strains* acts as antibiofilm, antioxidant, and stimulating factor for immune system. It was found that EPS of Lactobacillus strains was antagonistic toward Gram-negative (*S.typhimurium* and *P.aeruginosa*) as well as Gram-positive (*S.aureus* and *L.monocytogenes*) bacterial strains (Barzegari et al., 2020). Mechanisms for activity of bacteriocin against biofilm are not clear, but some bacteriocins induce a pore development on the cell surface of bacteria causing ATP leak, whereas some act as proteolytic enzymes (Okuda et al., 2013). Anti-QS effect of subtilosin was studied by Chikindas et al. against *E.coli O157:H7*, *Gardnerella vaginalis ATCC 14018*, *L.monocytogenes Scott A*. Subtilosin showed 60% inhibition in *E.coli*, 90% in *G.vaginalis*, and 80% in *L.monocytogenes* (Algburi et al., 2017). *Bacillus sonorensis MT93* produces sonorensin that can decrease *S. aureus* cell viability in biofilm, inhibition of biofilm attachment (Chopra et al., 2015). Biosurfactant produced by *Pediococcus acidilactici* and *L. plantarum* can obstruct attachment and biofilm development of *S.aureus CMCC 26003*. These biosurfactant affects regulation of biofilm related genes and interferes with AI-2 release in QS system (Yan et al., 2019).

Monocyte cells can be activated by *Streptococcus thermophilus* to release TNFα, IL-1β, IFN-γ, and IL-6, which help in activation of innate immunity to destroy pathogens. TNFα, CCL20, and IL-6 secretion can be activated by *L.paracasei DG* in monocytes leukemia cell line. Increased IFN-γ production and decreased IL-10 production can be achieved by *Lactobacillus* species to activate immunomodulatory effect against *S.mutans* in human cell lines (Barzegari et al., 2020). Inhibition of *V.parahaemolyticus* and *V.cholerae* biofilm was achieved by *Lactobacillus* species. Culture supernatant (CS) of seven *Lactobacillus* isolates showed antibiofilm activity on *V. cholerae* biofilm by dispersion activity (Kaur et al., 2018). Studies indicate that *Lactobacillus* and *Saccharomyces* could lower the danger of diarrhea associated with *C.difficile*. *Lactobacillus reuteri* and *Bifidobacterium infantis* biofilms could be used to delay growth of *L.monocytogenes* (Barzegari et al., 2020). Low levels of lactoferrin and human cationic host defense peptide LL-37 present around mucosal surfaces were used against *P.aeruginosa* biofilm which strongly inhibited the biofilm (Shakibaie, 2018). Nitric oxide in combination of SDS or antimicrobial showed enhanced activity against *P.aeruginosa* biofilms (Yoon et al., 2006).

As a strategy for prevention of biofilm formation on the medical tools, they are being coated with antimicrobial substances (Harrell et al., 2021). Disruption of the mature biofilm can be carried out using displacement strategy in which matrix degrading enzymes are used to penetrate the matrix to disrupt the biofilm structure. Adhesion of pathogenic planktonic cells is inhibited by using exclusion strategy in which the surface is precoated with probiotics so that adhesion of pathogenic cells is avoided, making it impossible to form biofilm. A competition strategy involves coculturing probiotic planktonic cells along with pathogenic planktonic cells. This mechanism depends on fight/competition of probiotics to get attached on the surface to prevent colonization of pathogenic bacteria. Probiotics might also be useful as they might compete for the limited nutrients (Carvalho et al., 2021).

The scope of research in antibiofilm is large and many important strategies might be introduced in coming future to control the detrimental biofilm formation in vivo. Yet as formation of biofilm is a natural way of life for microorganisms and is not a relevant indicator of disease, future strategies could aim at balancing host-biofilm homeostasis. As the biofilm niche is different in different region of GIT, the tactics should be region oriented (Motta et al., 2021).

Conclusion

Biofilm formation is a natural phenomenon which increases survival of microorganisms in harsh environmental conditions. Due to the structural integrity of biofilms, it has been a critical challenge in medical and food sectors as they are difficult to penetrate or remove. It has been an area of fascination to understand the mechanism used by microorganisms to form biofilms. Every microorganism has its own genetic regulation that aids in the attachment to any substratum. Knowledge about these mechanisms would be supportive in tactical aspect against biofilms. Probiotic biofilms have a beneficial role in human health and are most suitable natural remedy against pathogenic biofilms. Relying on antibiotics treatment as sole strategy is not an eminent way to tackle biofilms. We need to have multifactorial approach toward finding solution against biofilms which should include selection of sensitive and penetrative antibiotics in combination with anti-QS or biofilm dispersal supplements.

References

Abdallah, M., Benoliel, C., Drider, D., Dhulster, P., Chihib, N., 2014. Biofilm formation and persistence on abiotic surfaces in the context of food and medical environments. Arch. Microbiol. 196 (7), 453−472.

Alam, A., Kumar, A., Tripathi, P., Ehtesham, N.Z., Hasnain, S.E., 2019. Biofilms: a phenotypic mechanism of bacteria conferring tolerance against stress and antibiotics. Mycobacterium Tuberculosis: Molecular Infection Biology, Pathogenesis, Diagnostics and New Interventions. Springer, pp. 315−333.

Algburi, A., Zehm, S., Netrebov, V., Bren, A.B., Chistyakov, V., Chikindas, M.L., 2017. Subtilosin prevents biofilm formation by inhibiting bacterial quorum sensing. Probiotics Antimicrob. Proteins 9 (1), 81−90.

Barzegari, A., Kheyrolahzadeh, K., Khatibi, S.M.H., Sharifi, S., Memar, M.Y., Vahed, S.Z., 2020. The battle of probiotics and their derivatives against biofilms. Infect. Drug. Resist. 13, 659−672.

Béchon, N., Jiménez-Fernández, A., Witwinowski, J., Bierque, E., Taib, N., Cokelaer, T., et al., 2020. Autotransporters drive biofilm formation and autoaggregation in the diderm firmicute Veillonella parvula. J. Bacteriol. 202 (21), e00461-20.

Bjarnsholt, T., Jensen, P., Fiandaca, M., Pedersen, J., Hansen, C., Andersen, C., et al., 2009. Pseudomonas aeruginosa biofilms in the respiratory tract of cystic fibrosis patients. Pediatr. Pulmonol. 44 (6), 547−558.

Burne, R.A., Chen, Y.-Y.M., Penders, J.E., 1997. Analysis of gene expression in Streptococcus mutans in biofilms in vitro. Adv. Dental Res. 11 (1), 100−109.

Camargo, A.C., Woodward, J.J., Call, D.R., Nero, L.A, 2017. Listeria monocytogenes in food-processing facilities, food contamination, and human listeriosis: the Brazilian scenario. Foodborne Pathog. Dis. 14 (11), 623−636.

Carvalho, F.M., Teixeira-Santos, R., Mergulhão, F.J.M., Gomes, L.C., 2021. Targeting biofilms in medical devices using probiotic cells: a systematic review. AIMS Mater. Sci. 8 (4), 501−523.

Chao, Y., Zhang, T., 2012. Surface-enhanced Raman scattering (SERS) revealing chemical variation during biofilm formation: from initial attachment to mature biofilm. Anal. Bioanal. Chem. 404 (5), 1465−1475.

Chopra, L., Singh, G., Jena, K.K., Sahoo, D.K., 2015. Sonorensin: a new bacteriocin with potential of an anti-biofilm agent and a food biopreservative. Sci. Rep. 5 (1), 1−13.

Costerton, J.W., 1999. Introduction to biofilm. Int. J. Antimicrob. Agents 11 (3−4), 217−221.

Davies, D.G., Parsek, M.R., Pearson, J.P., Iglewski, B.H., Costerton, J.W., Greenberg, E.P., 1998. The involvement of cell-to-cell signals in the development of a bacterial biofilm. Science 280 (5361), 295−298.

Harrell, J.E., Hahn, M.M., D'Souza, S.J., Vasicek, E.M., Sandala, J.L., Gunn, J.S., et al., 2021. Salmonella biofilm formation, chronic infection, and immunity within the intestine and hepatobiliary tract. Front. Cell. Infect. Microbiol. 10, 910.

Hassett, D.J., Cuppoletti, J., Trapnell, B., Lymar, S.V., Rowe, J.J., Yoon, S.S., et al., 2002. Anaerobic metabolism and quorum sensing by Pseudomonas aeruginosa biofilms in chronically infected cystic fibrosis airways: rethinking antibiotic treatment strategies and drug targets. Adv. Drug. Delivery Rev. 54 (11), 1425−1443.

Hentzer, M., Teitzel, G.M., Balzer, G.J., Heydorn, A., Molin, S., Givskov, M., et al., 2001. Alginate overproduction affects Pseudomonas aeruginosa biofilm structure and function. J. Bacteriol. 183 (18), 5395−5401.

Hentzer, M., Givskov, M., Parsek, M.R., 2002. Targeting quorum sensing for treatment of chronic bacterial biofilm infections. Lab. Med. 33 (4), 295−306.

Hernández-Jiménez, E., Del Campo, R., Toledano, V., Vallejo-Cremades, M.T., Muñoz, A., Largo, C., et al., 2013. Biofilm vs. planktonic bacterial mode of growth: which do human macrophages prefer? Biochem. Biophys. Res. Commun. 441 (4), 947−952.

Joshua, G.W.P., Guthrie-Irons, C., Karlyshev, A.V., Wren, B.W., 2006. Biofilm formation in Campylobacter jejuni. Microbiology 152 (2), 387−396.

Kaur, S., Sharma, P., Kalia, N., Singh, J., Kaur, S., 2018. Anti-biofilm properties of the fecal probiotic lactobacilli against Vibrio spp. Front. Cell. Infect. Microbiol. 8, 120.

Latasa, C., Roux, A., Toledo-Arana, A., Ghigo, J., Gamazo, C., Penadés, J.R., et al., 2005. BapA, a large secreted protein required for biofilm formation and host colonization of Salmonella enterica serovar Enteritidis. Mol. Microbiol. 58 (5), 1322−1339.

Li, Y.-H., Lau, P.C.Y, Lee, J.H., Ellen, R.P., Cvitkovitch, D.G., 2001. Natural genetic transformation of Streptococcus mutans growing in biofilms. J. Bacteriol. 183 (3), 897−908.

Liu, Z., Li, L., Wang, Q., Sadiq, F.A., Lee, Y., Zhao, J., et al., 2021. Transcriptome analysis reveals the genes involved in Bifidobacterium longum FGSZY16M3 biofilm formation. Microorganisms 9 (2), 385.

Loo, C., Corliss, D., Ganeshkumar, N., 2000. Streptococcus gordonii biofilm formation: identification of genes that code for biofilm phenotypes. J. Bacteriol. 182 (5), 1374−1382.

Melo, R.T., Mendonça, E.P., Monteiro, G.P., Siqueira, M.C., Pereira, C.B., Peres, P.A.B.M., et al., 2017. Intrinsic and extrinsic aspects on Campylobacter jejuni biofilms. Frontiers in Microbiology. 8, 1332.

Mizan, M.F.R., Jahid, I.K., Ha, S.-D., 2015. Microbial biofilms in seafood: a food-hygiene challenge. Food Microbiol. 49, 41−55.

Motta, J.-P., Wallace, J.L., Buret, A.G., Deraison, C., Vergnolle, N., 2021. Gastrointestinal biofilms in health and disease. Nat. Rev. Gastroenterol. Hepatol. 18 (5), 314−334.

Nikolaev, Y.A., Plakunov, V., 2007. Biofilm—"City of microbes" or an analogue of multicellular organisms? Microbiology 76 (2), 125−138.

Okuda, K.-i, Zendo, T., Sugimoto, S., Iwase, T., Tajima, A., Yamada, S., et al., 2013. Effects of bacteriocins on methicillin-resistant Staphylococcus aureus biofilm. Antimicrob. Agents Chemother. 57 (11), 5572−5579.

Pamp, S.N.J., Tolker-Nielsen, T., 2007. Multiple roles of biosurfactants in structural biofilm development by Pseudomonas aeruginosa. J. Bacteriol. 189 (6), 2531−2539.

Schembri, M.A., Kjærgaard, K., Klemm, P., 2003. Global gene expression in Escherichia coli biofilms. Mol. Microbiol. 48 (1), 253−267.

Shakibaie, M.R., 2018. Bacterial biofilm and its clinical implications. Ann. Microbiol. Res. 2.

Sherlock, O., Vejborg, R.M., Klemm, P., 2005. The TibA adhesin/invasin from enterotoxigenic Escherichia coli is self recognizing and induces bacterial aggregation and biofilm formation. Infect. Immun. 73 (4), 1954−1963.

Thakur, A., Mikkelsen, H., Jungersen, G., 2019. Intracellular pathogens: host immunity and microbial persistence strategies. J. Immunol. Res. 2019.

Vélez, M.P., Petrova, M.I., Lebeer, S., Verhoeven, T.L.A., Claes, I., Lambrichts, I., et al., 2010. Characterization of MabA, a modulator of Lactobacillus rhamnosus GG adhesion and biofilm formation. FEMS Immunol. Med. Microbiol. 59 (3), 386−398.

Wagner, V.E., Iglewski, B.H., 2008. P. aeruginosa biofilms in CF infection. Clin. Rev. Allergy Immunol. 35 (3), 124−134.

Webb, J.S., Givskov, M., Kjelleberg, S., 2003. Bacterial biofilms: prokaryotic adventures in multicellularity. Curr. Opin. Microbiol. 6 (6), 578−585.

Yan, X., Gu, S., Cui, X., Shi, Y., Wen, S., Chen, H., et al., 2019. Antimicrobial, anti-adhesive and anti-biofilm potential of biosurfactants isolated from Pediococcus acidilactici and Lactobacillus plantarum against Staphylococcus aureus CMCC26003. Microb. Pathogen. 127, 12−20.

Yang, L., Barken, K.B., Skindersoe, M.E., Christensen, A.B., Givskov, M., Tolker-Nielsen, T., 2007. Effects of iron on DNA release and biofilm development by Pseudomonas aeruginosa. Microbiology 153 (5), 1318−1328.

Yin, W., Wang, Y., Liu, L., He, J., 2019. Biofilms: the microbial "protective clothing" in extreme environments. Int. J. Mol. Sci. 20 (14), 3423.

Yoon, S.S., Coakley, R., Lau, G.W., Lymar, S.V., Gaston, B., Karabulut, A.C., et al., 2006. Anaerobic killing of mucoid Pseudomonas aeruginosa by acidified nitrite derivatives under cystic fibrosis airway conditions. J. Clin. Investig. 116 (2), 436−446.

Yoshida, A., Kuramitsu, H.K., 2002. Multiple Streptococcus mutans genes are involved in biofilm formation. Appl. Environ. Microbiol. 68 (12), 6283−6291.

Young, E.C., Baumgartner, J.T., Karatan, E., Kuhn, M.L, 2021. A mutagenic screen reveals NspS residues important for regulation of Vibrio cholerae biofilm formation. Microbiology 167 (3).

CHAPTER 11

Biofilm: a coordinated response of bacteria against stresses

Roopshali Rakshit[1], Aayush Bahl[1], Ashutosh Kumar[2], Deeksha Tripathi[1] and Saurabh Pandey[3]

[1]Microbial Pathogenesis and Microbiome Lab, Department of Microbiology, School of Life Sciences, Central University of Rajasthan, Ajmer, Rajasthan, India
[2]Department of Microbiology, Tripura University (A Central University), Agartala, Tripura, India
[3]Department of Biochemistry, School of Chemical and Life Sciences, Jamia Hamdard, New Delhi, Delhi, India

Introduction

Adaptation to the microenvironment is essential for pathogens to invade and colonize the host. This adaptation helps them to sustain possible challenges of unknown terrain—the temperature, pH, osmotic gradient, exposure to reactive oxygen and nitrogen species, nutrient availability, variety of immune challenges, and the composition of the normal microflora in the host. These are the primary weapons that the host's artillery boasts of while throwing off a pathogen invasion. But the coordinated response of the bacteria against these stresses helps them to sustain. The coordinated response as biofilm is predominant in known microhabitat and macrohabitat. We will further detail about stress adaptation mechanisms and pathological impact of bacterial communities.

Host-mounted stresses against bacteria

Physical status

Pathogens are subjected to elevated temperatures as they migrate into the host from an external environment. Consequently, many virulence factors in pathogens are thermally regulated. Temperature may impact the topology of the DNA, modify RNA structure, influence the metabolism, and alter protein activity and processing. It plays a role in the regulation of gene expression by influencing DNA supercoiling which, in turn, modifies the rate of transcription. Interactions between the mRNA and ribosome are affected by alterations in the secondary and tertiary structures of proteins, brought about by temperature changes. Temperature also affects thermo-sensitive domains of proteins, which in turn, has a significant impact on altering protein structure or their susceptibility to degradation by proteases.

Often, a single transcription factor regulates multiple virulence factors. If such a transcription factor is thermo-regulated, it can have a significant impact on the ability

Bacterial Survival in the Hostile Environment
DOI: https://doi.org/10.1016/B978-0-323-91806-0.00006-0

of a pathogen to adapt to host environment. For example, the transcriptional activator LcrF, a thermo-regulated global regulator in *Yersinia pseudotuberculosis*, monitors the expression of genes encoding a type III secretion system (T3SS) and secreted proteins, which are important components of pathogenesis. Similar thermo-regulated transcriptional factors include the PrfA in *Listeria monocytogenes* and the BvgAS two-component system in *Bordetella* spp.

Although temperature-triggered adaptations are usually considered to be different from heat shock responses, a considerable overlap between the two mechanisms is frequently observed.

The temperature has been also found to influence bacterial motility. Increased flagellar motility at higher temperatures is observed in *Campylobacter* spp. while, in *L. monocytogenes*, motility is limited to temperatures below 37°C. *Yersinia* is found to be nonmotile at 37°C. In *Escherichia coli*, swimming behavior is found to be influenced by changes in temperature. Unidirectional movement is usually observed in warmer conditions while cooler temperatures are witness to tumbling and a change in direction. This behavior is, however, found to be reversed at higher bacterial densities. Nutritional availability also plays a role in this reversal. When nutrients are limited, *E. coli* swims to cooler areas to conserve energy. Interestingly, there seems to be a link between thermotaxis and chemotaxis (Lam et al., 2014).

Invading pathogens get trapped in neutrophil extracellular traps (NETs) or are directly phagocytized by neutrophils which are the first to reach infected tissues. Heme enzyme myeloperoxidase (MPO) is brought into the phagosome with the help of reactive oxygen/nitrogen species (ROS/RNS) manufactured by the superoxide generating NADPH complex. Acids like hypochlorous acid (HOCl), hypobromous acid (HOBr), and hypothiocyanate (HOSCN) inside the phagosome is catalyzed by MPO.

HOCl and HOBr possess strong oxidizing and halogenating abilities. In comparison to peroxynitrite ($ONOO^-$) and H_2O_2, these acids have a significantly higher reactivity against biomolecules and thus, act as key protective elements of the body in the face of a pathogen attack.

Host-microbe interactions

Microbial colonization may be beneficial or detrimental to the host. A huge diversity of microorganisms constitute what is known as the natural microbiota of living beings. These microbes, residing in all living organisms, are essential for the survival and proper functioning of the host. The natural microbiota of the host plays a key role in maintaining a healthy metabolism. Infection by opportunistic pathogens is often found to be a result of changes in the microbiome composition. This also gives rise to resistant species. Nitrogen and carbon sources utilized by pathogens to fulfill their nutritional requirements and enhance their virulence are often derived from microbiota-derived sources (Bäumler and Sperandio, 2016).

Mucins constituting the gut epithelium are known to be rich in various carbohydrates like mannose, galactose, fucose, sialic acid, N-acetylglucosamine (GlcNAc), and N-acetylgalactosamine (GalNAc). Bacteroidales and other saccharolytic members of the microbiota harvest these sugars, making them accessible to other members of the microbiota lacking this ability. Unfortunately, this also makes these sugars accessible to pathogenic bacteria in the gut which then utilize these to stimulate their growth and virulence. Microbiota-produced succinate has been known to influence gene expression in pathogenic bacteria like *Clostridium difficile*. The variations in composition and abundance of several short-chain fatty acids produced by the host-microbiome have been reported to aid pathogens in niche recognition in the gut. For instance, acetate concentration in the ileum tends to be around 30 mM. This acetate concentration has been reported to increase the expression of the *S. Typhimurium Salmonella* pathogenicity island 1 (SPI-1)-encoded T3SS (T3SS-1) is a contributing factor for colonization of gut by *S. typhimurium* (Bäumler and Sperandio, 2016).

Host inflammatory responses result in a decrease in the availability of trace elements like iron, which leads to increase in competition between the pathogenic invaders and members of the host microbiota. For instance, the release of lipocalin-2 from the gut epithelia of mice and rhesus macaques is induced by IL-22. This antimicrobial protein binds to a low-molecular-weight siderophore called enterobactin produced by *Enterobacteriaceae*. This leads to a decline in iron levels in the gut. Unlike the commensal *Enterobacteriacea*, which is solely dependent on enterobactin for iron acquisition, *S. typhimurium* can acquire iron through the secretion of salmochelin. Salmochelin is a glycosylated derivative of enterobactin that does not bind to lipocalin-2 (Behnsen et al., 2014; Flo et al., 2004; Goetz et al., 2002). Iron limitation induces competition between *Enterobacteriaceae* which utilize protein-based toxins called colicins (Bäumler and Sperandio, 2016).

Enhanced susceptibility to enteric pathogenic bacteria like *Shigella flexneri*, *Citrobacter rodentium*, *L. monocytogenes*, and *Salmonella enterica* was observed in mice bred under sterile conditions or the effect of antibiotics (Ferreira et al., 2011; Sprinz et al., 1961; Zachar and Savage, 1979; Kamada et al., 2012).

The status of the host's immune system and the adaptation strategies employed by the invading microorganisms decide the impact of microbial colonization on the host. Adaptation factors or virulence factors help microorganisms to adapt to the environment within the host and also facilitate the transmission of microbes to a new host. Located on mobile genetic elements like plasmids and genomic islands, these factors can be easily transmitted from one bacteria to another. *Mycobacterium* spp., however, forms an important exception to this phenomenon. Virulence factors assist bacteria in attaching to host cell surfaces and/or the extracellular matrix, penetrating the surface epithelia, invading the intracellular compartments, iron acquisition, evading defense strategies employed by the host, and transmitting to a new host (Medzhitov, 2007).

Host immune challenges and coordinated microbial response

A wide range of host defense mechanisms has evolved to combat pathogen attacks. In jawed vertebrates, the defense mechanisms may be categorized as innate or acquired, based on the type of receptors used to recognize the pathogen. While innate immune recognition is facilitated by pattern recognition receptors (PRRs), acquired immune recognition is mediated by antigen receptors. Acquired immune recognition is more specific in nature when compared to that mediated by PRRs.

Innate immune system

The innate immune system identifies molecular structures and patterns exclusive to microorganisms. PRRs have broad specificity. They can bind to a wide range of molecules with structural similarities. These molecules which serve as the targets for PRRs are called pathogen-associated molecular patterns (PAMPs), present in both pathogenic and nonpathogenic species. Usually playing critical roles in microbial physiology, PAMPs are products of metabolic pathways found solely in microbes. Microorganisms of a particular class tend to have identical PAMPs. These features make these molecules perfect targets for the innate immune system as they can be easily distinguished from the eukaryotic host cells and are rarely subject to evolution. Bacterial PAMPS are made up of cell wall components like lipoteichoic acids, lipopolysaccharides (LPS), lipoproteins, and peptidoglycan. Toll-like receptors (TLRs) constitute one of the well-known classes of PRRs. It is to be noted that the ligands of PRRs are not unique to pathogenic strains. Hence, pathogenic strains cannot be differentiated from nonpathogenic variants based on PRR alone (Medzhitov, 2007).

Adaptive immune system

Receptors of T- and B-lymphocytes constitute important elements of the adaptive immune response. Both innate like and conventional receptors express antigen receptors. Unlike innate-like lymphocytes (B1 cells, marginal-zone B-cells, natural-killer T-cells, and subsets of $\gamma\delta$T cells), conventional lymphocytes (B2 cells and most $\alpha\beta$T cells) lack a predetermined specificity for microbial antigens. To enhance their specificity toward microbial antigens, conventional lymphocytes are further differentiated into different effector cell types. These include the two types of $\alpha\beta$t cells: T-helper cells (T_H cells) and the cytotoxic T-cells, which identify the antigens linked to the major histocompatibility complex (MHC) class II and class I, respectively. The coreceptor, CD4, is expressed on the cell surface of T_H cells. Cytotoxic T-cells are marked by the coreceptor CD8. Conventional B-cells, on the other hand, can recognize any antigen associated with an epitope. These lymphocytes tend to move through the lymph nodes and spleen until they meet their specific antigen, which is brought in by the antigen-presenting cells. Cytokines and chemokines of the innate immune system bring about the differentiation of the conventional lymphocytes and carry them to the infection site.

The antigen receptors of innate-like lymphocytes are gathered in a more organized fashion. The location and function of these lymphocytes are usually predetermined (Medzhitov, 2007).

Complement and coagulation system

The complement and coagulation systems in the host help in detecting the presence of invading pathogens. Three major pathways constitute the complement system. These are—lectin pathway, the alternative pathway, and the classical pathway. The products of the complement system are known to assist in cellular immune responses in generating indirect stress on pathogens. These molecules may also directly disintegrate bacterial membranes. The lectin pathway contains recognition molecules like collectin-11, mannose-binding lectin (MBL), ficolins (Ficolin-1, Ficolin-2, and Ficolin-3). The recognition molecules of the complement systems bind to specific molecules on the bacterial surface. These molecules are unique to bacteria and are believed to have been conserved through the evolutionary process. The binding of recognition molecules to the target activates serine proteases. The C3 convertase enzyme is then produced. This enzyme cleaves C3 to mark the target surface with C3b. Similarly, the C5 convertase enzyme cleaves C5 to form C5a and C5b. C5b produces the membrane attack complex (MAC).

MAC is an sodium dodecyl sulfate (SDS)-stable multiprotein complex. It is constituted of single copies of C5b, C6, C7, and C8, along with 12—18 copies of C9. While MAC was known to destroy Gram-negative bacteria, it was long believed that the thick PG layer of Gram-positive bacteria prevents the insertion of MAC into the membrane, thus acting as a protective barrier. However, it has been recently reported that MAC collects on numerous Gram-positive bacteria as SDS-stable polymeric C9 structures.

Based on their cell wall composition, bacteria are classified as Gram-positive and Gram-negative. Peptidoglycan, composed of alternating strands of GlcNAc and N-acetylmuramic acid (MurNAc), connected by short peptide bridges, is an integral component of the bacterial cell wall. It is responsible for maintaining cellular integrity and is found in abundance in Gram-positive bacteria. In contrast, Gram-negative bacteria, possess a much thinner peptidoglycan layer, encased in an outer membrane.

In vitro studies have found isolated peptidoglycan to be a potent activator of the complement system (Ma et al., 2004; Verbrugh et al., 1980). However, the peptidoglycan layer is well shielded in living bacteria, making it imperative for the complement system to be able to recognize other conserved bacterial structures.

MBL and collectin-11 bind to carbohydrate and glycoconjugate assemblies on microorganisms, with the help of C-type carbohydrate recognition domain (CRD) (Jack et al., 2001). LPS, present in Gram-negative bacteria and wall teichoic acid (WTA), found in Gram-positive bacteria, are considered to be the principal bacterial targets of MBL. The CRD in ficolins can identify both sugar motifs and acetylated groups present in GlcNAc and GalNAc. Binding to GlcNAc facilitates the binding to Gram-positive bacteria. The C1q molecule activates the classical pathway. It binds to the Fc regions of IgG or IgM antibodies on the microbial surface. C1q binds bacteria with the help of plasma proteins called pentraxins, which can detect nonself antigens and self-altered ligands. The classical pathway may also be triggered into action by the direct binding of the C1q to the bacterial surface. The protein molecule, properdin, has been recently reported to function as a pattern recognition receptor for bacteria, allowing the alternative pathway to directly recognize invading bacteria in the host.

The coagulation cascade can be triggered by two pathways: the intrinsic pathway (contact system) and extrinsic (tissue factor) pathway. The coagulation factor XII (FXII)-dependent intrinsic pathway has been reported to activate microbe-specific coagulation. Bradykinin (BK), a pro-inflammatory peptide is produced and clotting is facilitated by the activation of Factor XII. BK recruits innate immune cells at the infection site. It instigates the release of chemoattractants with the help of macrophages and stimulates vascular leakage. Additionally, inflammation is initiated by the cleavage of protease-activated receptors (PARs) expressed by various immune cells, by coagulation proteases (thrombin and FXa). Chemoattractants like fibrinopeptides A and B are released following the activation of fibrinogen.

Considerable crosstalk has been observed between the complement and coagulation systems. The formation of fibrin clots at the terminal stages of coagulation is facilitated by complement proteases. Pattern recognition activates MASPs, found on bacterial membranes, in association with MBLs or ficolins. These proteases have been reported to promote fibrin clot formation by cleaving prothrombin into thrombin. Conversely, coagulation proteases, like thrombin, FXa, FIXa, FXIa have been found to cleave the central complement proteins C3 and C5, thus activating the complement system (Berends et al., 2014).

Antimicrobial peptides

Production of antimicrobial peptides (AMPs) is brought about by the activation of the coagulation cascade. Neutrophil-derived proteases cleave coagulation proteins to form peptides with antimicrobial activity. For example, upon incubation with neutrophil elastase, human plasma or fibrin clots produced thrombin-derived C-terminal peptides

(TCPs). TCPs were found to lyse microbial membranes and have been reported to exhibit antimicrobial properties against Gram-negative *E. coli* and *Pseudomonas aeruginosa* and Gram-positive *Staphylococcus aureus*. GHR28, a fibrinogen-derived peptide, has also been reported to display antimicrobial properties against *Streptococcus agalactiae* (GBS) and *S. aureus* but not against *Enterococcus faecalis* and *E. coli*.

Biofilms

Biofilms are the troupes of microbial cells with the aggregates of different biomolecules that are latched on a robust surface in a predestined fashion (Di Somma et al., 2020). The biomolecules in the matrix include polysaccharides, proteins, teichoic acid, and DNA which protects bacteria from stress conditions. Extrachromosomal DNA is a vital component of the biofilm because it underwrites the structural integrity of biofilm as well as provides resistance against antibiotics due to horizontal gene transfer from the environment. Moreover, application of any such antibiotics or disinfectants during the growth phase can induce the formation of biofilm.

Stages of biofilm formation are as follows:
1. Primary changeable attachment
2. Irrevocable attachment
3. Microcolonial stage
4. Development
5. Disengagement then diffusion to different areas

Note—exodus is a newly discovered detachment phase mediated by the expression of nucleases.

Pathogenic bacteria after the creation of biofilm deploy both resistance and tolerance against antimicrobial challenges. The bacteria which come under the planktonic category can stick to the nonliving and the living surfaces by the virtue of proteins expressed on its surface. The proteins responsible for the process can adhere to the surface of the host proteins or they can release signals for auto-aggregation (Campoccia et al., 2019). A sessile bacterial colony marks the commencement of the first stage of biofilm formation. At this stage, the colony produces extracellular polymeric substances (EPS) which alleviates biofilm stability. Extracellular DNA (eDNA) is also an important component in the biofilm stabilization process. This is confirmed by the susceptibility of the newly formed biofilms to DNase enzymes and on the other hand, mature biofilms are stabilized after the expressions of EPS (Whitchurch et al., 2002). The final stage involves the amalgamation of bacteria to form a multilayered biofilm around the colonies. As time passes, the bacterial colonies widens in volume and the thickness can reach up to 10 μm (Gupta et al., 2016). The final stage is the detachment which can be achieved by three different strategies—desorption, detachment, and dispersion (Petrova and Sauer, 2016).

Quorum sensing

Quorum sensing (QS) is a cell—cell communication method utilized by the microorganisms to change the outflow of specific aggregates once the population thickness beginning is reached. QS can handle the outflow of virulence factors, biofilm development, swarming and can likewise shield the microorganisms from the hostile immune system of the host. QS prevents lymphocyte propagation and cytokine formation by macrophages in *P. aeruginosa* (Telford et al., 1998). The biofilms also prevent phagocytosis and free radical bursts by neutrophils (Bjarnsholt et al., 2005). The virulence factors controlled by the QS signaling like elastase, rhamnolipids, and alkaline protease hinder the immunological response system of the host (Laarman et al., 2012). QS systems are controlled by the production, release, build-up, and group-wide recognition of molecules called autoinducers (Fig. 11.1).

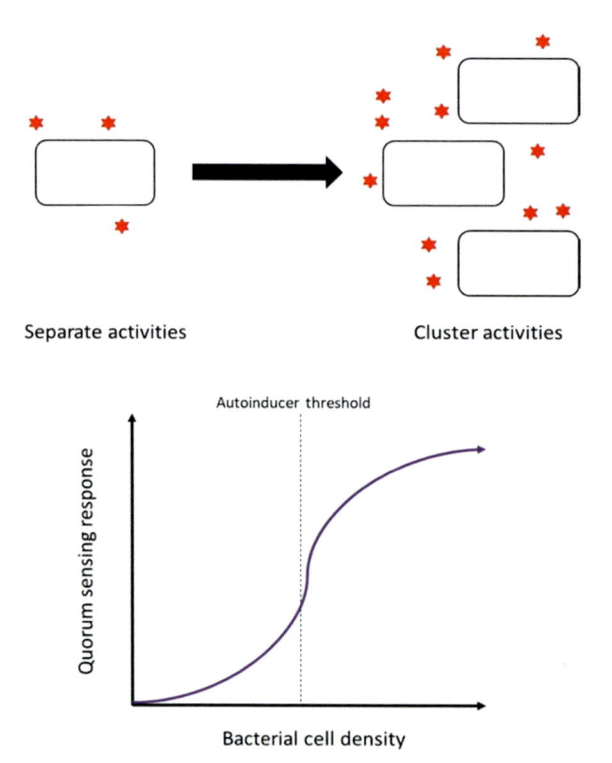

Figure 11.1 *Heterogeneity in quorum sensing.* Heterogeneity in quorum sensing is observed in bacteria which means that not all the bacteria in a population show any response to the autoinducer signals. The autoinducer threshold level is represented using a dashed line, and the curve line shows the quorum sensing response to variable levels of autoinducer signals. Cells can react to different levels of autoinducers. The cells that can detect the autoinducer level below the threshold value can reduce the production levels of the molecules and vice versa.

Expression of toxins, specialized secretion systems

Many bacteria live inside various communities which can create a competitive state. As a result, bacteria evolve an array of mechanisms that can be used for their benefit. Previous studies have confirmed that Gram-negative bacteria release toxins that can hinder colony development in the surroundings. The bacteria have to transport the toxin from inside to the surface, this is performed by a process called Esx pathway, and the toxin released is called LXG protein. To protect from the toxin, the bacteria also produce an antidote. The LXG protein interferes with the ability of the bacteria to form cell membrane and also inhibits other metabolic activities (Whitney et al., 2017). Gram-negative bacteria also possess a secretion system called type VI secretion system (T6SS). The secretion system acts like a nano-syringe to deliver the effector protein to neighboring cells. Some other examples may include the pneumococcus which secretes virulence factors such as pneumolysin (PLY), adhesins, pili proteins, and exoglycosidases (Subramanian et al., 2019). Virulence factors in *Helicobacter pylori* are categorized into three broad categories. The first type includes the cytotoxin-associated gene pathogenicity island (*cag*PAI) which is responsible for the expression of T4SS. The second group contains the genes that can be altered in the life duration of the bacteria to ensure its survival. The last group contains the genes that are particular to a specific strain like the vacuolating cytotoxin gene (*vacA* gene) (Roesler et al., 2014). Similarly, toxin−antitoxin (TA) systems are found in all organisms. They may be present on plasmid or chromosome and can be absent or present abundantly in a particular bacterium. Typically, the TA system codes for a toxin and an antitoxin to counter its effect. The toxin is a protein but the antitoxin can be a protein or RNA and TA systems are broadly classified into VI categories. TA systems perform various functions from affecting the metabolic pathways to causing infections in the host (Sierra et al., 2019).

Responses from extracellular bacteria

Bacteria continuously discharge extracellular substances to seize nutrients from the surroundings, hydrolase solid resources, and construct impeccable biofilm communities. In the community, the producers synthesize some substances that are used up by the nonproducers called public goods (Griffin et al., 2004). Important functions of bacteria as a community are controlled by QS to survive sustainably. One such example is seen in *Vibrio cholerae* biofilm formed on solid surface chitin in Fig. 11.2.

Bacterial extracellular vesicles (EVs) are part of intracellular communication and are conserved among species. The composition of EVs and their respective functions vary across the species but their secretion pathway is conserved (Jurkoshek et al., 2016). The EV secretion by *Mycobacterium tuberculosis* (*Mtb*) effects on system with

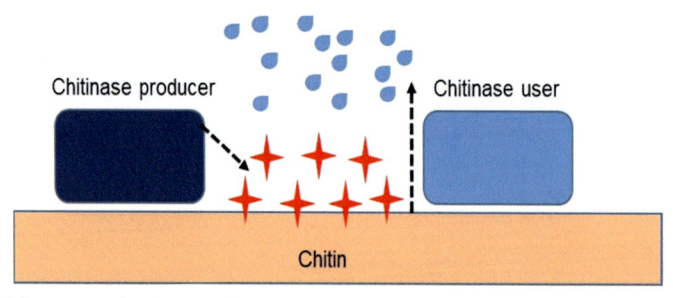

Figure 11.2 *Chitinase production and quorum sensing.* Chitinase producers synthesize chitinase enzyme (star-shaped) and hydrolase the chitin present on the surface to form N-acetylglucosamine which is taken up by both the producers and the consumers. This process is controlled by QS systems and the whole community survives in a coordinated manner to grow thick biofilms.

both constructive and destructive manners. To induce an infection, *Mtb* releases multiple EVs to suppress antimycobacterial function of the host macrophages. These EVs prevent the lysosome—phagosome fusion by lipoarabinomannan (LAM) secretion. This allows the bacteria to reside within the infected cell for a long time. Other immune invasion mechanisms like secretion of TLR2 agonists induce production of immunosuppressive cytokines, thus inhibiting MHC class II presentation (see Fig. 11.3). As a result, the CD4 + T-cells continue the *Mtb* within granulomas but are not able to eradicate the pathogen.

Responses from intracellular bacteria

Host-generated stresses reprogram the bacterial gene expression profile. *L. monocytogenes*, a Gram-positive, facultative intracellular pathogen, which causes listeriosis in humans and livestock, gain entry into the host cell by phagocytosis or receptor-mediated endocytosis. Subsequently, it gets encapsulated in a host vacuole. Then it breaks out of the vacuole by secreting listeriolysin O (LLO)—a cholesterol-dependent pore-forming toxin and the phospholipases PlcA and PlcB. Following its release into the cytosol, cell-to-cell spread and motility of *L. monocytogenes* are facilitated by the protein, ActA. *S. typhimurium* is located in the cytosol of macrophages but moves into the cytosol to replicate. This implies that bacterial multiplication and survival inside the host are more efficiently thwarted in some cells. It has also been observed that certain metabolic pathways in *L. monocytogenes* and *Francisella* sp. play key roles in cytosolic replication and the overall survival of the pathogens inside the host.

Nutrient acquisition inside host cells is a major challenge for intracellular pathogens. *L. monocytogenes* reduces metabolic disturbances detectable by the host cell and makes use of diverse carbohydrate and nitrogen sources to fulfill its nutritional

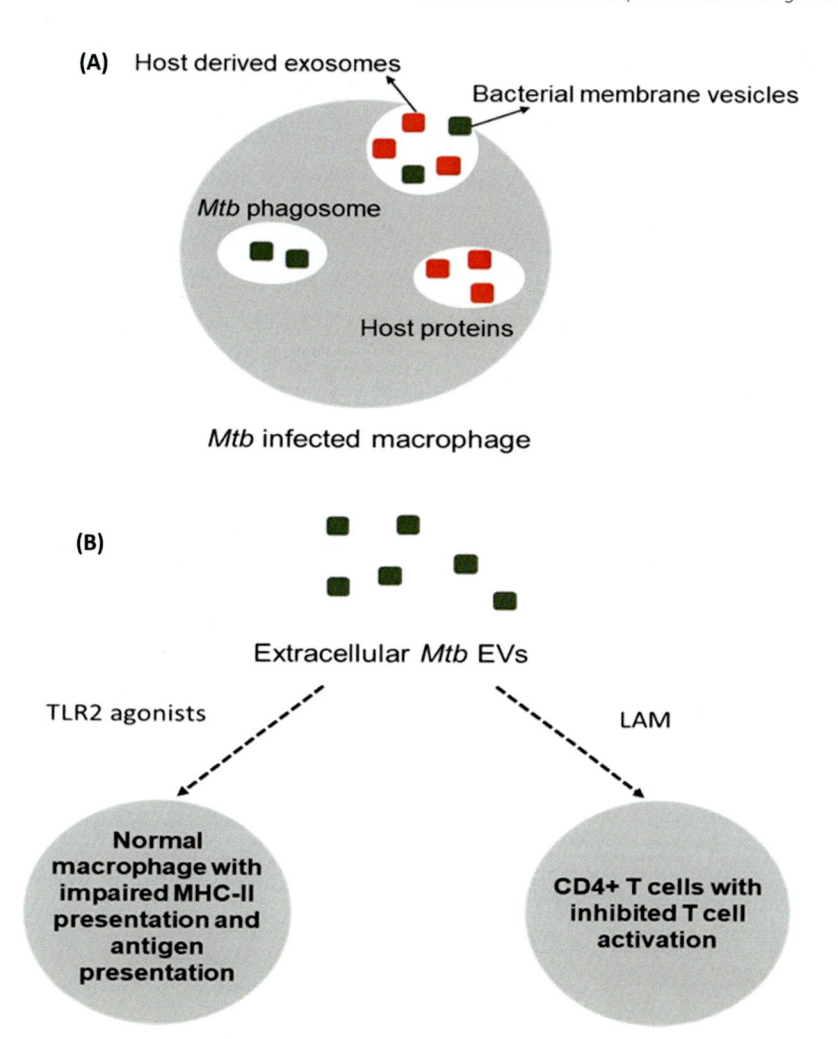

Figure 11.3 *EVs secretion by the Mtb-infected host cells.* (A) The model of EV trafficking in the host macrophage is presented here. The EVs released inside the cell suppress the effector function of the cell and are also released in the environment to infect other cells. (B) EVs secreted by the bacteria negatively regulate the host immune system.

requirements inside the cytosol. *Francisella* sp. uses γ glutamyl peptides of the host to produce amino acid cysteine, essential for its survival. Despite acquiring nutrients from the cytosol of the host to enable replication, *L. monocytogenes* and *S. flexneri* can effectively evade host innate immune responses.

Interactions between α proteobacteria and their host are mediated by IV secretion system. The type IV secretion system is vital for the virulence of *Anaplasma phagocytophilum* and *Ehrlichia chaffeensis*, closely related to the pathogenic Rickettsia (Chen et al., 2017; O'Callaghan and Stebbins, 2010).

Salmonella uses toxin-like proteins and adhesive structures to survive inside host cells (Wagner and Hensel, 2011).

Mtb uses multiple mechanisms for successful persistence inside host cells. *Mtb* utilizes superoxide dismutases (SODs), heme-dependent and thiol-dependent peroxidases, truncated hemoglobins and various low molecular weight (LMW) reductants which work in concert against host-derived oxidants. *Mtb* relies on Mycothiol (MSH) (1-D-myoinosityl-l-2-N-acetylcysteinyl-amido-2-deoxy-α-D-glucopyranoside) as its principal thiol instead of glutathione (GSH), commonly used by eukaryotes and bacteria in the maintenance of redox homeostasis inside the cell. A decrease in MSH content in *Mtb* has been linked to increased susceptibility to acidic conditions, drugs, and oxidants. MSH, ergothioneine (2-mercaptohistidine trimethylbetaine), and gamma-glutamylcysteine protect the pathogen from reactive species. MSH is kept in its reduced form by mycothiol disulfide reductase (MshR). MshR is thus crucial for the survival of *Mtb*. Rv2466c, a novel MSH-dependent reductase enhanced bacterial survival in high concentrations of H_2O_2. This enzyme also activate a thieno-pyrimidine derivative, TP052, known to have a fatal impact on *Mtb* (Piacenza et al., 2019). In addition to protecting against reactive oxygen intermediate (ROI) and reactive nitrogen intermediate, the SecA2 protein in *Mtb* plays a role in suppressing macrophage immune responses in the host, thus assisting in intracellular growth. Furthermore, macrophages infected with ΔsecA2 mutants enhance production of tumor necrosis factor α, interleukin 6, RNI, and major histocompatibility complex class II (induced by γ-interferons) (Kurtz et al., 2006).

Conclusion

Inside the host, pathogens face myriad factors such as changes in temperature, pH, the action of antimicrobial peptides, host immune responses, and competition with the natural microbiota, which act as threats to their survival. Pathogens employ various adaptive mechanisms to overcome these challenges and thrive in the host environment. Bacteria work in concert to generate a coordinated response against host-imposed threats. Under stress, bacteria form heterogeneous communities often called biofilms, which function as a single organism. Within a biofilm, bacteria communicate with the help of QS to express a certain phenotype essential for their survival within the host. The coordinated response generated by biofilms makes them a formidable opponent. Pathogens are continuously evolving to generate novel adaptive mechanisms in response to host-generated stress. Gaining a better understanding of these mechanisms will help us develop innovative combat strategies, which is the need of the hour. In combination with more conventional approaches, this will put us one step forward in the perpetual war against pathogenic microorganisms.

References

Bäumler, A.J., Sperandio, V., 2016. Interactions between the microbiota and pathogenic bacteria in the gut. Nature 535 (7610), 85–93.

Behnsen, J., Jellbauer, S., Wong, C.P., Edwards, R.A., George, M.D., Ouyang, W., et al., 2014. The cytokine IL-22 promotes pathogen colonization by suppressing related commensal bacteria. Immunity 40 (2), 262–273.

Berends, E.T., Kuipers, A., Ravesloot, M.M., Urbanus, R.T., Rooijakkers, S.H., 2014. Bacteria under stress by complement and coagulation. FEMS Microbiol. Rev. 38 (6), 1146–1171.

Bjarnsholt, T., Jensen, P.Ø., Burmølle, M., Hentzer, M., Haagensen, J., Hougen, H.P., et al., 2005. Pseudomonas aeruginosa tolerance to tobramycin, hydrogen peroxide and polymorphonuclear leukocytes is quorum-sensing dependent. Microbiology 151 (2), 373–383.

Campoccia, D., Mirzaei, R., Montanaro, L., Arciola, C.R., 2019. Hijacking of immune defences by biofilms: a multifront strategy. Biofouling 35 (10), 1055–1074.

Chen, G.Y., Pensinger, D.A., Sauer, J.D., 2017. Listeria monocytogenes cytosolic metabolism promotes replication, survival, and evasion of innate immunity. Cell Microbiol. 19 (10).

Di Somma, A., Moretta, A., Canè, C., Cirillo, A., Duilio, A., 2020. Antimicrobial and antibiofilm peptides. Biomolecules 10 (4).

Ferreira, R.B., Gill, N., Willing, B.P., Antunes, L.C., Russell, S.L., Croxen, M.A., 2011. The intestinal microbiota plays a role in Salmonella-induced colitis independent of pathogen colonization. PLoS One 6 (5), e20338.

Flo, T.H., Smith, K.D., Sato, S., Rodriguez, D.J., Holmes, M.A., Strong, R.K., 2004. Lipocalin 2 mediates an innate immune response to bacterial infection by sequestrating iron. Nature 432 (7019), 917–921.

Goetz, D.H., Holmes, M.A., Borregaard, N., Bluhm, M.E., Raymond, K.N., Strong, R.K., 2002. The neutrophil lipocalin NGAL is a bacteriostatic agent that interferes with siderophore-mediated iron acquisition. Mol. Cell 10 (5), 1033–1043.

Griffin, A.S., West, S.A., Buckling, A.J.N., 2004. Cooperation and competition in pathogenic bacteria. Nature 430 (7003), 1024–1027.

Gupta, P., Sarkar, S., Das, B., Bhattacharjee, S., Tribedi, P., 2016. Biofilm, pathogenesis and prevention—a journey to break the wall: a review. Arch. Microbiol. 198 (1), 1–15.

Jack, D.L., Klein, N.J., Turner, M.W., 2001. Mannose-binding lectin: targeting the microbial world for complement attack and opsonophagocytosis. Immunol. Rev. 180, 86–99.

Jurkoshek, K.S., Wang, Y., Athman, J.J., Barton, M.R., Wearsch, P.A., 2016. Interspecies communication between pathogens and immune cells via bacterial membrane vesicles. Front. Cell Dev. Biol. 4, 125.

Kamada, N., Kim, Y.G., Sham, H.P., Vallance, B.A., Puente, J.L., Martens, E.C., 2012. Regulated virulence controls the ability of a pathogen to compete with the gut microbiota. Science 336 (6086), 1325–1329.

Kurtz, S., McKinnon, K.P., Runge, M.S., Ting, J.P., Braunstein, M., 2006. The SecA2 secretion factor of Mycobacterium tuberculosis promotes growth in macrophages and inhibits the host immune response. Infect. Immun. 74 (12), 6855–6864.

Laarman, A.J., Bardoel, B.W., Ruyken, M., Fernie, J., Milder, F.J., van Strijp, J.A., et al., 2012. Pseudomonas aeruginosa alkaline protease blocks complement activation via the classical and lectin pathways. J. Immunol. 188 (1), 386–393.

Lam, O., Wheeler, J., Tang, C.M., 2014. Thermal control of virulence factors in bacteria: a hot topic. Virulence 5 (8), 852–862.

Ma, Y.G., Cho, M.Y., Zhao, M., Park, J.W., Matsushita, M., Fujita, T., et al., 2004. Human mannose-binding lectin and L-ficolin function as specific pattern recognition proteins in the lectin activation pathway of complement. J. Biol. Chem. 279 (24), 25307–25312.

Medzhitov, R., 2007. Recognition of microorganisms and activation of the immune response. Nature 449 (7164), 819–826.

O'Callaghan, D., Stebbins, C.E., 2010. Host-microbe interactions: bacteria. Curr. Opin. Microbiol. 13 (1), 1–3.

Petrova, O.E., Sauer, K., 2016. Escaping the biofilm in more than one way: desorption, detachment or dispersion. Curr. Opin. Microbiol. 30, 67–78.

Piacenza, L., Trujillo, M., Radi, R., 2019. Reactive species and pathogen antioxidant networks during phagocytosis. J. Exp. Med. 216 (3), 501–516.

Roesler, B.M., Rabelo-Gonçalves, E.M.A., Zeitune, J.M.R., 2014. Virulence factors of Helicobacter pylori: a review. Clin. Med. Insights Gastroenterol. 7, CGast.S13760.

Sierra, R., Viollier, P., Renzoni, A., 2019. Linking toxin-antitoxin systems with phenotypes: A Staphylococcus aureus viewpoint. Biochim. Biophys. Acta Gene Regul. Mech. 1862 (7), 742–751.

Sprinz, H., Kundel, D.W., Dammin, G.J., Horowitz, R.E., Schneider, H., Formal, S.B., 1961. The response of the germfree guinea pig to oral bacterial challenge with Escherichia coli and Shigella flexneri. Am. J. Pathol. 39 (6), 681–695.

Subramanian, K., Henriques-Normark, B., Normark, S., 2019. Emerging concepts in the pathogenesis of the Streptococcus pneumoniae: from nasopharyngeal colonizer to intracellular pathogen. Cell. Microbiol. 21 (11), e13077.

Telford, G., Wheeler, D., Williams, P., Tomkins, P.T., Appleby, P., Sewell, H., et al., 1998. The Pseudomonas aeruginosa quorum-sensing signal molecule N-(3- oxododecanoyl)-L-homoserine lactone has immunomodulatory activity. Infect. Immun. 66 (1), 36–42.

Verbrugh, H.A., van Dijk, W.C., Peters, R., van Erne, M.E., Daha, M.R., Peterson, P.K., et al., 1980. Opsonic recognition of staphylococci mediated by cell wall peptidoglycan: antibody-independent activation of human complement and opsonic activity of peptidoglycan antibodies. J. Immunol. 124 (3), 1167–1173.

Wagner, C., Hensel, M., 2011. Adhesive mechanisms of Salmonella enterica. Adv. Exp. Med. Biol. 715, 17–34.

Whitchurch, C.B., Tolker-Nielsen, T., Ragas, P.C., Mattick, J.S., 2002. Extracellular DNA required for bacterial biofilm formation. Science 295 (5559), 1487.

Whitney, J.C., Peterson, S.B., Kim, J., Pazos, M., Verster, A.J., Radey, M.C., et al., 2017. A broadly distributed toxin family mediates contact-dependent antagonism between gram-positive bacteria. Elife 6, e26938.

Zachar, Z., Savage, D.C., 1979. Microbial interference and colonization of the murine gastrointestinal tract by Listeria monocytogenes. Infect. Immun. 23 (1), 168–174.

CHAPTER 12

The bacterial communication system and its interference as an antivirulence strategy

Suruchi Aggarwal[1], Pallavi Mahajan[1], Payal Gupta[1], Alka Yadav[2], Gagan Dhawan[3], Uma Dhawan[2] and Amit Kumar Yadav[1]

[1]Translational Health Science and Technology Institute, NCR Biotech Science Cluster, Faridabad, Haryana, India
[2]Department of Biomedical Science, Bhaskaracharya College of Applied Sciences, University of Delhi, New Delhi, Delhi, India
[3]Department of Biomedical Science, Acharya Narendra Dev College, University of Delhi, New Delhi, Delhi, India

Introduction

Bacteria talk—and they talk a lot. Depending on the type of chatter—cooperative or competitive, community compositions are shaped up, and coordinated behaviors akin to multicellular organisms are impersonated to ensure community well-being, survival, or niche colonization. This cooperativity demands a chemical language for signal synthesis, secretion, relay, and transmission. This coordinated, cooperative chatter called *quorum sensing* (QS) gathers information about surrounding bacteria and colony density, a primitive form of multicellular behavior. This density-dependent chemical signal detection phenomenon that allows coordinated behavior (synthesis of virulence factors, biofilms, proteases, etc.) between microbes is called *quorum sensing*. This language also has dialects with built-in promiscuity, transcending the barriers of species, genera, and even kingdoms. By fine-tuning these chemical molecules, different bacteria enjoy various degrees of private and public communications with varying degrees of specificity (Defoirdt, 2018; Mukherjee and Bassler, 2019). Bacteria are also known to harbor more than one QS system which are often interconnected, like in the case of *Pseudomonas aeruginosa*. While these complex networks of regulatory circuits appear to carry unwarranted redundancy for regulating QS-mediated functions, these provide robustness to sudden fluctuations in QS signals. Using acyl-homoserine lactone (AHL) autoinducers in gram-negative bacteria, peptide lactones in gram-positive bacteria, and autoinducers-2 (AI-2) mediated signals used by both; bacteria can regulate their specific and generic communication. Bacteria have evolved a global language (autoinducers-2 or AI-2) for broader trade among diverse genera with common goals and specialized local dialects for trading with their nearby kin. Bacteria need to constantly assess the composition, structure, and population dynamics of their surrounding environment. Activities with higher energy costs such as the production of biofilms, toxins, and virulence factors, etc. are entrusted to the QS system, which activates the genes only when an optimal population density is reached. Bacteria can sense the density and the composition of the species present in their surroundings by "listening to" the various quorum signals. In bacteria, about 4%–10% of the transcriptome and >20% of the proteome is dedicated to the quorum-sensing genes.

At low concentrations of autoinducers, the bacteria live individually because the other members of the community are not discernable. When community-contributed autoinducer molecules accumulate to a higher concentration, the bacteria can sense their kin and act as social beings. Their conversation decides several aspects of their life—viz. foraging for a carbon source, virulence, bioluminescence, metabolism, toxin production, symbiosis, competence, antibiotic synthesis, biofilm formation, etc. The QS inhibition (QSI) or quorum quenching (QQ) by natural or synthetic molecules is being explored as a viable therapeutic strategy.

The antivirulence strategy aims to disarm, instead of killing the pathogen, by interfering with the production of virulence factors. Bacteria cause virulence through various factors like toxins, adhesins, invasins, siderophores, and specialized secretion systems (Defoirdt, 2018). The antibiotics were developed to kill the bacteria by interfering with their growth and development. Since the discovery of antibiotics, the host-pathogen war has intensified due to the development of antimicrobial resistance. The bacteria evolve to evade the deleterious effects of antibiotics by developing resistance, which may be natural or acquired (Kandpal et al., 2017). The intense selection pressure is an unavoidable cost of using antibiotics and a major cause of resistance. Drug resistance can evolve through several mechanisms like efflux pumps, kinome-reprogramming, mutations, horizontal gene transfer, etc. (Kandpal et al., 2017; Aggarwal et al., 2017). The resistance to antibiotics is reported soon after their first introduction. This leads to huge losses of human lives and creates economic liability. As the resistance to an antibiotic makes the economic lifespan of the drug short, there is low monetary interest for the pharmaceutical companies to invest in the development of new antibiotics. Only five new antibiotics have entered the market between 2000 and 2015 as the discovery and regulatory approvals are lengthy, expensive, and challenging. To make matters worse, the antibiotics cannot be marketed too expensively, the infectious diseases they treat are there for a limited period and thus less profitable, and antibiotics are beneficial only transiently until resistance develops due to their indiscriminate use. So, the euphoria around antibiotics has long subdued, and the battle continues to search for novel strategies to fight the pathogenic bacteria (Dickey et al., 2017). Since it is imminent that all antibiotics will eventually fall prey to resistance, there is an urgent need to search for alternate strategies for combating infectious diseases. Antivirulence strategies aimed at disarming the pathogen by countering their virulence factors instead of killing them is gaining popularity as a possible alternative. This includes immunotherapy (monoclonal antibodies) and QSI or alternatively, QQ, in pathogens in which the QS system controls the virulence (Dickey et al., 2017; Allen et al., 2014). These strategies are believed to posit lesser selection pressure on pathogenic bacteria to develop resistance since these do not inhibit the normal growth or function. This antivirulence strategy can potentially provide better alternatives to traditional antibiotics with better control of pathogens and aimed at better return-on-investment for pharmaceutical companies. This can also spur drug development efforts to counter the infectious disease and resistance burden.

Quorum-sensing and quorum quenching

Quorum-sensing phenomenon was discovered in *Vibrio fischeri*, where their symbiotic relationship with Hawaiian bobtail squids was studied based on bioluminescence. The bioluminescence was controlled by quorum sensing and it has been a subject of interest for biologists to understand the mechanism underlying the process. Bacteria have devoted around 4%—10% of their genomes and >20% of proteomes for QS, which regulates a large number of genes and regulons. This suggests the role of QS in regulating critical signaling networks in manifesting virulence. A coordinated communication network is critical for pathogenic bacteria to mount an attack or colonize a niche effectively. The outcome of tussle between infectious bacteria and the human host hinges on effective exploitation of these communication systems. Virulence in many bacteria is under QS regulation. Several competing organisms eavesdrop on interspecies communication and produce QS inhibitors as signal mimics to confuse the competitor by interrupting the communication channel. The consequence of this signal war decides the species composition and dominance stratification of the microbial community. While bacteria jam the interspecies signals, they inform their kin of their presence by enhancing their own QS signals. This helps the bacteria decide on the behavioral switch between living a solitary life at low cellular densities (LCD) versus a multicellular social life at higher cell densities (HCD) (Mukherjee and Bassler, 2019).

The bacterial communication varies from one species to another. In gram-negative bacteria, the QS signals are autoinducer molecules (AHLs) synthesized from S-adenosyl-methionine (SAM), which can freely diffuse through the membrane. These can bind to membrane-bound receptors and regulate hundreds of genes covering numerous biological processes (Papenfort and Bassler, 2016). The acyl-chain lengths may differ for different variants/homologs of the common LuxI/R QS systems in different species that impart specificity to the intraspecies bacterial dialog. Sometimes bacteria may harbor multiple QS systems interconnected to provide functional redundancy to these circuits like—LasI/R, RhlI/R, PQS, IQS—systems in *P. aeruginosa* (Mukherjee and Bassler, 2019; Papenfort and Bassler, 2016). In gram-positive bacteria, the signals are oligopeptides that are synthesized and modified in the cell and secreted out of the cell using ABC transporters. These peptides are detected by specific two-component sensory apparatus (histidine-kinase two-component system) that receives and transduces the signal to the response regulator. The response regulator controls the expression of genes under QS control. ComX peptide autoinducer in *Bacillus subtilis* is sensed by ComP/ComA system to activate ComK that controls the competence state of the bacterium. Some bacteria have evolved multiple ways of communication, for example, *Vibrio harveyi*, which communicates between different species via AI-2, and within the same species via AI-1. Some of these systems have been represented in Fig. 12.1.

QQ is defined as the natural obstruction of quorum-sensing systems to prevent one species from taking over the niche where multiple species competitively coexist. The term *quorum-sensing inhibition* (QSI) is also frequently used to describe the same phenomenon and both terms are often used interchangeably. Not all QS-related communications in bacteria control virulence and, therefore, do not always cause disease. For example, the LuxI/LuxR QS system controls bioluminescence in *V. fischeri*. However, the bacterial communication that regulates virulence in species like *Vibrio cholerae*, *P. aeruginosa*, *Agrobacterium tumefaciens*, etc. can lead to debilitating human

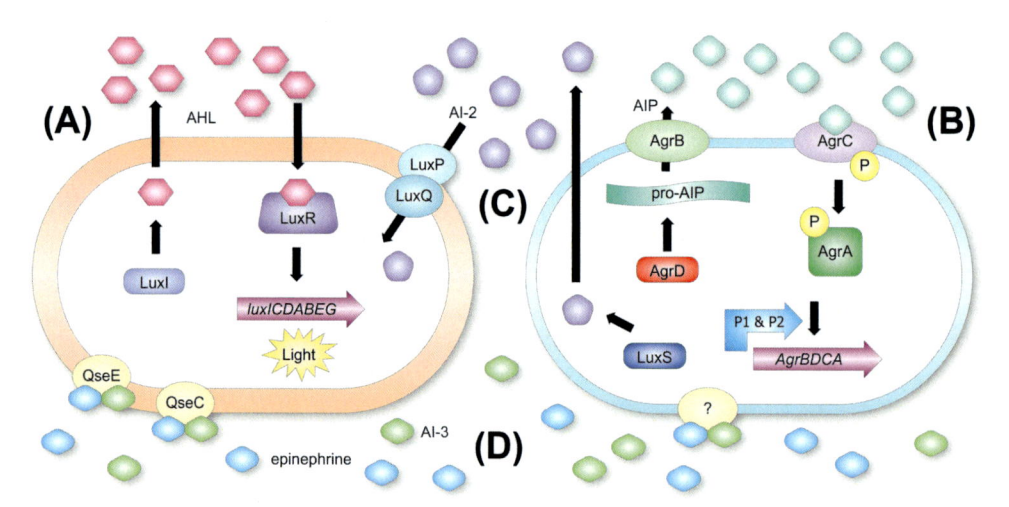

Figure 12.1 The typical quorum-sensing systems in gram-negative (left) and gram-positive (right) bacteria. Autoinducers—acyl-homoserine lactones (AHLs) (A) and autoinducer peptides (AIPs) (B) regulate the quorum sensing in gram-negative and gram-positive bacteria, respectively. Both types of bacteria can also communicate between them through AI-2 mediated quorum sensing (C). The bacteria can also crosstalk across kingdoms in which AI-3 mediated signals are used by bacteria, while the host uses epinephrine or other small hormones. These hormones are signal mimics of AI-3 molecules and the bacterial receptors can use either AI-3 or hormone for activation of quorum sensing regulated gene expression.

diseases. A growing body of literature states that even the eukaryote hosts communicate and produce signals that mimic AHLs and cause QQ. Still the levels are not enough to mount a considerable quenching effect. These may also act as immunomodulatory signals, for example, AI-3 has been known to respond to epinephrine and norepinephrine hormones. There has been evidence that cancer metastasis in humans is associated with quorum sensing with the gut microbiome. There have been numerous discoveries of natural QQ compounds and researchers are also evaluating several synthetic QQ compound to fight the bacterial infection. Although many of these QQ compounds do not boast of successful clinical trials, these molecules are providing hope for the future as antivirulence compounds (see Table 12.1) and many are being tested to counter the ever-growing resistance to traditional antibiotics. Here, we explore the basics of QS systems and the mechanisms affecting QQ systems. We then guide the reader through how the primitive multicellularity of the pathogen might be exploited to avoid resistance and develop better therapeutics for infectious diseases—in a quest for the development of the so-called *evolution-proof* drugs (Allen et al., 2014). Even though the status of QQ mechanisms as being impervious to resistance is gradually being challenged, it offers significantly better hope than antibiotics.

The major difference between the QQ compounds and antibiotics lies in the strength of the selection pressure that is applicable, owing to the proposed nonessentiality of such QS systems. The QSI strategy targets only to disarm the pathogen and not affect its fitness. Accumulating evidence suggests that these circuits are more complex than were initially thought, and some amount of selection pressure may be applicable due to the interconnectedness of QS systems and signaling pathways that either synthesize the signal molecules, relay, or receive the signals. Understanding these QS systems in more detail and insights into the network architectures of the associated biomolecules can aid in better antivirulence strategies. By allowing the QSI approach to gain full traction—either by itself or in a synergistic combination therapy with low doses of antibiotics or other prophylactics, it is possible to deal with the ominous antibiotic resistance problem. The QQ resistance reported so far is still debatable. It is only likely to happen if the QS systems are somehow interconnected with bacterial survival and fitness. If we understand the circuits involved in QS more deeply at the systems-level, it will be easier to predict the effects and propose better alternate QQ compounds or synthesize compounds with a novel mechanism of action.

The architecture of quorum-sensing circuits

Cell-to-cell communication systems run on chemical signals and signaling networks to bring about the desired biological effect. The QS systems are regulated at several points which are amenable to exploitation by QS inhibitors. We briefly enumerate some of the most studied bacteria in context of their QSS architecture. The list is not

Table 12.1 Quorum-sensing inhibitors and their mechanism of action.

Inhibitor	Organism	Target	Mechanism of action	Further development stage	Indication	References
C-30 (Halogenated Furanone)	*P. aeruginosa*	LasR	Efflux pump	Preclinical	Urinary tract, burn, wound infection	Hentzer et al. (2003)
Gallium	*P. aeruginosa*	hitA	Ferric citrate iron transport system	Preclinical	Urinary tract, burn, wound infection	Garcia-Contreras et al. (2013b)
5-Fu	*P. aeruginosa*	AHL receptor	Pyocyanin, elastase production	-	Urinary tract, burn, wound infection	Garcia-Contreras et al. (2013a)
ZnO	*P. aeruginosa*	PQS, CzcR	Zinc cation efflux pump	-	Urinary tract, burn, wound infection	Garcia-Lara et al. (2015)
Hamamelitannin	*S. aureus*	Agr	RNAIII production	-	Sepsis, abscess, keratitis	Kiran et al. (2008)
RIP	*S. aureus*	Agr	traP phosphorylation	-	Sepsis, abscess, keratitis	Balaban et al. (2007)
Savirin (3-(4-Propan-2-ylphenly) sulfonyl-1H-triazolo[1,5-a]quinazolin-5-one)	*S. aureus*	AgrA	Efflux pump and DNA binding	Preclinical	Sepsis, abscess, keratitis	Sully et al. (2014)

E22 (Phenoxyacetyl homoserine lactone)	P. aeruginosa	RhlR	Receptor destabilization, affinity for transcription factors	Preclinical	Urinary tract, burn, wound infection	Eibergen et al. (2015)
mBTL (Halogenated thiolactone)	P. aeruginosa	RhlR	Pyocyanin production and biofilm formation	Preclinical	Urinary tract, burn, wound infection	O'Loughlin et al. (2013)
Itc–12 and itc–13 (Isothiocyanate-AHL)	P. aeruginosa	LasR	Pyocyanin production and biofilm formation	Preclinical	Urinary tract, burn, wound infection	Amara et al. (2009)
ZBzl-YAA5911 (Cyclic peptide)	E. faecalis	FsrC	Inhibit gelatinase and secreted serine protease	Preclinical	Endocarditis, meningitis	Nakayama et al. (2013)
Solonomide A, B	S. aureus	AgrC	Virulence Factors	–	Sepsis, abscess, keratitis	Mansson et al. (2011)
Ajoene	P. aeruginosa	LasR	Rhamnolipid production	–	Urinary tract, burn, wound infection	Jakobsen et al. (2012a)
Iberin	P. aeruginosa	LasR, RhlR	Rhamnolipid production	–	Urinary tract, burn, wound infection	Jakobsen et al. (2012b)
Catechin	P. aeruginosa	LasR	Pyocyanin, elastase and biofilm production	–	Urinary tract, burn, wound infection	Vandeputte et al. (2010)
Furanone C-2	D. pulchra	LuxR, LuxS	Biofilm formation	–	Defense against epiphytes and herbivores	de Nys et al. (1993)

(Continued)

Table 12.1 (Continued)

Inhibitor	Organism	Target	Mechanism of action	Further development stage	Indication	References
Pyrogallol	*V. harveyi*	AI-2 analog	Bioluminescence	–	Vasculitis in shrimp	Ni et al. (2008)
Farnesol	*P. aeruginosa*	PQS	Transcriptional regulator PqsR with pqsA promoter	–	Urinary tract, burn, wound infection	Cugini et al. (2007)
pCIPhT-DADMe-ImmA	*E. coli*	MTAN	Inhibit polyamine biosynthesis and salvage pathways for adenine and methionine	–	Gastroenteritis, urinary tract infections	Singh et al. (2005)
CTL, CL	*C. violaceum*	CviR	Inhibit DNA binding	–	Skin lesions, sepsis	Swem et al. (2009)
J8-C8	*B. glumae*	Tofl	Block QS receptor	–	Bacterial panicle blight of rice	Chung et al. (2011)
Compound 10	*B. subtilis*	LuxS	Binding site of the LuxS	–	Fungicide, food poisoning	Shen et al. (2006)
TP 5	*P. aeruginosa*	LasR	PON activity	–	Urinary tract, burn, wound infection	Muh et al. (2006)
Compounds 19 and 20	*P. aeruginosa*	PqsR	Inhibit pyocyanin production	–	Urinary tract, burn, wound infection	Lu et al. (2012)
ML370	*V. cholera*	AI-2 (LuxO)	Inhibit ATPase activity	–	Cholera	Faloon et al. (2010)

exhaustive but aimed at highlighting the signal circuits at play during the QS process. The QS systems canonically comprise the following parts:

1. Signal synthesis
2. Signal secretion
3. Signal reception
4. Response regulators

The proteins from each of the above components form various QS systems and diverge among different bacterial species. The diversity in the arrangement of the components can give rise to several types of systems but the basic architecture remains similar. We briefly present some well-studied QS systems to demonstrate the regulation of density-dependent phenomenon like virulence, biofilms, or bioluminescence, etc. Fig. 12.1 shows an overview of some of the QS systems in gram-negative and gram-positive bacteria.

Bacteria synthesize, disseminate, and integrate the information in the form of small chemical signals called autoinducers to regulate genes under QS control in a density-dependent manner. The autoinducer signals accumulate in the environment after passive diffusion or active transport outside the cell. Bacteria keep monitoring the signal concentration to track their population density which, after a certain critical threshold, activates QS-controlled gene expression that allows the bacteria to modulate their collective behavior. The virulence factors are costly to produce for a single bacterium but advantageous and effective in manifesting infection as a group behavior (Papenfort and Bassler, 2016).

Quorum-sensing systems in bacteria

The bacterial quorum-sensing systems in gram-negative and gram-positive bacteria are regulated through autoinducers—AHLs and autoinducer peptides (AIPs) (Fig. 12.1A and B). The different type of bacteria can communicate through AI-2 between species (Fig. 12.1). The bacteria through AI-3 signals and human hormones can even communicate between kingdoms (Fig. 12.1D). Some examples of QS system circuits in few of the most studied organisms are discussed in the following sections.

Vibrio fischeri

In gram-negative bacteria, the QS systems depend on a prototypical LuxI/R system for the synthesis and recognition of autoinducer signals, respectively. In *V. fischeri*, the luxI enzyme synthesizes N-(3-oxohexanoyl) homoserine lactone (C6HSL) that can easily diffuse in and out of the cell. At low cell densities (LCD), a small amount of C6HSL autoinducer is present. When it increases to a higher cell density (HCD) and leads to accumulation of autoinducer beyond the threshold, it diffuses in, binds to LuxR to form a complex and binds to DNA at the *luxICDABEG* operon promoter, to transcribe luciferase and other genes under QS control. LuxR receptors are prone to degradation in the absence of their cognate autoinducers, which when present, help them dimerize and bind to DNA at the so-called "lux boxes" upstream of QS regulated genes. The other two QS systems, AinS-AinR and LuxS-LuxP/Q, also regulate LuxR through a complex phospho-relay pathway. This phospho-relay pathway consists of two sensor kinases, AinR and LuxP/Q and downstream regulators, LuxU and LuxO. Extracellular matrix with limiting threshold of C8-HSL and AI-2 leads to the phosphorylation of their receptors, that is, AinR and LuxP/Q, respectively. These two sensors then donate phosphate to LuxU, followed by the phosphate transfer to LuxO. Eventually, it activates the expression of sRNA qrr1 in cooperation with σ54. This qrr1 inhibits the translation of the mRNA litR in the presence of chaperon protein Hfq. Inhibition of litR leads to the inhibition of LuxR activation. When high concentrations of C8-HSL and AI-2 are present, the two sensor kinases act as phosphatases and remove the phosphoryl group from LuxU and LuxO leading to LitR translation. This further activates LuxR resulting in bioluminescence and colonization. Some studies have shown that AinS/R branch is more important than LuxS-LuxP/Q branch. In addition to bioluminescence, the three interconnected QS networks control the acetate utilization, motility, and colonization in *V. fischeri*. By varying the lengths of acyl chains, cognate receptors maintain species-specific signaling.

Pseudomonas aeruginosa

In *P. aeruginosa*, LuxI/LuxR homologs that control QS are LasI/LasR and RhlI/RhlR systems that help in expression of virulent factors like exotoxinA, pyocyanin, rhamnolipids, and proteases like elastase, etc. LasR binds to 3-oxo-C12-homoserine lactone (3O-C12-HSL) and RhlR binds to butanoyl-homoserine lactone (C4HSL) autoinducers, respectively. These signals lead to autoinduction of feed-forward loops, as they activate their own synthesis besides the virulence factors. Apart from these, the bacteria also produce another signal, 2-heptyl-3-hydroxy-4-quinolone (PQS or *Psuedomonas* quinolone signaling system), produced by PqsA, PqsB, PqsC, PqsD, and PqsH. It is detected by PqsR, also called MvfR. The lasR-HSL complex also activates the pqsH and pqsR that expresses PQS and also RhlI/RhlR system by activating the *pqsABCDH*. RhlR-C4HSL binding causes the repression of pqsABCD and pqsR genes, creating a feedback loop that regulates the QS-controlled genes by LasR and RhlR. QscR is an orphan or solo LuxR receptor homolog with no LuxI homolog counterpart. QscR can bind to C8-HSL, C10-HSL, 3-oxo-C10-HSL, C12-HSL, and C14HSL. It is suggested that QscR detects cohabitating species like *Burkholderia cepacia* in a mixed biofilm. There is another QS control system for the chemical signal 2-(2-hyroxylphenyl)-thiazole-4-carbaldehyde (IQS), which under low phosphate conditions, activate a gene cluster *ambBCDE* that controls LasR independent regulation of RhlR- and PqsR-dependent genes for virulence and biofilm formation. This hierarchical system of multiple QSSs in *Pseudomonas* demonstrates the complex networks of the QS circuits and the extensive overlap between them. Several other factors feed these QS circuits for robust integration of information to produce virulence factors independent of signal fluctuations as the virulence factor production is a costly affair for the pathogen. The pathogen QS systems ensure that no premature commitment to making virulence factor ensues due to a signal with low autoinducer concentration. Autoinducers do not activate the full QS regulon in the log growth phase. Only when the bacteria are in the stationary phase with a signal depicting robust cell counts of neighbors, the full QS expression takes over.

Vibrio cholerae

In another gram-negative bacterium, *V. cholerae*, responsible for causing diarrheal cholera, multiple QSSs are present to control the complete virulence expression, Type-VI secretion, competence, and biofilm formation (Jung et al., 2015). The (S)-3-hydroxy-triecan-4-one autoinducer (CAI-1) is synthesized by CAI-1 synthase, CqsA. Using S-adenosyl-methionine (SAM) and decanoyl-coA, CqsA produces amino-CAI-1 which converts into CAI-1, probably naturally. It is believed to be an intragenus communication signal, as it is found in several *Vibrio* species. *Vibrio* spp. have different affinities for CAI-1 signals other than their own cognate one, due to varying acyl-chain moieties synthesized by their respective CqsA/CqsS systems. The luxS/LuxPQ two-component system synthesizes and senses autoinducer-2 (S-TMHF borate) signal. It is believed to be an intergenera language based on the observation in diverse sets of gram-positive and gram-negative bacteria. At LCD, both CqsS and LuxQ act as kinases to phosphorylate LuxU protein through histidine-kinase two-component system. The phosphorylated LuxU activates LuxO, the response regulator, to transcribe four regulatory small RNAs- Qrr1−4. These sRNAs are assisted by Hfq chaperon to activate the translation of AphA, which inhibits translation of HapR regulator. At HCD, the receptors CqsS and LuxQ bind to the accumulated cognate signals and their kinase activities are inhibited, leading to reversal of phosphate flow in the circuit. Kinases act at phosphatases activating a dephosphorylating cascade that causes dephosphorylation-mediated inhibition of LuxO. The qrr RNAs are not formed and HapR is expressed while AphA is inhibited (Jung et al., 2015). Interestingly, both AphA and HapR inhibit each other which accentuates the role of these master regulators in committing completely to one pathway. Unlike other bacteria, *V. cholerae* expresses its virulent factors at low cell density, producing cholera toxin through transcription factor ToxT. Biofilm formation is also induced at low cell density. After attaining high cell density, the bacteria tries to exit and disseminate to other hosts when it senses that the infection titer is high enough.

It was noticed that both CqsA and LuxQ mutants were still able to maintain virulence, and this led to the discovery of hybrid histidine-kinase VpsS that caused LuxO-dependent upregulation of biofilm production. Another histidine-kinase VC1831, later named CqsR (cholera quorum-sensing receptor) along with VpsS could be phosphorylated by LuxU in vitro. These two new kinases work in parallel with the two previously explained canonical QS systems and any single receptor out of these four can induce virulence through the QS pathways converging at LuxO. This makes the bacteria insensitive to perturbation in signal variations through any single receptor (Jung et al., 2015).

Recently, a new autoinducer, 3,5-dimethylpyrazin-2-ol (DPO) was discovered that binds to VqmA receptor, controlling *vqmR* that transcribes VqmR sRNA regulating biofilm formation (Papenfort et al., 2017). The autoinducer is a product of threonine catabolism and requires threonine and alanine for its biosynthesis. This expands the repertoire of QS networks in *V. cholerae* and suggests that many more autoinducers-receptor pairs are waiting to be discovered (Papenfort et al., 2017).

Vibrio harveyi

V. harveyi, a free-living marine bacterium, is a pathogen to several marine animals. It controls bioluminescence by LuxI/R type QS similar to *V. cholerae*, yet both harbor differences in their QSSs, probably displaying their evolutionary adaptation to different niches. Their LuxI/R systems are unlike other gram-negative bacteria but quite similar to each other in some respects. *V. harveyi* was the first bacterium that was discovered to use three types of autoinducers for intraspecies, interspecies, and intergenera communication mediated via an autoinducer-3-OH-C4HSL (HAI-1), (Z)-3-amino-undec-2-en-4-one (CAI-1) and the set of interconverting autoinducer-2 molecules derived from 4,5-dihydroxy-2,3-pentanedione (DPD), respectively. HAI-1 autoinducer is synthesized by LuxM synthase, CAI-1 by CqsA, and boron-containing AI-2 (S-THMF-borate) by LuxS, which is the DPD synthase. In this bacteria, the HAI-1 signal is stronger than AI-2, which is stronger than CAI-1, which is the reverse in *V. cholerae*, that is, CAI-1 is the strongest signal followed by AI-2. This indicates that despite some similarities, there are notable differences in the QS systems of these two species.

LuxN is the receptor for HAI-1, LuxP/Q for AI-2, and CqsS for CAI-1 in *V. harveyi* (Papenfort and Bassler, 2016). These cognate receptors are highly specific for AHL and CAI signals and longer chain acyl lengths act as antagonists. The bacterium senses autoinducers from other organisms to shut down its quorum sensing to stop other bacteria from exploiting the *"public goods"* it produces. LuxPQ forms a heterotetramer, which when binds to AI-2, undergoes a significant conformational change to accommodate the boron in AI-2. Localizing the receptors LuxPQ to the membrane effectively prevent unwanted response to endogenous AI-2. The LuxN, CqsS, and LuxPQ act as kinases when cognate autoinducer is absent and transfer the phosphoryl group to LuxU, which in turn passes it on to LuxO. In concert with σ^{54}, the LuxO response regulator forms five regulatory sRNAs called Qrr 1−5. These are HfQ-dependent sRNAs that regulate master regulators LuxR negatively and AphA positively. AphA controls genes involved in individual behavior (biofilm formation, virulence, etc.), while LuxR controls the expression of genes involved in group behavior (bioluminescence). The Qrr sRNAs also repress *luxMN* coding for LuxM synthase and LuxN receptor. At HCD, however, the kinases start acting as phosphatases and the phosphoryl group flow is reversed deactivating the expression of Qrr 1−5 by inactivating LuxO. This keeps AphA expression in check while LuxR is synthesized, leading to the expression of bioluminescence phenotype. These multiple circuits and interconnections ensure fidelity of signal transduction to maintain cooperative or conflicting behavior as per cell density or the presence of kin.

Staphylococcus aureus

The gram-positive bacterial QS systems have genetically encoded peptide autoinducers, thereby enabling each species to have a unique peptide sequence. *Staphylococcus aureus* is a gram-positive opportunistic bacterial pathogen causing nosocomial infections like sepsis and pneumonia. There is a rapid rise in antimicrobial resistance in *S. aureus*, including methicillin-resistant *S. aureus* (MRSA) and others with a broad range of β-lactams. The accessory gene regulator (Agr) is the QS system controlling the virulence. The two promoters P2 and P3, respectively, transcribe RNAII and RNAIII transcripts. The RNAII transcript encodes for *AgrBDCA* locus that controls the feed-forward loop for expression of quorum-sensing autoinducer peptide. AgrD synthesizes the 45−47 residue pro-AIP, which is processed to 7−9 residue AIP and secreted out of the cell by AgrB transporter. This processing truncates the pro-AIP and cyclizes a five-residue peptide through a thiolactone bond between C-terminal and cysteine residue. The AIP accumulates when other bacteria are in the milieu and triggers the QS system by binding to the Agr C membrane-bound histidine-kinase (Fig. 12.1B). A conserved histidine is auto-phosphorylated and the phosphoryl group is transferred to an aspartate residue on AgrA, the response regulator protein kinase. The AgrA induces the *agr* operon by binding upstream of P2 promotor, thus activating the feed-forward loop. AgrA also controls PIII promoter that synthesizes RNAIII expressing *hld* gene coding for virulence factor δ-hemolysin. RNAIII expresses α-toxin on one hand and represses *rot* factor (repressor of toxins) on the other. This QS regulatory cascade expresses secretion of virulence factors and inhibition of factors that repress the toxin production. The QS system also controls biofilm formation at LCD and represses it at HCD. This ensures that the bacteria grow inside the biofilm until HCD is achieved, for dissemination and dispersal to other hosts.

There are four subtypes of *S. aureus*, each with its own AIP molecules based on the hypervariability within the *agrD* gene and a part of *agrB* gene. This leads to the synthesis of four types of AIP molecules based on the subtype, called specificity groups I−IV. In the *agrC* gene, a corresponding hypervariability leads to each AIP being specifically recognized by its coevolved cognate AgrC receptor. The binding of incorrect

AIP can cause inhibition of the QS system of other subtypes leading to the establishment of the infection by the subtype that first expresses its QS system.

There is the presence of another QS signal produced by LuxS called AI-2. AI-2 is present both in gram-positive and gram-negative bacteria. Not much is known about AI-2 signaling in *S. aureus*, although recently it has been shown that its absence increases biofilm formation and higher polysaccharide intercellular adhesion (PIA) production (Ma et al., 2017). LuxS produces AI-2 but the corresponding receptor for the AI-2 is not known and it does not actively form a part of quorum sensing but plays a metabolic role in the activated methyl cycle (AMC). The AI-2 signaling controls the production of capsular polysaccharide using a two-component system KdpDE phosphorylation, whose role is not clear. The regulation occurs by binding of the phosphorylated form of KdpDE that binds to the *cap* promoter. AI-2 system negatively controls the signaling of rbf transcription, which is a positive regulator of PIA-dependent biofilm formation.

Connecting the dots between interspecies and interkingdom circuits

AI-2 deserves a special mention since this signal is primarily a universal bacterial language, unlike the AHLs and AIPs. Both the gram-negative and the gram-positive bacteria are commonly governed by AI-2 mediated QS systems, and over 55 species of bacteria have been found to quorate via AI-2. Therefore AI-2 is the global language of QS systems that is common to both gram-positive and gram-negative bacteria and the connecting link of communication systems between the two. This signal is mediated by receptors and inducer mechanisms that were already in place during the evolutionary path from simple prokaryotes to multicellular organisms. These QS signals may be the remnants of progression from a single cell to multicellular behavior.

The LuxS gene controls the formation of a protein that generates AI-2. The AI-2 inducer is a furanosyl borate di-ester, formed during the AMC, where the homocysteine formed by the oxaloacetate or sulfate is converted to S-adenosyl-methionine (SAM). This product is used in RNA, DNA, metabolite, and protein methylations, where SAM is an active methyl donor and is reduced to S-adenosyl homocysteine (SAH) during the reaction. This molecule in eukaryotes can directly be converted back to homocysteine using SAH hydrolase (SAHh) without AI-2 production. However, in prokaryotes, it may be converted to S-ribosyl-L-homocysteine (SRH) by the enzyme 5′-methylthioadenosine/S-adenosylhomocysteine (MTA/SAH) nucleosidase (Pfs). Eukaryotes can replace this two-step process. Subsequently, LuxS can form homocysteine and DPD from SRH. Here, DPD is an inter-converting set of molecules of AI-2 and can produce R-THMF or S-THMF, based on the organism it occurs in, to be used as an inducer (Ross-Gillespie et al., 2014).

While the QS phenomenon is mostly studied in bacterial communities, the communication also exists between the bacteria and their human hosts, facilitated by aromatic autoinducer-3 signals (AI-3). While the bacteria talk through AI-3, the hosts communicate through various hormones. Epidermal growth factor (EGF), adrenaline, noradrenaline, steroid hormones, and dynorphin are the host hormonal factors that induce QS and virulence expression in bacteria. The pathogens hijack the host signals for their growth and to activate the virulence gene expression, toxin production, or other factors for host colonization. In enterohemorrhagic *Escherichia coli* (EHEC), the two-component signal systems, QseBC and QseEF, can sense host adrenaline or noradrenaline through sensor QseC to get activated. This controls the regulation of nearly 400 genes including toxins, biofilms, motility, type-III secretion systems, and stress response genes. To counter such attacks, the virulence receptors like QseC can be targets for therapy.

Quorum quenching

Due to their global control of virulence among other biological functions, QS regulation has been looked at as an alternative therapy to curb the antibiotic resistance problem. It was not surprising when natural quorum sensing inhibiting compounds were discovered. As bacteria have different languages for intraspecies, interspecies, and intergenus communication, both cooperative and conflicting roles can be played by quorum sensing. It is expected to observe positive selection of features enhancing the fitness in both types of communication. The role of QS has been extensively seen in supportive behaviors in single as well as multispecies communities, for example, pathogenic bacteria in biofilms. *P. aeruginosa* biofilms have been observed to harbor *B. cepacia* in lungs of cystic fibrosis patients (Tomlin et al., 2001). Similarly, conflicting behaviors in which species fight to establish their community over other competitors are also seen frequently. For example, *S. aureus* has four pathogenic subtypes depending on the QS system to express virulence. The AIPs of these four subtypes are different in sequence and highly specific. Using their specificity, they correctly bind to their cognate receptors and inhibit all other subtypes. This helps them in establishing their infection along with blocking out other competitors by the autoinducer peptide molecule. There have been cases where higher eukaryotes have been discovered to inhibit the quorum-sensing mechanisms of the bacteria to prevent harmful effects of colony behavior like virulence and biofouling. This is also seen in many organisms wherein the bacteria secrete molecules to confuse the other bacterial species by messing with their QS signals and establish themselves at their expense. Therefore the promiscuity and specificity in QS systems also define the roles played by autoinducers for disrupting QS in other bacteria. While maintaining specificity, these are also used to confuse other bacteria by antagonizing their QS receptors, or feeding on AI-2 signals from other species to prevent them from communicating. It is also used to count the density of kin and nonrelated species to make decisions on QS, virulence, and biofilm formation, based on their ecological supremacy as a community.

These observations demonstrate that QSI or QQ by natural compounds is an ancient, naturally occurring phenomenon. Exploiting QSI to attenuate pathogenic bacterial populations can be an innovative antivirulence approach, since QSI is not believed to exert much selection pressure, thereby circumventing the problem of resistance (Dickey et al., 2017; Allen et al., 2014). There are two major ways to disrupt bacterial QS—by interfering with signal detection and inactivating the signal. A red marine algae *Delisea pulchra* that produces halogenated furanones is able to disrupt the AHL QS systems as well as the parallel QS systems in *Vibrio* spp. by interacting with the transcriptional regulators of the QS systems and AI synthesizing enzyme LuxS. Besides natural inhibitors, there are several synthetic QS inhibiting compounds or antagonists that are analogs of AHLs and furanones (Muh et al., 2006). Defoirdt et al. were the first to argue that the resistance is possible against QS disruption in the light of evolutionary selection pressure (Defoirdt et al., 2010). The authors methodically enumerated the factors and synthesized a compelling hypothesis to explain how the factors could lead to resistance. It was later reported that the lack of controls in screening for QS inhibitors leads to erroneous QSI observed by many compounds. In such screening systems, the authors argued to control reporter strains with constitutive or inducible promotors. This can enhance the specificity of the screen and the inhibitors thus found will have better efficiency.

While testing the QSI approach, several in vitro evolutionary experiments reported the development of resistance to QQ compounds. The discovery is still not proven unequivocally. The said resistance is still debatable in a clinical or real-world setting, as the reporter strains used in the studies are not ideal method to measure resistance, and the QS regulation was not completely independent of bacterial survival (Defoirdt, 2018). Views suggesting resistance for QSI as well as their limitations exist in literature (Defoirdt et al., 2010; Maeda et al., 2012; Garcia-Contreras et al., 2013a).

Antivirulence strategies based on quorum quenching

The interference to QS can be achieved by interrupting/degrading signals, their receptors, their synthesis, or uptake. It can also be brought about by using signal mimics that will outcompete the QS signal molecules. Antivirulence is the most sought after and promising strategy to combat the problem of antibiotic resistance (Dickey et al., 2017). Numerous studies have reported and reviewed the status of QSI-mediated strategy for pathogen control. A comprehensive list of some QSI molecules and their mechanism is shown in Table 12.1. Here, we provide selective examples of novel strategies utilizing QSI. These strategies are also summarized in (Fig. 12.2).

Controlling biofilms

Bacterial biofilms are complex and structured social aggregates that form a layer at the interfaces of liquid—liquid, liquid—air, solid—liquid, and solid—air in diverse ecological

Figure 12.2 Various antivirulence strategies using quorum-sensing inhibition/quorum quenching.

niches. These are bacterial survival strategies in harsh environments like host mucosal surfaces, medical devices, wounds, and complex environments in plant-microbe interfaces and other ecological niches with biological competition for resources. Biofilms are structures to help bacteria persist in such environments and are major cause for pathogens reappearing even after antibiotic treatments as these contain extensive mechanisms like efflux pumps for keeping the drugs out. These are under QS system control and bacteria start producing biofilm once the microbes reach high density. Already approved drugs like 5-fluorouracil have QSI properties and have been reported to control biofilm formation on catheters (Walz et al., 2010). In *P. aeruginosa*, *meta*-bromo–thiolactone (mBTL) was found to be the most active inhibitor for biofilm formation and pyocyanin production (O'Loughlin et al., 2013).

Prophylactic use

In hospital settings, the pathogens pose great danger to patient health and quality of healthcare service by affecting medicinal devices and implants—like catheters and prosthetic grafts. The prophylactic use of several QSI compounds can help curb the nosocomial infections by interrupting adhesion and biofilm formation by pathogens. The prophylactic use of RNA III-inhibiting peptide (RIP) to coat Dacron vascular prosthetic graft lowers the infection from vancomycin-resistant *Staphylococcus epidermidis* and *S. aureus*. This is an effective way to prevent infection of postsurgery wounds (Balaban et al., 2003). A garlic compound, ajoene, is used for prophylactic prevention of *P. aeruginosa* infections by reducing rhamnolipid production, a major virulence factor (Jakobsen et al., 2012a). A fungal QSI compound, Apicidin, was found to be effective as a prophylactic in skin infection by preventing agr signaling in methicillin-resistant *S. aureus* (MRSA). The prophylactic use prevented necrosis of the skin due to MRSA (Parlet et al., 2019).

Cocktail of quorum-sensing inhibitors (quorum-sensing inhibitor combination therapy)

As the QS depends heavily on the nutrients, species composition, kin density, etc., the immediate environment decides how virulence is manifested. For example, in *P. aeruginosa*, virulence is generally regulated by las QS system. However, in iron or phosphate depleted media, pqs and Rhl QS systems take over bypassing las to produce pyocyanin and rhamnolipids. In cystic fibrosis patients, lasR inhibitors can accomplish inhibition of QS. In phosphate and iron-deficient environment in immunocompromised intestine, a cocktail of QSI inhibitors to tackle LasR, RhlR, and Pqs circuits together is more beneficial (Markou and Apidianakis, 2014).

Combination of quorum-sensing inhibitors with antibiotics

It is postulated that a synergistic combination therapy of QQ compounds with antibiotics can tackle the antibiotic resistance problem to a great extent. While the QSI controls virulence, the sensitivity to antibiotics would increase, thereby allowing the antibiotic to eradicate the pathogen. It has been observed in *S. aureus* infections that a combination of farnesol and hamamelitannin reduces the bacterial virulence and pathogenicity. The therapy also induces beta-lactam antibiotic sensitivity in the bacteria. Similar twofold effects are observed in using the QQ compounds, FS10 and epigallocatechin-3-gallate, which were used with the antibiotic tetracyclin (Simonetti et al., 2016). In *P. aeruginosa* infections, a combination of furanone C-30, ajoene, or horseradish extract was used as QQ compounds with the antibiotic tobramycin (Christensen et al., 2012). The QQ compound GW5072 (which is a 4-hydroxybenzylidene indolinone compound) has been reported to be useful in down-regulation of histidine-kinase sensor and response regulator, in combination with antibiotics. It was shown to be useful for treating MRSA and vancomycin-resistant *Enterococcus faecalis* (VRE) by resensitizing them to the antibiotic after QSI controls the quorum-mediated virulence (Opoku-Temeng et al., 2019). The antivirulence effect of QSI combined with the antibiotics to remove the pathogen is a promising application of the QQ strategy to tackle and control the resistance problem, but it is not yet in commercial use.

Immunotherapy as quorum-quenching agent

The key QSI strategy hinges on effectively disrupting the quorum-sensing signal, by degrading or sequestering the signal molecule to stop the bacterial communication. Immunomodulatory therapy consisting of monoclonal or polyclonal antibodies may also target the quorum-sensing acting as inhibitors against the autoinducers. These can mimic the natural autoinducers of varying lengths to reduce the quorum signals and thus control virulence. This has been reported in *S. aureus* virulence control by using AP4−24 H11, a monoclonal antibody (Mab) that acted as a bait receptor for confiscating AIP-4 autoinducer, thereby attenuating the virulence in a murine model (Park et al., 2007). Using immunoglobulin Y from egg yolk exposed to *P. aeruginosa*, passive immunization was observed for a murine infection model. Prophylactic use of the IgY antibodies is suggested for cystic fibrosis patients (Thomsen et al., 2016). Virulence in *P. aeruginosa* is also reported to be attenuated by another Mab, RS2−1G9. This Mab protects bone marrow macrophages from autoinducer-3-oxo-C12-HSL mediated cytotoxicity (Kaufmann et al., 2008). Several such immunotherapy trials are underway that target antivirulence therapy. The challenge is to circumvent redundancy in QS systems to target antibodies to specific autoinducers that bring about cytotoxic effects.

Quorum-sensing inhibitor molecules as nutritional supplements

Another strategy to utilize QSI is to integrate them in dietary supplements, like N-acetylglucosamine (NAG) is used as a QQ compound that downregulates AHL synthesis and is reported to be used as a dietary supplement. Since it is nontoxic and safe, it has been reported as an additive with milk, beer, and wine as a dietary supplement (Chen et al., 2010). Its further development and supplemental application as a QS-related virulence regulator can be highly advantageous. Similarly, plant extracts are a known source of antibacterial properties. In a study testing 14 herbs and spices, Sedanenolide, an active compound from celery extract, was highly effective against gram-negative bacteria that produce violacein (Cosa et al., 2019). For *P. aeruginosa* infections, ajoene from garlic and iberin from horseradish have been shown to have antivirulence properties (Jakobsen et al., 2012a). The probiotic release of secreted fengyncin lipopeptides from *B. subtilis* (β-OH-C17-fengyncin) can eradicate *S. aureus* colony. This QQ compound is structurally similar to Agr and attenuates virulence by competing for AgrC receptor binding, thereby inhibiting Agr expression (Piewngam et al., 2018). In another example, *Lactobacillus acidophilus* taken as a probiotic supplement inhibited the AI-2 activity of EHEC in pig intestine and led to boosting of immune system by increasing the IgG levels (Kim et al., 2018). The probiotic and dietary use of QSI molecules to counter pathogens is promising but requires a lot more work before their QQ effects can be devised into an effective antivirulence therapy.

Social cheaters as therapy

The bacterial communities use quorum sensing to regulate public goods like virulence factors, exoproteases, siderophores, and other such products of quorum sensing related gene expression. Even though the autoinducer synthesis and release is a costly affair for a single bacterium, the cost is necessary as it allows the bacteria to thrive and succeed as a colony. Usually the public goods are useful for the whole community and every bacterial cell produces it (cooperators), except for some mutants that are incapable of quorum sensing but can consume the public goods. Such bacteria are called social cheaters which thrive on publicly available consumables (Mukherjee and Bassler, 2019; Sandoz et al., 2007). The cooperators cause virulence by active quorum sensing, while the quorum-sensing cheaters can consume the costly public goods and cause the collapse of the colony structure. In a mixed culture of cooperators and cheaters, the wild type alone produces the QS signal 3-oxo-dodecanoyl-homoserine lactone, which the ΔlasI mutants can consume and grow (Mukherjee and Bassler, 2019; Sandoz et al., 2007). The nutrient media like adenosine or casein as sole carbon sources can determine which mutants act as cheaters. The ΔlasI acts as noncheater in the first case (adenosine source) while ΔlasR mutants act as noncheaters in the latter (casein source). This can be exploited as an effective QQ strategy by utilizing cheater mutants to collapse the community structure of wild-type pathogens (Ozkaya et al., 2018). Genes can be introduced in these mutants which can then make the pathogen vulnerable to antibiotics or can release antibiotics against them. While this seems like a rewarding strategy, cheater-based therapy requires further development, as the microbial communities have structure and dynamics for social policing to weed out cheaters.

Is there resistance to quorum-sensing inhibitors?

Conceptually, the QSI resistance is similar to antibiotic resistance and therefore the mechanisms remain nearly the same—efflux pumps, mutations, horizontal gene transfers, etc. Since QS exerts less selection pressure, few specific mechanisms of developing resistance may also be involved, although not yet proven in clinical settings. Here we explain the routes or mechanisms by which QQ resistance may arise, during in vitro or clinical settings. The rerouting of circuits and cross-connections between social or molecular players turns out to be the underlying theme by which resistance to QQ phenomenon may develop. Some of these are also covered in Table 12.1.

Efflux pumps in gram-positive or negative bacteria eject toxic substances, drugs, antibiotics, and quorum-sensing molecules from the cells usually by active transport to maintain the cellular homeostasis. Efflux pumps that generate antibiotic resistance can also contribute toward resistance to QSI. Efflux pumps which prevent metabolic toxicity within the biofilm and thereby support a high cell density. Maeda et al. reported that resistant in the opportunistic pathogen *P. aeruginosa* arises rapidly to the so far best characterized QS inhibitor C-30, a brominated furanone. Growth of *P. aeruginosa* on sole carbon source adenosine requires active quorum sensing and addition of C-30 masks QS pathways that eventually impaired the growth of the pathogen (Maeda et al., 2012). Further, resistance to C-30 was observed in vitro using reporter strains within four sequential dilutions after transposon mutagenesis. The authors observed enhanced efflux of C-30 due to mutation in *mexR* and *nalC*, which encode *mexAB-oprM* multidrug resistance operon. *MexAB-OprM* complex exports a number of diverse antibiotics and autoinducer 3OC12-HSL of Las QS system. The *mexR* mutant provides resistance to QQ compound C-30 (Maeda et al., 2012). It was reported that the inhibition of *BpeAB-OprB* could result in virulence attenuation via the inhibition of quorum sensing in *B. pseudomallei*, although later it was argued that the drug efflux system does not necessarily target the QS but rather is a broad-spectrum efflux system (Mima and Schweizer, 2010). The clinical isolates of *P. aeruginosa* were studied to compare the effect of quorum quencher C-30 and observed variable results. Out of eight clinical isolates, three were highly resistant to C-30 while others showed small-scale resistance. Resistance to C-30 was mediated by efflux *mexAB-oprM*. These results provide the clinical evidence for the presence of QQ resistant strains present before the introduction of C-30 in *P. aeruginosa* (Garcia-Contreras et al., 2013a).

Some QS inhibitors work by blocking the AHL-producing genes to prevent the synthesis of AHL. Others attach themselves to the receptor or signal molecule to block their reception. The latter are generally mimics of AHL molecules (like brominated furanones discussed above) that allow the binding but do not induce the QS activity. In both the cases, resistance to QSI may arise due to mutation changes that affect the binding of QQ drug. In Vibrios, LuxI/LuxR systems mainly control the QS responses, while *P. aeruginosa* harbors LasI/LasR along with RhlI/RhlR systems. Mutations in these QS systems may arise that can render these systems inaccessible to the QQ compounds due to changes in structural properties or binding sites. Reduced virulence in animal models of infection has been observed due to mutations in LasI/LasR and RhlI/RhlR genes. The biofilm formation was also observed to be reduced, supporting the role of AHLs in disease. Mutation in the *pfm* gene affects the biofilm formation in *P. aeruginosa* (Mou et al., 2011). Further, a non-AHL QS system based on 4-hydroxy-2-alkylquinolines (HAQs), synthesis of which is controlled by transcriptional regulator *MvfR* has been identified in *P. aeruginosa*. In the gram-positive bacterium *S. aureus*, the mutants of *agr* QS system have been reported to be attenuated in various animal models of disease compared to the wild type (Plata et al., 2011). It is also reported that mutations in luxS gene of *S. mutans* resulted in structural changes in biofilm formation, bacteriocin, and mutacin I production (Ng et al., 2011) along with variation in inter-species behavior.

The most common QS regulation occurs by AHL autoinducers produced by LuxI family. There are also many reported orphan LuxR homologs in several organisms that may lead to redundancy in QS circuits. One such homolog PpoR, in *P. putida* KT2440, regulates the fitness and biofilm surface motility independent of AHLs (Fernandez-Pinar et al., 2011). Several accessory regulators of *agr* QS system besides the P2 and P3 regulators help *S. aureus* regulate the QS system optimally under various conditions. Some of these are—ArlS, CcpA, CodY, σ^B, SarA, SarR, SarT, SarU, SarX, SarZ, Rsr, SrrA/SrrB, etc. Under extracellular stress, σ^B regulates surface proteins and production of pigment, while repressing the formation of virulence factors. As the synthesis of virulence factors is energetically costly for the bacteria, this ensures their survival in stress condition. CodY responds to isoleucine starvation conditions and it has binding sites in *agrC* gene. It is believed to delay the QS responses until a nutrient-rich environment is found. SrrA/SrrB responds to an anaerobic environment and binds to P2 and P3 promoters via SrrB binding. Their overexpression decreases virulence in similar ways to other accessory systems explained earlier. This is to ensure that QS is expressed in ideal environments only. The multiple signal transduction cascades can help the bacteria to conserve the valuable resources by modulating accessory signals. When the QS inhibitors target a particular pathway, the organisms use other QS systems for signaling and crosstalk and activate its virulence genes. In starvation-induced circumstances, the selection *of P. aeruginosa* lasR mutants leads to the expression of virulence through the quorum-sensing system Rhl. In the absence of acetyllactone with multiple QSS, the transcriptional regulators may bind to each other to form a heterodimer. This heterodimer may bind to a promoter region to express various genes leading the bacteria to sense the generated stress in the environment and rewire the QS circuit. *P. aeruginosa* consists of multiple QSS for activating its virulence factors like pyocyanin, protease like elastase, etc. Out of three QSS LasI/LasR and RhlI/RhlR systems are LuxI/LuxR homologs that control QSS and the third is the PQS system playing its role in pathogenesis. In the absence of acyl-HSLs, a luxR type protein, QscR, forms heterodimer with LuxR or RhlR to modulate expression of hydrogen cyanide, pyocyanin, and elastase (Mukherjee and Bassler, 2019). Multiple QS systems found in *V. cholerae* are CAI-1/CqsS, AI-2/LuxP/Q, and VarS/VarA. The kinase activity of CqsS and LuxPQ is inversely proportional to CAI-1 and AI-2 concentrations, and thus the disruption of these signaling pathways is impossible as they have undergone mutations and are able to regulate their gene expression. At low cell density, AI level is low, the kinase activities of CqsS, LuxPQ, VpsS, and CqsR increase, and they via phosphor-relay reaction activate the response regulator LuxO.

This promotes the transcription, thus activating the translation of AphA and inhibiting the production of HapR. At high cell density, AI level is high and thus dephosphory-lation rendering the LuxO inactive, thus repressing the transcription. AphA is not produced but the HapR is activated. Inverse expression of these signals help in biofilm formation and the production of virulence factors. Because of this multiplicity of QS systems, it has become challenging to use a narrow spectrum quencher for pathogenic bacteria. A stressed environment like QSI might lead the bacteria to explore and exploit the parallel QSS resulting in QSI-resistant mutants (Mukherjee and Bassler, 2019).

The immediate microenvironment of a bacterial cell is crucial for deciding the cell fate and the response to its surroundings. Quorum sensing is one of the mechanisms for bacteria to ascertain if the microenvironment is conducive for its growth. Antibiotic resistance occurs when a cell bears natural survival mechanism during the antibiotic invasion. In case of such antibiotic assault, only the resistant cells remain, multiply, and may survive to cause infection. The antivirulence strategy like QSI meet the same fate when it has a fitness cost for the pathogen. Certain cells exist naturally in the bacterial population, which are not capable of quorum sensing. Due to the absence of QS, they do not contribute to the synthesis of costly virulence factors or other "public goods" for community growth and survival. However, they utilize the public goods secreted by other bacteria capable of QS. These cells are known as "Social Cheaters" in that community and will be naturally resistant to an attack from any QQ compound. Since these cells are deficient in quorum-sensing activity, any QSI in the medium will not affect the growth and health of these cells. The inhibitor causes grad-ual death of QS sensitive cells while the cheaters can evolve unaffected and are selected by evolution due to their inherent fitness. The cheater cells can either evolve to produce the new signaling molecules unaffected by the quenching compound or utilize the public goods for their growth. A QSI-resistant population thus evolves due to the community dynamics between cheaters and providers.

There is no broadly established clinical proof of antivirulence resistance. Since the observed antivirulence compounds are mostly at preclinical stage or at best reached clinical trials, no data on resistance are available in a natural population.

Conclusion

Though QSI has been known for decades, no molecule has yet reached the clinical stage of development (Allen et al., 2014). There could be diverse reasons for this delay but the picture is not as gloomy as it seems since several such molecules are in a pre-clinical stage where proof of concepts have been demonstrated and their effectiveness is encouraging. Several challenges, however, remain.

Several potent signal mimics like halogenated furanones are not very specific. The overarching reasons are the circuit complexity and redundancy we do not yet fully comprehend, as some of the autoinducers are used as cross-species communication systems. This redundancy and cross-species communication can be manipulated for signal interference by sequestering the autoinducers. These can have broader effects than anticipated and may also harm gut microbiota of the patients.

Since the QSI compounds only render the bacteria harmless, an effective immune system is necessary to clear out the pathogen from the site of infection and should not be ignored. Current and future drugs should focus on combinatorial therapy that disrupts long-range interspecies crosstalk between different bacteria and even with the host. While QSI seems a very lucrative option to thwart infectious diseases, its real-world application is not so simple and straightforward. Drugs are administered (antibiotic or QSI) after disease symptoms are manifested and diagnosed. There seems to be a practical challenge of disrupting quorum sensing before the disease is diagnosed, as that is when the disruption can have any positive effect. After the pathogen has occupied its niche in the host and expressed the virulence factors, it is already late for any QSI mechanism to work effectively, albeit in some cases where it can restore the homeostatic balance in favor of antivirulence. The timing is, therefore, an utmost and often neglected criterion for QSI research. The supplemental intake of QQ compounds as nutritional additives is still in infancy. Therefore even if there is no selection for resistance against QSI inhibitors, only those compounds might be useful which can bring the phenotype back to original state *after* the disease has manifested itself. There is some hope, as it has been reported in some cases that antivirulence therapy helps even after infection (Allen et al., 2014).

Antivirulence drugs are narrow spectrum and specific, thereby allowing the host microbiome to remain unperturbed and hopefully allow convalescence period to be lesser and patient recovery faster due to maintenance of potentially helpful gut microbiota. Also, the use of QQ compounds in prophylactic use is possible. When used alone, immune cells must clear the bacterial pathogen load. So, the role for QSI therapy in immunocompromised individuals is not clear. Tipping the balance in favor of the host can help the host immunity to take over and clear the pathogen, instead of the drug killing the pathogen. However, the economic viability of developing narrow spectrum inhibitors is less.

Another challenge is the reliable proof of molecules as QSI agents. Extracellular quenching of siderophore pyoverdine with gallium has been shown to be more robust against the rise of QSI resistance as compared to antibiotics in an evolutionary experiment (Ross-Gillespie et al., 2014). However, it was reported that some assumptions were left untested, and certain experiments could have been performed to make the claims robust that Ga acted mainly on pyoverdine quenching and it worked better than antibiotics in caterpillar hosts. In reporter strains, the identifications appear easy, but the confirmation of QS activity requires more controls using constitutive or inducible promoters. Specific quorum-sensing disruption activity, A_{QSI}, was suggested to improve this aspect of QSI discovery, measured as the ratio of percentage inhibition of QS activity in reporter strain phenotype when controlled by QS, to the percentage inhibition when phenotype was independent of QS control.

Despite these challenges, many QS systems and corresponding QSI compounds have been discovered. Many more are still being discovered. Hopefully, enough QS information will be present in the near future to evaluate the evidence for their widespread and effective application. For now, this antivirulence strategy appears to be one of the best bet against the rising challenge of antimicrobial resistance, which may have been made further worse by the COVID-19 pandemic, in which there was indiscriminate and widespread use of antibiotics than usual or necessary.

Acknowledgments

AKY is supported by DBT-Big Data Initiative grant (BT/PR16456/BID/7/624/2016), Translational Research Program (TRP) at THSTI funded by DBT also supports AKY and SA. AKY is also supported by THSTI intramural grant (2021-2023). UD is supported by SERB-DST grant (ECR/2017/000605).

Conflict of interest

The authors declare no competing interests.

Author contributions

SA, PM, PG, AY, GD, UD, and AKY contributed to writing and editing the manuscript.

References

Aggarwal, S., Kandpal, M., Asthana, S., Yadav, A.K., 2017. Perturbed signaling and role of posttranslational modifications in cancer drug resistance. In: Arora, G., Sajid, A., Kalia, V.C. (Eds.), Drug Resistance in Bacteria, Fungi, Malaria, and Cancer. Springer International Publishing, Cham, pp. 483−510. Available from: http://doi.org/10.1007/978-3-319-48683-3_22.

Allen, R.C., Popat, R., Diggle, S.P., Brown, S.P., 2014. Targeting virulence: can we make evolution-proof drugs? Nat. Rev. Microbiol. 12 (4), 300−308. Available from: https://doi.org/10.1038/nrmicro3232.

Amara, N., Mashiach, R., Amar, D., Krief, P., Spieser, S.A., Bottomley, M.J., et al., 2009. Covalent inhibition of bacterial quorum sensing. J. Am. Chem. Soc. 131 (30), 10610−10619. Available from: https://doi.org/10.1021/ja903292v.

Balaban, N., Giacometti, A., Cirioni, O., Gov, Y., Ghiselli, R., Mocchegiani, F., et al., 2003. Use of the quorum-sensing inhibitor RNAIII-inhibiting peptide to prevent biofilm formation in vivo by drug-resistant *Staphylococcus epidermidis*. J. Infect. Dis. 187 (4), 625−630. Available from: https://doi.org/10.1086/345879.

Balaban, N., Cirioni, O., Giacometti, A., Ghiselli, R., Braunstein, J.B., Silvestri, C., et al., 2007. Treatment of *Staphylococcus aureus* biofilm infection by the quorum-sensing inhibitor RIP. Antimicrob. Agents Chemother. 51 (6), 2226–2229. Available from: https://doi.org/10.1128/AAC.01097-06.

Chen, J.K., Shen, C.R., Liu, C.L., 2010. N-acetylglucosamine: production and applications. Mar. Drugs 8 (9), 2493–2516. Available from: https://doi.org/10.3390/md8092493.

Christensen, L.D., van Gennip, M., Jakobsen, T.H., Alhede, M., Hougen, H.P., Hoiby, N., et al., 2012. Synergistic antibacterial efficacy of early combination treatment with tobramycin and quorum-sensing inhibitors against *Pseudomonas aeruginosa* in an intraperitoneal foreign-body infection mouse model. J. Antimicrob. Chemother. 67 (5), 1198–1206. Available from: https://doi.org/10.1093/jac/dks002.

Chung, J., Goo, E., Yu, S., Choi, O., Lee, J., Kim, J., et al., 2011. Small-molecule inhibitor binding to an N-acyl-homoserine lactone synthase. Proc. Natl. Acad. Sci. USA 108 (29), 12089–12094. Available from: https://doi.org/10.1073/pnas.1103165108.

Cosa, S., Chaudhary, S.K., Chen, W., Combrinck, S., Viljoen, A., 2019. Exploring common culinary herbs and spices as potential anti-quorum sensing agents. Nutrients 11 (4). Available from: https://doi.org/10.3390/nu11040739.

Cugini, C., Calfee, M.W., Farrow 3rd, J.M., Morales, D.K., Pesci, E.C., Hogan, D.A., 2007. Farnesol, a common sesquiterpene, inhibits PQS production in *Pseudomonas aeruginosa*. Mol. Microbiol. 65 (4), 896–906. Available from: https://doi.org/10.1111/j.1365-2958.2007.05840.x.

de Nys, R., Wright, A.D., König, G.M., Sticher, O., 1993. New halogenated furanones from the marine alga *Delisea pulchra* (cf. fimbriata). Tetrahedron 49 (48), 11213–11220. Available from: https://doi.org/10.1016/S0040-4020(01)81808-1.

Defoirdt, T., 2018. Quorum-sensing systems as targets for antivirulence therapy. Trends Microbiol. 26 (4), 313–328. Available from: https://doi.org/10.1016/j.tim.2017.10.005.

Defoirdt, T., Boon, N., Bossier, P., 2010. Can bacteria evolve resistance to quorum sensing disruption? PLoS Pathog. 6 (7), e1000989. Available from: https://doi.org/10.1371/journal.ppat.1000989.

Dickey, S.W., Cheung, G.Y.C., Otto, M., 2017. Different drugs for bad bugs: antivirulence strategies in the age of antibiotic resistance. Nat. Rev. Drug. Discov. 16 (7), 457–471. Available from: https://doi.org/10.1038/nrd.2017.23.

Eibergen, N.R., Moore, J.D., Mattmann, M.E., Blackwell, H.E., 2015. Potent and selective modulation of the RhlR quorum sensing receptor by using non-native ligands: an emerging target for virulence control in *Pseudomonas aeruginosa*. Chembiochem 16 (16), 2348–2356. Available from: https://doi.org/10.1002/cbic.201500357.

Faloon, P., Weiner, W.S., Matharu, D.S., Neuenswander, B., Porubsky, P., Youngsaye, W., et al., 2010. Discovery of ML370, an inhibitor of *Vibrio cholerae* quorum sensing acting via the LuxO response regulator. Probe Reports from the NIH Molecular Libraries Program. Bethesda (MD).

Fernandez-Pinar, R., Camara, M., Soriano, M.I., Dubern, J.F., Heeb, S., Ramos, J.L., et al., 2011. PpoR, an orphan LuxR-family protein of *Pseudomonas putida* KT2440, modulates competitive fitness and surface motility independently of N-acylhomoserine lactones. Environ. Microbiol. Rep. 3 (1), 79–85. Available from: https://doi.org/10.1111/j.1758-2229.2010.00190.x.

Garcia-Contreras, R., Martinez-Vazquez, M., Velazquez Guadarrama, N., Villegas Paneda, A.G., Hashimoto, T., Maeda, T., et al., 2013a. Resistance to the quorum-quenching compounds brominated furanone C-30 and 5-fluorouracil in *Pseudomonas aeruginosa* clinical isolates. Pathog. Dis. 68 (1), 8–11. Available from: https://doi.org/10.1111/2049-632X.12039.

Garcia-Contreras, R., Lira-Silva, E., Jasso-Chavez, R., Hernandez-Gonzalez, I.L., Maeda, T., Hashimoto, T., et al., 2013b. Isolation and characterization of gallium resistant *Pseudomonas aeruginosa* mutants. Int. J. Med. Microbiol. 303 (8), 574–582. Available from: https://doi.org/10.1016/j.ijmm.2013.07.009.

Garcia-Lara, B., Saucedo-Mora, M.A., Roldan-Sanchez, J.A., Perez-Eretza, B., Ramasamy, M., Lee, J., et al., 2015. Inhibition of quorum-sensing-dependent virulence factors and biofilm formation of clinical and environmental *Pseudomonas aeruginosa* strains by ZnO nanoparticles. Lett. Appl. Microbiol. 61 (3), 299–305. Available from: https://doi.org/10.1111/lam.12456.

Hentzer, M., Wu, H., Andersen, J.B., Riedel, K., Rasmussen, T.B., Bagge, N., et al., 2003. Attenuation of *Pseudomonas aeruginosa* virulence by quorum sensing inhibitors. EMBO J. 22 (15), 3803–3815. Available from: https://doi.org/10.1093/emboj/cdg366.

Jakobsen, T.H., van Gennip, M., Phipps, R.K., Shanmugham, M.S., Christensen, L.D., Alhede, M., et al., 2012a. Ajoene, a sulfur-rich molecule from garlic, inhibits genes controlled by quorum sensing. Antimicrob. Agents Chemother. 56 (5), 2314−2325. Available from: https://doi.org/10.1128/AAC.05919-11.

Jakobsen, T.H., Bragason, S.K., Phipps, R.K., Christensen, L.D., van Gennip, M., Alhede, M., et al., 2012b. Food as a source for quorum sensing inhibitors: iberin from horseradish revealed as a quorum sensing inhibitor of *Pseudomonas aeruginosa*. Appl. Environ. Microbiol. 78 (7), 2410−2421. Available from: https://doi.org/10.1128/AEM.05992-11.

Jung, S.A., Chapman, C.A., Ng, W.L., 2015. Quadruple quorum-sensing inputs control *Vibrio cholerae* virulence and maintain system robustness. PLoS Pathog. 11 (4), e1004837. Available from: https://doi.org/10.1371/journal.ppat.1004837.

Kandpal, M., Aggarwal, S., Jamwal, S., Yadav, A.K., 2017. Emergence of drug resistance in mycobacterium and other bacterial pathogens: the posttranslational modification perspective. In: Arora, G., Sajid, A., Kalia, V.C. (Eds.), Drug Resistance in Bacteria, Fungi, Malaria, and Cancer. Springer International Publishing, Cham, pp. 209−231. Available from: http://doi.org/10.1007/978-3-319-48683-3_9.

Kaufmann, G.F., Park, J., Mee, J.M., Ulevitch, R.J., Janda, K.D., 2008. The quorum quenching antibody RS2-1G9 protects macrophages from the cytotoxic effects of the *Pseudomonas aeruginosa* quorum sensing signalling molecule N-3-oxo-dodecanoyl-homoserine lactone. Mol. Immunol. 45 (9), 2710−2714. Available from: https://doi.org/10.1016/j.molimm.2008.01.010.

Kim, J., Kim, J., Kim, Y., Oh, S., Song, M., Choe, J.H., et al., 2018. Influences of quorum-quenching probiotic bacteria on the gut microbial community and immune function in weaning pigs. Anim. Sci. J. 89 (2), 412−422. Available from: https://doi.org/10.1111/asj.12954.

Kiran, M.D., Adikesavan, N.V., Cirioni, O., Giacometti, A., Silvestri, C., Scalise, G., et al., 2008. Discovery of a quorum-sensing inhibitor of drug-resistant staphylococcal infections by structure-based virtual screening. Mol. Pharmacol. 73 (5), 1578−1586. Available from: https://doi.org/10.1124/mol.107.044164.

Lu, C., Kirsch, B., Zimmer, C., de Jong, J.C., Henn, C., Maurer, C.K., et al., 2012. Discovery of antagonists of PqsR, a key player in 2-alkyl-4-quinolone-dependent quorum sensing in *Pseudomonas aeruginosa*. Chem. Biol. 19 (3), 381−390. Available from: https://doi.org/10.1016/j.chembiol.2012.01.015.

Ma, R., Qiu, S., Jiang, Q., Sun, H., Xue, T., Cai, G., et al., 2017. AI-2 quorum sensing negatively regulates rbf expression and biofilm formation in *Staphylococcus aureus*. Int. J. Med. Microbiol. 307 (4-5), 257−267. Available from: https://doi.org/10.1016/j.ijmm.2017.03.003.

Maeda, T., Garcia-Contreras, R., Pu, M., Sheng, L., Garcia, L.R., Tomas, M., et al., 2012. Quorum quenching quandary: resistance to antivirulence compounds. ISME J. 6 (3), 493−501. Available from: https://doi.org/10.1038/ismej.2011.122.

Mansson, M., Nielsen, A., Kjaerulff, L., Gotfredsen, C.H., Wietz, M., Ingmer, H., et al., 2011. Inhibition of virulence gene expression in *Staphylococcus aureus* by novel depsipeptides from a marine photobacterium. Mar. Drugs 9 (12), 2537−2552. Available from: https://doi.org/10.3390/md9122537.

Markou, P., Apidianakis, Y., 2014. Pathogenesis of intestinal *Pseudomonas aeruginosa* infection in patients with cancer. Front. Cell. Infect. Microbiol. 3, 115. Available from: https://doi.org/10.3389/fcimb.2013.00115.

Mima, T., Schweizer, H.P., 2010. The BpeAB-OprB efflux pump of *Burkholderia pseudomallei* 1026b does not play a role in quorum sensing, virulence factor production, or extrusion of aminoglycosides but is a broad-spectrum drug efflux system. Antimicrob. Agents Chemother. 54 (8), 3113−3120. Available from: https://doi.org/10.1128/AAC.01803-09.

Mou, R., Bai, F., Duan, Q., Wang, X., Xu, H., Bai, Y., et al., 2011. Mutation of pfm affects the adherence of *Pseudomonas aeruginosa* to host cells and the quorum sensing system. FEMS Microbiol. Lett. 324 (2), 173−180. Available from: https://doi.org/10.1111/j.1574-6968.2011.02401.x.

Muh, U., Hare, B.J., Duerkop, B.A., Schuster, M., Hanzelka, B.L., Heim, R., et al., 2006. A structurally unrelated mimic of a *Pseudomonas aeruginosa* acyl-homoserine lactone quorum-sensing signal. Proc. Natl. Acad. Sci. USA 103 (45), 16948−16952. Available from: https://doi.org/10.1073/pnas.0608348103.

Mukherjee, S., Bassler, B.L., 2019. Bacterial quorum sensing in complex and dynamically changing environments. Nat. Rev. Microbiol. 17 (6), 371–382. Available from: https://doi.org/10.1038/s41579-019-0186-5.

Nakayama, J., Yokohata, R., Sato, M., Suzuki, T., Matsufuji, T., Nishiguchi, K., et al., 2013. Development of a peptide antagonist against fsr quorum sensing of *Enterococcus faecalis*. ACS Chem. Biol. 8 (4), 804–811. Available from: https://doi.org/10.1021/cb300717f.

Ng, W.L., Perez, L.J., Wei, Y., Kraml, C., Semmelhack, M.F., Bassler, B.L., 2011. Signal production and detection specificity in Vibrio CqsA/CqsS quorum-sensing systems. Mol. Microbiol. 79 (6), 1407–1417. Available from: https://doi.org/10.1111/j.1365-2958.2011.07548.x.

Ni, N., Choudhary, G., Li, M., Wang, B., 2008. Pyrogallol and its analogs can antagonize bacterial quorum sensing in *Vibrio harveyi*. Bioorganic Med. Chem. Lett. 18 (5), 1567–1572. Available from: https://doi.org/10.1016/j.bmcl.2008.01.081.

O'Loughlin, C.T., Miller, L.C., Siryaporn, A., Drescher, K., Semmelhack, M.F., Bassler, B.L., 2013. A quorum-sensing inhibitor blocks *Pseudomonas aeruginosa* virulence and biofilm formation. Proc. Natl. Acad. Sci. USA 110 (44), 17981–17986. Available from: https://doi.org/10.1073/pnas.1316981110.

Opoku-Temeng, C., Onyedibe, K.I., Aryal, U.K., Sintim, H.O., 2019. Proteomic analysis of bacterial response to a 4-hydroxybenzylidene indolinone compound, which re-sensitizes bacteria to traditional antibiotics. J. Proteom. 202, 103368. Available from: https://doi.org/10.1016/j.jprot.2019.04.018.

Ozkaya, O., Balbontin, R., Gordo, I., Xavier, K.B., 2018. Cheating on cheaters stabilizes cooperation in *Pseudomonas aeruginosa*. Curr. Biol. 28 (13), 2070–2080. Available from: https://doi.org/10.1016/j.cub.2018.04.093. e2076.

Papenfort, K., Bassler, B.L., 2016. Quorum sensing signal-response systems in Gram-negative bacteria. Nat. Rev. Microbiol. 14 (9), 576–588. Available from: https://doi.org/10.1038/nrmicro.2016.89.

Papenfort, K., Silpe, J.E., Schramma, K.R., Cong, J.P., Seyedsayamdost, M.R., Bassler, B.L., 2017. A *Vibrio cholerae* autoinducer-receptor pair that controls biofilm formation. Nat. Chem. Biol. 13 (5), 551–557. Available from: https://doi.org/10.1038/nchembio.2336.

Park, J., Jagasia, R., Kaufmann, G.F., Mathison, J.C., Ruiz, D.I., Moss, J.A., et al., 2007. Infection control by antibody disruption of bacterial quorum sensing signaling. Chem. Biol. 14 (10), 1119–1127. Available from: https://doi.org/10.1016/j.chembiol.2007.08.013.

Parlet, C.P., Kavanaugh, J.S., Crosby, H.A., Raja, H.A., El-Elimat, T., Todd, D.A., et al., 2019. Apicidin attenuates MRSA virulence through quorum-sensing inhibition and enhanced host defense. Cell Rep. 27 (1), 187–198. Available from: https://doi.org/10.1016/j.celrep.2019.03.018. e186.

Piewngam, P., Zheng, Y., Nguyen, T.H., Dickey, S.W., Joo, H.S., Villaruz, A.E., et al., 2018. Pathogen elimination by probiotic *Bacillus* via signalling interference. Nature 562 (7728), 532–537. Available from: https://doi.org/10.1038/s41586-018-0616-y.

Plata, K.B., Rosato, R.R., Rosato, A.E., 2011. Fate of mutation rate depends on agr locus expression during oxacillin-mediated heterogeneous-homogeneous selection in methicillin-resistant *Staphylococcus aureus* clinical strains. Antimicrob. Agents Chemother. 55 (7), 3176–3186. Available from: https://doi.org/10.1128/AAC.01119-09.

Ross-Gillespie, A., Weigert, M., Brown, S.P., Kummerli, R., 2014. Gallium-mediated siderophore quenching as an evolutionarily robust antibacterial treatment. Evol. Med. Public Health 2014 (1), 18–29. Available from: https://doi.org/10.1093/emph/eou003.

Sandoz, K.M., Mitzimberg, S.M., Schuster, M., 2007. Social cheating in *Pseudomonas aeruginosa* quorum sensing. Proc. Natl. Acad. Sci. USA 104 (40), 15876–15881. Available from: https://doi.org/10.1073/pnas.0705653104.

Shen, G., Rajan, R., Zhu, J., Bell, C.E., Pei, D., 2006. Design and synthesis of substrate and intermediate analogue inhibitors of S-ribosylhomocysteinase. J. Med. Chem. 49 (10), 3003–3011. Available from: https://doi.org/10.1021/jm060047g.

Simonetti, O., Cirioni, O., Cacciatore, I., Baldassarre, L., Orlando, F., Pierpaoli, E., et al., 2016. Efficacy of the quorum sensing inhibitor FS10 alone and in combination with tigecycline in an animal model of staphylococcal infected wound. PLoS One 11 (6), e0151956. Available from: https://doi.org/10.1371/journal.pone.0151956.

Singh, V., Evans, G.B., Lenz, D.H., Mason, J.M., Clinch, K., Mee, S., et al., 2005. Femtomolar transition state analogue inhibitors of 5'-methylthioadenosine/S-adenosylhomocysteine nucleosidase from *Escherichia coli*. J. Biol. Chem. 280 (18), 18265−18273. Available from: https://doi.org/10.1074/jbc.M414472200.

Sully, E.K., Malachowa, N., Elmore, B.O., Alexander, S.M., Femling, J.K., Gray, B.M., et al., 2014. Selective chemical inhibition of agr quorum sensing in *Staphylococcus aureus* promotes host defense with minimal impact on resistance. PLoS Pathog. 10 (6), e1004174. Available from: https://doi.org/10.1371/journal.ppat.1004174.

Swem, L.R., Swem, D.L., O'Loughlin, C.T., Gatmaitan, R., Zhao, B., Ulrich, S.M., et al., 2009. A quorum-sensing antagonist targets both membrane-bound and cytoplasmic receptors and controls bacterial pathogenicity. Mol. Cell 35 (2), 143−153. Available from: https://doi.org/10.1016/j.molcel.2009.05.029.

Thomsen, K., Christophersen, L., Bjarnsholt, T., Jensen, P.O., Moser, C., Hoiby, N., 2016. Anti-*Pseudomonas aeruginosa* IgY antibodies augment bacterial clearance in a murine pneumonia model. J. Cyst. Fibros. 15 (2), 171−178. Available from: https://doi.org/10.1016/j.jcf.2015.08.002.

Tomlin, K.L., Coll, O.P., Ceri, H., 2001. Interspecies biofilms of *Pseudomonas aeruginosa* and *Burkholderia cepacia*. Can. J. Microbiol. 47 (10), 949−954.

Vandeputte, O.M., Kiendrebeogo, M., Rajaonson, S., Diallo, B., Mol, A., El Jaziri, M., et al., 2010. Identification of catechin as one of the flavonoids from Combretum albiflorum bark extract that reduces the production of quorum-sensing-controlled virulence factors in *Pseudomonas aeruginosa* PAO1. Appl. Environ. Microbiol. 76 (1), 243−253. Available from: https://doi.org/10.1128/AEM.01059-09.

Walz, J.M., Avelar, R.L., Longtine, K.J., Carter, K.L., Mermel, L.A., Heard, S.O., et al., 2010. Anti-infective external coating of central venous catheters: a randomized, noninferiority trial comparing 5-fluorouracil with chlorhexidine/silver sulfadiazine in preventing catheter colonization. Crit. Care Med. 38 (11), 2095−2102. Available from: https://doi.org/10.1097/CCM.0b013e3181f265ba.

CHAPTER 13

Microbial adaptations in extreme environmental conditions

Jayshree Sarma[1], Aveepsa Sengupta[1], Mani Kankana Laskar[1], Shatabdi Sengupta[1], Shivendra Tenguria[2] and Ashutosh Kumar[1]
[1]Department of Microbiology, Tripura University (A Central University), Agartala, Tripura, India
[2]Department of Pathology and Laboratory Medicine, Cedars-Sinai Medical Center (UCLA), Los Angeles, CA, United States

Introduction

The earth has a vast microbial diversity and the survival of each species on earth requires a certain set of environmental conditions such as the temperature, light, humidity, precipitation, wind, and atmosphere. The requirement of each species is distinct from another, and only when the conditions are optimum, they can thrive in that environment. Extreme temperature, pressure, salinity, pH, or drought disturb the molecular interactions of the biomolecules to remain functional and hence disrupts the cellular integrity of the cell (Rampelotto, 2010). Biologists in 1980s and 1990s, were able to figure out about the amazing flexibility of microbes surviving in some extreme environments that otherwise would not be tolerable by other complex organisms. This ability to survive in extreme conditions like extraordinarily hot or acidic ones placed some microbes in the 'extremophile' category. The organisms which survive in the extreme conditions are known as extremophiles. Most extremophiles are prokaryotic in nature with few of them being eukaryotes (Gupta et al., 2014). The state of conditions that these extremophiles survive in are known to be hostile and evidently, they have developed different molecular mechanisms to sustain themselves in these hostile environments. Extremophiles, for many years, had been terra incognita, as environments which made their survival possible were regarded as a dead zone. It was found a long time ago that some fungi can thrive in mildly acidic conditions. The first ever obligatory acidophilic bacteria identified was *Acidithiobacillus ferrooxidans*. Later on, lithotrophic thermophilic acidophiles were discovered and hyperacidophilic organisms of genus *Picrophilus* growing at extreme lower pH were identified in 1996. The word "psychrophile" is introduced in 1902 by Schmidt-Nielsen for describing bacteria growing at $0°C$. Although Arctic diatoms were already studied years ago without using the term "psychrophile" (Pikuta et al., 2007). Extremophiles such as the thermophiles or psychrophiles are more difficult to study as they need to be cultured at extraneous temperatures that they survive in. The astonishing feature of these extremophiles is the ability of their proteins to remain stable and operational. One of the interesting components of these extremophiles is their enzymes known as extremozyme. Extremozymes and their ability to be more

Bacterial Survival in the Hostile Environment
DOI: https://doi.org/10.1016/B978-0-323-91806-0.00007-2

stable naturally make them very crucial for industrial works, compared to the enzymes isolated from other mesophilic organisms (Kohli et al., 2020). Cultivating extremophilic microbes is difficult as they require a certain type of media preparation with various safety protocols and due to this they have been explored much less in comparison with mesophiles. The diligence required for culturing these microbes has also led to the implementation of culture-independent techniques such as metaproteomics, metagenomics, and metatranscriptomics (Charlesworth & Burns, 2016). These techniques have been advantageous in exploring many nonviable microorganisms including extremophiles which would rather remain unexplored due to the rigors of its culturing techniques. The research in these field has escalated over the past decade with more genome sequenced, filing of patents, and many funding programs launched such as the National Science Foundation (US) and programs in NASA on Life in extreme environments, astrobiology, and exobiology (Rothschild & Mancinelli, 2001). The distinct metabolic processes and biological functions attributed by their enzymes and proteins make these extremophiles crucial in biotechnological fields. The fact that these extremophiles can withstand extreme conditions also makes it possible that the enzymes secreted by these organisms will remain functional against denaturants such as chemicals, and can be used in polymer degradation such as cellulose and chitin which requires enzymes resistant to high temperatures (Van Den Burg, 2003; Gupta et al., 2014). Extremozymes can catalyze many chemical reactions in severe conditions like those seen in industrial processes that were otherwise not suitable for enzymatic activity (Sarmiento et al., 2015).

Mechanism of adaptation of extremophiles

Thermophiles

Thermophiles are extremophiles that grow at higher temperatures and are found in heated soils and deep-sea hydrothermal vents (Nakagawa and Takai, 2006). At high temperatures, membrane fluidity is increased, membrane proteins and enzymes needed for cellular processes denature which results into disturbed metabolism of these thermophiles. Protein thermostability is one of the core aspects of their survival. These organisms, at high temperatures, evolve and adapt to increase the thermostability and functional activity of membrane proteins and enzymes by undergoing molecular changes such as more number of hydrogen bonds, enhanced Van der Waal interactions, shortening of surface loops, ionic interactions, increased hydrophobicity, strengthened packing density of proteins and enzymes (Coker, 2019). Hydrophobicity plays a prominent role in protein stabilization of thermophiles. DNA at high temperatures gets denatured and undergoes modification chemically but the DNA in hyperthermophiles, for example, *Pyrococcus furiosus* are more stable compared to a mesophile like *Escherichia coli* (Rothschild & Mancinelli, 2001). Fluidity and permeability of membrane is raised with increase in temperature and have adverse effects on motility and function of the membrane proteins, cell division, and nutrients diffusion. Homeostasis is maintained physiologically and the maintenance of membrane integrity is achieved by a mechanism known as "homeoviscous adaptation" It was observed in *E. coli*, where membrane fluidity remains constant at relatively high temperatures (Sinensky, 1974; Siliakus et al., 2017). When the temperature goes above the optimum, the lipid acquires a fluid phase and then a nonlamellar phase which resembles the hexagonal or the cubic structure (Escribá, 2006; Siliakus et al., 2017). Thermophiles maintain their fluidity by various methods, they increase their content of saturated fatty acids (Oshima and Miyagawa, 1974), branched chain iso-fatty acids (Sinensky, 1971, 1974; Patel et al., 1991), polar carotenoid content (Ray et al., 1971; Yokoyama et al., 1996), and long chain fatty acids (Siliakus et al., 2017). Protein stability can also be increased by chaperones as they help to repair the denatured proteins. The nucleic acids stability is improved by certain divalent and monovalent salts that neutralize the negative charges of phosphate groups, and $MgCl_2$ and KCl prevents the DNA from undergoing hydrolysis and depurination (Rampelotto, 2010). DNA is also stabilized by increasing the number of G-C base pairs in the certain regions. Histones such as H_3, H_2A/B of eukaryotes are known to be related to histones of archaeal thermophiles. The melting point of DNA has been shown to increase due to binding of such histones (Coker, 2019).

Psychrophiles

Psychrophiles exist at lower to extreme low temperatures, ranging from 5°C to 15°C or lower. It can grow even up to the freezing point of water. They are found in freezing environments, such as glaciers, deep oceans, and polar regions. Psychrophilic bacteria have been known to inhabit Permafrost which comprises 24% of total land cover, glaciers (10% land cover), deep ocean (90% ocean volume), lakes, freezing deserts, snow, sea ice, and most areas with low temperature zones (Margesin and Miteva, 2011). Extreme low temperature disrupt membrane integrity, hence psychrophiles are adapted to regulate membrane fluidity by increasing the number of unsaturated fatty acids, or by shrinking the length of fatty-acyl chains. The increase in the fatty acids disturb the phospholipid bilayer and decreases the packing density, which leads to transferring from liquid to gel phase conversion temperature and maintaining the liquid bilayers at the lowest temperatures (Collins & Margesin, 2019). At low temperature, the enzymes stiffen, frozen water can penetrate the cell membrane and destroy the cellular integrity. Certain antifreeze proteins found in these psychrophiles combine with ice crystals by a large complementary surface and does not allow the ice crystals to penetrate the cell membrane. Certain molecular chaperones also help in the protein refolding affecting the rate of protein synthesis. Some osmoprotectants also act as cryoprotectors which helps in decreasing the freezing point inside the cytoplasm and prevents denaturation of their proteins, and enhance membrane stability. These also leads to scavenging free radicles in low temperature conditions (Coker, 2019). The enzymes produced at these low temperatures enable transcription and translation. These enzymes are adapted to act at low temperatures with modified properties such as having lower activation energy and higher catalytic efficiency (Gupta et al., 2020). These enzymes also attain less thermal stability at room temperature. High catalytic efficiency in psychrophiles make them important by allowing energy saving in cellular processes by heating or enabling these processes in cold temperatures. These characteristics of these enzymes makes them useful for commercial use. The psychrophilic enzymes have higher catalytic efficiency compared to the mesophiles or thermophiles by removing certain residues in loops enclosing active site or by replacing the heavy side chains by smaller groups, hence making ligands more attainable to the active site. This phenomenon of having better catalytic activity is assumed to be a modified structural adaptability of psychrophiles (Kohli et.al., 2020). Some extreme psychrophiles show the presence of cold acclimation proteins (CAPs) that balance the cell cycle and also regulates proper growth. Reverse gyrase, a type I DNA topoisomerase causes positive supertwists giving thermal stabilization to double helix DNA (Kikuchi and Asai, 1984). Apart from these above-mentioned mechanisms, psychrophilic organisms also undergo other adaptive changes including seasonal dormancy, selective permeability of ions, and polymerization of microtubules (Feller and Gerday, 1997).

Acidophiles

Acidophiles are organisms that grow in acidic environment (at low pH). Geothermal regions like the Yellowstone national park is home to sulphur reducing microorganisms of the archaea genera that oxidize sulphur to produce sulphuric acid and thrive at pH below 3, making it evident that many heat loving bacteria also happen to be acidophllic (Satyanarayana et.al., 2005). They have proton transporter systems in their membrane which helps to balance cytosolic pH levels in acidic range by transporting protons across the cell membrane. These organisms have a special property of using proton pumps to balance a neutral pH. This supports a steady supply of protons to facilitate the proton motive force and formation of ATP (adenosine triphosphate) via the electron transport chain. In acidophiles, this mechanism to balance acidic in environment using proton pump is very well regulated, if remained unchecked, the higher concentrations of incoming protons can drastically reduce the cellular pH to highly acidic and eventually may lead to cell death. These organisms have evolved certain mechanisms over the time to tightly regulate acid balance. In these acidophiles, cell membrane, impenetrable to protons, have tetraether lipids, a heavy isoprenoid core, variability in head-group of lipids. However, ester linkages of eukarya and bacteria domains of life are more vulnerable to acid hydrolysis compared to the ether linkages of archaea domain (Coker, 2019). Extreme acidity can cause denaturation of the proteins. Alkaliphiles include more neutral amino acids, thereby protecting their proteins (Rampelotto, 2010). An extremophilic organism named *Ferroplasma acidarmanus* is an *Eubacterium* which can grow at pH 0. This species has a cell membrane which separates its cytoplasm from the concentrated sulfuric acid which have higher concentrations of zinc, copper, cadmium in the surrounding environment. They have no cell membrane and cell wall is the only protective barrier. An archaea *Picrophilus oshimae* defied all records by showing that it can survive even at minimum pH of 0.2. This organism was discovered in Japan near a hydrothermal spring comprising of solfataric (sulfur-rich) gases. Metals may accumulate in the cells of acidophiles which can have a toxic effect by forming bonds with anions which can disrupt the functional enzymes and damage membrane integrity, inhibiting the functioning of different transport systems by displacing essential metals from their binding site. Mechanisms developed by these acidophiles to show resistance against different metals include impermeable barrier for separation, enzymatic conversion, outpouring the metal out of the cell, and reducing sensitivity of different cellular targets (Pikuta et al., 2007). In acidophiles, membrane associated antiporters aid in uptake of secondary proton. Other mechanisms include selective permeability and up-regulation of H^+ exporters (Gupta et al., 2014). Acidophilic proteins adapt to the low pH environment as the fluctuation in pH can affect the charges on polar amino acid residues. The polar-charged residues get protonated in low temperatures, which results in the change of the charges. These changes can contribute in unfolding of the protein leading to destabilizing of many structural interactions. These proteins have acidic amino acids on the surfaces (Reed et al., 2013).

Halophiles

These organisms that grow in high salt concentrations are known as halophiles. They can tolerate the ionic stress imparted by high salinity. These organisms are seen in all the three domains of life: bacteria, archaea and eukarya. They are found in environments with various concentrations of salinity mostly in aquatic habitats (salt marshes and subterranean salt lakes). They attain high concentration of solutes in the cytoplasm for maintaining their osmotic balance. The solubility, conformation, and stability of a protein is significantly affected by salts which eventually affect functional ability (Reed et al., 2013). Some halophiles synthesize osmolytes such as ectoine, which are small organic molecules and maintain the integrity of cells. These organic molecules do not intervene in enzymatic activity. Some halophiles scavenge osmoprotectants or compatible solutes from other organisms by using transporter systems as they do not produce their own compatible solutes. For example, *Halobacterium salinarum* has been known to show chemotactic activities toward osmoprotectants (Charlesworth & Burns, 2016). Haloarchaea possess C-5O carotenoids and hence they have red or orange cell vesicle (Fendrihan et al., 2006). Biofilm formation is another adaptive strategy acquired by halophiles to resist the pressure exerted on cells by extreme saline conditions. *F. acidarmanus* is one such example which forms a multilayered biofilm when grown under saline conditions. *F. acidarmanus* also shows their presence in filamentous biofilms which were observed to be linked with pyritic deposits in acid mine drainage streams. A multilayered, flat biofilm structures comprising of *Ferroplasma* spp. were also found with other thick biofilms populating the acid mine drainage system. Enzymatic degradation of the extracellular polymeric matrices is vital for the dispersal of biofilms to populate new habitat (Fröls, 2013). Water is less available to proteins at higher salt concentrations as most of the water forms a layer surrounding the salt in an ionic lattice. Therefore, low availability of water to hydrophobic amino acids in a protein causes dehydration of these amino acids and accumulate. This is how higher concentration of salt strengthen the hydrophobic interactions in a protein (Reed et al., 2013).

Alkaliphiles

These microorganisms grow at a pH of 8 or higher. Though the obligate once can grow at neutral pH they grow optimally between 9-10 pH. There is another category called alkalitolerant. These are the ones that grow optimally at neutral to acidic pH and are capable of growth at basic pH (Jones et al. 1994). Alkaline environments are mostly found in areas where high rate of Ca^{2+} is produced by serpentinites (a metamorphic rock mostly composed of one or more minerals of the serpentinite group) minerals such as silica. Some of the examples of these are hyperalkaline water found in Jordan and in the soda deserts and lakes, the East African Rift valley, Plateaus of Mongolia, Tibet, and West of the United States deserts (Rampelotto, 2010). Cyanobacteria are alkaliphilic organisms that grow at a pH of 12–13. Alkaliphiles generally grow in environments with high concentrations of sodium ions. This makes them similar to halophiles. Alkaliphilic organisms contain compatible solutes to provide resistance against different osmotic stresses. Alkaliphiles balance their intracellular pH in response to high external pH, and hence facilitating different pathways for transporting ions (Pikuta et al., 2007). One of the distinct properties of alkaliphiles is its variability of lipids. Bis-monoacylglycero-phosphate lipids, CLs (cardiolipins) are found in high quantities in some species (Krulwich, 2006). MUFAs (monounsaturated fatty acids) are mostly found in high concentrations, along with BCFAs (branched chain fatty acids) which are found on the membranes of alkaliphiles (Siliakus et al., 2017). Alkaliphiles also show presence of a secondary acidic wall comprised of polyglutamic acid and teichurono-peptide (Coker, 2019). Another adaptation for low/high pH organisms is the rigid membrane with less permeability to water (Siliakus et al., 2017).

Piezophiles

Piezophiles grow in high hydrostatic pressure surrounding the seafloor and deep in the Earth's crust. Piezophiles are adjusted to survive at pressure 40–110 MPa (Rampelotto, 2010). Pressure affects hydration and tertiary structures of protein. High pressure constricts the packaging of membrane lipids and finally affects its fluidity. To overcome this, piezophiles have adapted their cell membranes by incorporating monounsaturated fatty acids and phosphatidylcholine rather than phosphatidylethanolamine and polyunsaturated fatty acids (Coker, 2019). They have a dense hydrophobic core and higher multimerization. TET3 peptidase, a piezophilic protein, found in *Pyrococcus horikoshii* shows formation of a dodecamer that is very crucial for making compact layers around the cells so that the water cannot penetrate easily in high pressure (Reed et al., 2013).

Radiophiles

Microorganisms survive under high radiation such as ultraviolet (UV) radiation, gamma rays and radio waves are called as radiophiles or radio-resistants. Radiation can damage the nucleic acids and proteins. *Deinococcus radiodurans* is known to be resistant toward high radiation such as supra-lethal ionizing radiation and UV radiation. The high mobility group N1 (HMGN1) protein has evolved in response to alterations in chromatin due to the radiation-induced damage of DNA (Singh & Gabani, 2011). *D. radiodurans* show resistance to UV which is 33-fold times higher than *E. coli*. The desiccation resistance of *Deinococcus* is attributed by the homologs of plant desiccation resistance—associated proteins and the expanded nudix hydrolase superfamily (Pikuta et al., 2007). Since *D. radiodurans* can survive under extreme radiation, dehydration, cold, vacuum, and acidic environments, it is a good example of polyextremophile. Formation of cyclobutene pyrimidine dimers in DNA due to UV radiation hinders DNA replication and accounts for lesions in DNA. The DNA lesions are repaired using different DNA repairing mechanisms; excision repair of nucleotide (rad, xpf, uvrABCD), photoreactivation genes, homologous recombination (radA/51 and recA), and base excision repair (nth and mutY). These organisms also develop photoprotection mechanisms which protect them from the frequent exposure to UV radiation (Coker, 2019).

Xerophiles

Microorganisms that thrive in xeric (desert or inadequate supply of water) conditions are known as xerophiles. Water is crucial in all organisms, it maintains the structure and shape, balances the osmotic concentration in cells, and dissolves molecules. Lack of water can cause xeric stress where organisms cannot grow. Xerophiles are found in cracks and pores of rocks which protect them from different abiotic stresses. Some of these xerotolerant microorganisms are also found in food items, for example, *Salmonella enterica*. The cell membrane of xerotolerant bacteria have a slightly different composition of phospholipid, an increased ratio of both unsaturated to saturated fatty acids and monoenoic fatty acids to cyclopropane fatty acids conversion (Mutnuri et al. 2005; Halverson and Firestone, 2000). This ratio of fatty acids helps in preserving the membrane during moderate desiccation when it is in a liquid crystalline phase and during extreme desiccation provides the temperature for the transition of membrane from liquid phase to a disordered hexagonal phase (Halverson and Firestone, 2000; Van de Mortel and Halverson, 2004). It has been demonstrated that different tolerant pathways such as secreting extracellular polymeric substances and trehalose biosynthesis is a revital for the xerophilic organisms (Lebre et al., 2017). Osmoregulatory adaptive mechanisms help microbes to adapt in xeric stress which leads to deposition of solutes. Many cyanobacteria synthesize extracellular polysaccharides to adapt in low water conditions. There are responses such as upregulation of chaperone and heat shock proteins which are involved in protein stability to encounter oxidative stresses (LeBlanc et al., 2008). Xerophiles can also retain a state of dormancy. Species like *Bacillus* and *Clostridium*, undergo differentiation to form spores which are resistant to not only desiccation but also radiation, pressure, temperature, etc. (Rittershaus, 2013; Setlow, 2014).

Applications of extremophiles/extremozymes

Extremophiles have great biotechnological importance which comes from their unique capabilities of survival in extreme environmental conditions. They are used in various fields including bioremediation and biomass productivity. The extremozymes have great economic value because of their enhanced performances in otherwise adverse conditions found in the industrial processes. Extremophiles found in marine environments are an important source of hydrolase (Dalmaso et al., 2015). Acidophiles are being engineered to be used in biomining enabling extraction of metals (Gumulya et al., 2018). Since the extremophiles survive in diverse conditions, enzymes from these extremophiles have enhanced performances in otherwise adverse conditions found in the industrial processes. Starch hydrolyzing enzymes (glucoamylase, alpha amylase, pullulanase) from thermophilic and alkaliphilic organisms are used to make high-fructose syrup and ethanol, lactose hydrolysis in dairy products by cold-active beta-galactosidase. Certain other products from the extremophiles such as sugars (trehalose) and derivatives of amino acid (ectoine) are used to stabilize antibodies and for skin care products or vaccines (Coker, 2019).

Psychrophilic extremozymes have various applications in industries which is attributed by their higher specific activity at low temperature. Organic contaminants and their biodegradation at low temperature in various ecosystems is due to the degradation ability of the psychrophilic microbes. Psychrophiles are capable of mineralizing organic pollutants into less harmful and nonhazardous substances which are then incorporated into biogeochemical cycles. At the industrial level, enzymes used in the detergents constitutes 30%−40% of extremozymes. Subtilisin, a protease produced by *Bacillus* species, is needed for proper washing, at normal temperature with fresh water. Subtilisin found in psychrophiles (Antarctic *Bacillus* species) is properly characterized to abide by this requirement. In baking industry, psychrophilic xylanase has been useful to enhance bread quality and dough properties. Cold-active lactase has been known for its capability of hydrolyzing lactose during storage of milk at lower temperatures (Margesin & Feller, 2010).

Alkaliphiles proteases have major applications in industries ranging from detergent formulations, production of cheese, meat processing, and silver recovery from photographic films. Use of these enzymes leads to a gradual decrease in generation of toxic waste. To become a component in formulation of detergents, it is important that proteases function effectively at alkaline pH and higher temperature. The alkaline enzymes are useful in hide-dehairing process which is done at pH 8−10. It is also reported that alkaline protease is used in decomposing gelatinous coating of X-ray films and silver recovery from it. A novel alkaline pullulanase produced by *Bacillus* sp. has been a good candidate to be used as an additive in detergents. Pectic lyase produced by *Bacillus* sp. is used in enhancing Japanese paper production. Xylanase and pectinase have the ability of degumming ramie fibers (Horikoshi, 1999). Alkaline keratinases are useful in degrading feathers which are an unwanted by-product of some industrial processes. Cyclomaltodextrin glucotransferases are used in production of cyclodextrins which is an important component in food industry, pharmaceuticals, and other chemical reactions. Alkaliphiles are also deployed in microbial fuel cells, for example, *Pseudomonas alcaliphila* (releases phenazine-1-carboxylic acid). Also, the alkaliphiles produce important metabolites including antibiotics (Preiss et al., 2015; Sarethy et al., 2011).

Halophiles have been engaged in many centuries-old processes, like yielding of solar salt from the seawater and in fermented food production. Different processes where halophilic enzymes are applicable include environmental bioremediation, food processing, and biosynthetic processes. It is crucial to know that enzymatic use of the halophiles is not just limited to stability in higher concentrations of salts as they also show tolerance toward higher temperatures and remain stable in the presence of organic solvents. Halophilic amylases are used in starch saccharification, brewing, food, textile, and distilling industries (Moreno Amador et al., 2013). Bacteriorhodopsin (retinol protein pump of *Halobacterium*) is a potential candidate in photochemical processes and optoelectronic devices. Halophiles are used in treatment of hypersaline and saline wastewater and exopolysaccharide production. Nuclease H^+ from *Micrococcus varians* subsp. *halophilus* are used in producing flavoring agent 5'- guanylic acid (5'-GMP). It also helps in degradation of RNA at 60°C and 12% salt. *Halomonas boliviensis* are known to produce PHA (poly-β-hydroxyalkanoate), which contains β-hydroxyvalerate and β-hydroxybutyrate which is used in producing biodegradable plastics. Exopolysaccharides from halophiles have many applications as emulsifiers and gelling agents. β-Carotene produced by *Dunaliella* is potential source for biofuel production (Oren, 2010).

Thermophiles are considered more efficient in degradation of cellulose. Many aerobic thermophiles are involved in utilization of cellulose, for example, *Clostridium straminisolvens* and *Clostridium thermocellum*. Enzymes from hyperthermophiles and thermophiles enable preparation of catalytic modules with thermal stability and excellent selectivity. This thermal stability allows better flexibility in many in vitro bioconversion systems. Thermophilic strains isolated from paper pulp mills and brewery wastewater are known to produce ethanol from D-xylose. Thermophiles have been useful in microbial degradation of environmental pollutants and crude oil (Mehta et al., 2016). Taq polymerase is an extremozyme, extracted from *Thermus aquaticus* which is a thermophilic bacteria, discovered in hot springs of Yellowstone National Park, Wyoming, United States in 1969 (Brock, 1997) and being used in polymerase chain reactions.

Radiophiles are microorganisms that survive under different types of radiation as they possess adaptive mechanisms attributed by their metabolites (Table 13.1). Till date, many UVR-protective compounds have been discovered from radiophiles including sphaerophorin (involved in apoptotic activity), bacterioruberin (DNA repairing), mycosporine-like amino acids (useful in treating erythema), and melanin (involved in immune stimulatory activity). The therapeutic roles of radiophiles have a promising future in protecting skin from damage caused by radiations. Many other MAAs are known to have photoprotective and antioxidant roles which include palythinol, palythine, and asterina. Radiation-resistant extremolytes are also known to have the ability of nuclear waste bioremediation. C-type cytochrome from *Geobacter sulfurreducens* are involved in reduction of uranium radioisotopes. Ectoine has been largely studied for their ability to protect our skin against desiccation, UV damage, and water loss (Gabani & Singh, 2013; Raddadi et al., 2015).

Table 13.1 Extremozymes and their economical applications.

Extremozyme	Source/organisms	Use	Reference
Acidic proteases (rennin)	Acidophiles	Cheese industry	Sharma et al. (2012)
Alcalase, natalase, lipolase	Psychrophiles	Low temperature washing	Cavicchioli et al. (2011)
Alkaline cellulases	*Bacillius* sp.	Detergent additives	Charlesworth and Burns (2016)
Alkaline phosphatase	Psychrophiles	Molecular biology	Gupta et al. (2014)
Amylases	Psychrophiles	Degradation of polymers in detergents	Rothschild and Mancinelli (2001)
Catalase	*Vibrio* sp.	Textile, research, cosmetic applications	Sarmiento et al. (2015), Karan et al. (2012)
Cellulases	Alkaliphiles	Low temperature washing ($\geq 20°C$)	Cavicchioli et al. (2011)
Chitinase	*Aeromonas* sp., *Bacillus licheniformis*	Agriculture industry, food industry	Kohli et al. (2020)
Cold-adapted lipase	*Candida antarctica*	Cosmetic and pharmaceutical industries Flavor compounds	Collins and Margesin (2019), Kirk and Christensen (2002)
Dehydrogenases	Halophiles	Biocatalysis in organic media	Van Den Burg (2003)
Dehydrogenases	Psychrophiles	As biosensors	Rothschild and Mancinelli (2001)
Endoglucanases	Acidophiles	Paper industry	Blanco et al. (1998), Sharma et al. (2012)
Esterases	Psychrophiles	To increase the potency and spectrum of drugs by chiral resolution	Cavicchioli et al. (2011)
Feruloyl esterase	*P. haloplanktis*	Photoprotectant (ferulic acid), antioxidant	Aurilia et al. (2008)
Glucoamylase	Acidophiles	In starch processing	Pikuta et al. (2007)
Glucoamylase	Acidophile, Thermoacidophile (*S. solfataricus*)	Starch processing, dextrose and fructose syrups	Dumorné et al. (2017), Sharma et al. (2012)
Glucosidases	Thermophiles	Starch hydrolysis	Van Den Burg (2003)
Halophilic lipase	*Acinetobacter*	Production of food and biodiesel	Dumorné et al. (2017)
Lipase	*Thermosyntropha lipolytica*	Removing and degrading xenobiotics and other toxic compounds, bioremediation	Cavicchioli et al. (2011), Kohli et al. (2020)

Enzyme	Source	Application	Reference
Mannanase	Psychrophiles	Degrading of gum or mannan	Sarmiento et al. (2015)
Nucleoside phosphorylase	*Aeropyrum pernix K1*	Nucleoside analogs synthesis used in antiviral therapies	Raddadi et al. (2015)
Pectate lyases	Psychrophiles	Pectin-stain removal activity	Sarmiento et al. (2015)
Pretaq protease	Thermophiles	Cleaning up DNA before PCR amplification	Raddadi et al. (2015)
Protease	*Bacillus stearothermophilus*	Meat tenderizing	Sarmiento et al. (2015), Kohli et al. (2020)
Serine protease	*Bacillus* sp.	Leather industry, pharmaceutical industry	Raddadi et al. (2015)
Taq polymerase	*Thermus aquaticus*	Useful in PCR	Eichler (2001)
Thermolysin	*Bacillus thermoproteolyticus*	Dipeptide synthesis	Bruins et al. (2001), Jin et al. (2019)
Uracil-DNA N-glycosylase	Atlantic cod (*Gadus morhua*)	Control the carry-over contamination in RTPCR and PCR, SNP genotyping and in site-directed mutagenesis	Sarmiento et al. (2015), Lanes et al. (2002)
Urease	*Bacillus* sp. strain TB-90, *Campylobacter sputorumbiovarparaureolyticus*	Agriculture industry, chemical industry	Kohli et al. (2020)
Xylanase	Thermophiles	Enzyme associated paper bleaching, baking industry	Eichler (2001)
α–Amylase	*Geobacillus stearothermophilus*	Textile industry	Kohli et al. (2020)
β –Glucosidase	*Thermoanaerobacter brockii*	Chemical industry	Kohli et al. (2020)
β –Glycanase	Psychrophiles	Detergent industry	Eichler (2001)
β –Lactamases	*Pseudomonas fluorescens*	Antibiotic resistance	Michaux et al. (2008)
β–Galactosidase	*Bacillus* sp., *Arthrobacter psychrolactophilus*, *Bacillus subtilis KL88*	Brewing industry, biofuel, chemical industry	Kohli et al. (2020)

Conclusion

The extremophilic microorganisms have their unique survival strategy and that can be threatened by a mere change in their environment. The earthly existence of extremophiles and beyond have opened up a new set of possibilities in terms of astrobiological science and its biotechnological applications. The colonization of microbial diversity in vast regions of earth and their specialized adaptation mechanisms make them capable of having enzymes with unique biological properties. The extreme environments and the survival strategies lead to modifications in different cellular components such as polysaccharides, nucleic acids, and protein. Different extremozymes have specific uses in industrial fields, which is attributed by their distinct metabolic pathways. Many of the extremophiles are able to show resistance against antibiotics.

References

Aurilia, V., Parracino, A., D'Auria, S., 2008. Microbial carbohydrate esterases in cold adapted environments. Gene 410 (2), 234−240.

Blanco, A., Diaz, P., Martinez, J., Vidal, T., Torres, A.L., Pastor, F.I.J., 1998. Cloning of a new endoglucanase gene from *Bacillus* sp. BP-23 and characterisation of the enzyme. Performance in paper manufacture from cereal straw. Appl. Microbiol. Biotechnol. 50 (1), 48−54.

Brock, T.D., 1997. The value of basic research: discovery of *Thermus aquaticus* and other extreme thermophiles. Genetics 146 (4), 1207.

Bruins, M.E., Janssen, A.E., Boom, R.M., 2001. Thermozymes and their applications. Appl. Biochem. Biotechnol. 90 (2), 155−186.

Cavicchioli, R., Charlton, T., Ertan, H., Omar, S.M., Siddiqui, K.S., Williams, T., 2011. Biotechnological uses of enzymes from psychrophiles. Microb. Biotechnol. 4 (4), 449−460.

Charlesworth, J., Burns, B.P., 2016. Extremophilic adaptations and biotechnological applications in diverse environments. AIMS Microbiol. 2 (3), 251−261.

Coker, J.A., 2019. Recent advances in understanding extremophiles. F1000Research 8.

Collins, T., Margesin, R., 2019. Psychrophilic lifestyles: mechanisms of adaptation and biotechnological tools. Appl. Microbiol. Biotechnol. 103 (7), 2857−2871.

Dalmaso, G.Z.L., Ferreira, D., Vermelho, A.B., 2015. Marine extremophiles: a source of hydrolases for biotechnological applications. Marine Drugs 13 (4), 1925−1965.

Dumorné, K., Córdova, D.C., Astorga-Eló, M., Renganathan, P., 2017. Extremozymes: a potential source for industrial applications.

Eichler, J., 2001. Biotechnological uses of archaeal extremozymes. Biotechnol. Adv. 19 (4), 261−278.

Escribá, P.V., 2006. Membrane-lipid therapy: a new approach in molecular medicine. Trends Mol. Med. 12 (1), 34−43.

Feller, G., Gerday, C., 1997. Psychrophilic enzymes: molecular basis of cold adaptation. Cell. Mol. Life Sci. CMLS 53 (10), 830−841.

Fendrihan, S., Legat, A., Pfaffenhuemer, M., Gruber, C., Weidler, G., Gerbl, F., Stan-Lotter, H., 2006. Extremely halophilic archaea and the issue of long-term microbial survival. Life in Extreme Environments. Springer, Dordrecht, pp. 125−140.

Fröls, S., 2013. Archaeal biofilms: widespread and complex. Biochem. Soc. Trans. 41 (1), 393−398.

Gabani, P., Singh, O.V., 2013. Radiation-resistant extremophiles and their potential in biotechnology and therapeutics. Appl. Microbiol. Biotechnol. 97 (3), 993−1004.

Gumulya, Y., Boxall, N.J., Khaleque, H.N., Santala, V., Carlson, R.P., Kaksonen, A.H., 2018. In a quest for engineering acidophiles for biomining applications: challenges and opportunities. Genes 9 (2), 116.

Gupta, G.N., Srivastava, S., Khare, S.K., Prakash, V., 2014. Extremophiles: an overview of microorganism from extreme environment. Int. J. Agriculture Environ. Biotechnol. 7 (2), 371−380.

Gupta, S.K., Kataki, S., Chatterjee, S., Prasad, R.K., Datta, S., Vairale, M.G., et al., 2020. Cold adaptation in bacteria with special focus on cellulase production and its potential application. J. Clean. Prod. 258, 120351.

Halverson, L.J., Firestone, M.K., 2000. Differential effects of permeating and nonpermeating solutes on the fatty acid composition of *Pseudomonas putida*. Appl. Environ. Microbiol. 66, 2414−2421.

Horikoshi, K., 1999. Alkaliphiles: some applications of their products for biotechnology. Microbiol. Mol. Biol. Rev. 63 (4), 735−750.

Jin, M., Gai, Y., Guo, X., Hou, Y., Zeng, R., 2019. Properties and applications of extremozymes from deep-sea extremophilic microorganisms: a mini review. Mar. Drugs 17 (12), 656.

Jones, B.E., Grant, W.D., Collins, N.C., Mwatha, W.E., 1994. Alkaliphiles diversity and identification. In: Priest, F.G., Ramos Cormenzana, A., Tindall, B.J. (Eds.), Bacterial Diversity and Systematics.

Karan, R., Capes, M.D., DasSarma, S., 2012. Function and biotechnology of extremophilic enzymes in low water activity. Aquat. Biosyst. 8 (1), 1−15.

Kikuchi, A., Asai, K., 1984. Reverse gyrase—a topoisomerase which introduces positive superhelical turns into DNA. Nature 309 (5970), 677−681.

Kirk, O., Christensen, M.W., 2002. Lipases from *Candida antarctica*: unique biocatalysts from a unique origin. Org. Process. Res. Dev. 6 (4), 446−451.

Kohli, I., Joshi, N.C., Mohapatra, S., Varma, A., 2020. Extremophile—an adaptive strategy for extreme conditions and applications. Curr. Genomics 21 (2), 96−110.

Krulwich, T.A., 2006. Alkaliphilic prokaryotes. Prokaryotes 2, 283−308.

Lanes, O., Leiros, I., Smalås, A.O., Willassen, N., 2002. Identification, cloning, and expression of uracil-DNA glycosylase from Atlantic cod (*Gadus morhua*): characterization and homology modeling of the cold-active catalytic domain. Extremophiles 6 (1), 73−86.

LeBlanc, J.C., Gonçalves, E.R., Mohn, W.W., 2008. Global response to desiccation stress in the soil actinomycete Rhodococcus jostii RHA1. Appl. Environ. Microbiol. 74 (9), 2627−2636.

Lebre, P.H., De Maayer, P., Cowan, D.A., 2017. Xerotolerant bacteria: surviving through a dry spell. Nat. Rev. Microbiol. 15 (5), 285−296.

Margesin, R., Feller, G., 2010. Biotechnological applications of psychrophiles. Environ. Technol. 31 (8−9), 835−844.

Margesin, R., Miteva, V., 2011. Diversity and ecology of psychrophilic microorganisms. Res. Microbiol. 162 (3), 346−361.

Mehta, R., Singhal, P., Singh, H., Damle, D., Sharma, A.K., 2016. Insight into thermophiles and their wide-spectrum applications. 3 Biotech. 6 (1), 81.

Michaux, C., Massant, J., Kerff, F., Frère, J.M., Docquier, J.D., Vandenberghe, I., et al., 2008. Crystal structure of a cold-adapted class C β-lactamase. FEBS J. 275 (8), 1687−1697.

Moreno Amador, M.D.L., Pérez Gómez, D., García Gutiérrez, M.T., MelladoDurán, M.E., 2013. Halophilic Bacteria as a source of novel hydrolytic enzymes. Life 3, 38−51.

Mutnuri, S., Vasudevan, N., Kastner, M., Heipieper, H.J., 2005. Changes in fatty acid composition of *Chromobacter israelensis* with varying salt concentrations. Curr. Microbiol. 50, 151−154.

Nakagawa, S., Takai, K., 2006. The isolation of thermophiles from deep-sea hydrothermal environments. In: Rainey, F.A., Oren, A. (Eds.), Methods in Microbiology: Extremophiles. Elsevier, New York, NY, USA.

Oren, A., 2010. Industrial and environmental applications of halophilic microorganisms. Environ. Technol. 31 (8−9), 825−834.

Oshima, M., Miyagawa, A., 1974. Comparative studies on the fatty acid composition of moderately and extremely thermophilic bacteria. Lipids 9 (7), 476−480.

Patel, B.K.C., Skerratt, J.H., Nichols, P.D., 1991. The phospholipid ester-linked fatty acid composition of thermophilic bacteria. Syst. Appl. Microbiol. 14 (4), 311−316.

Pikuta, E.V., Hoover, R.B., Tang, J., 2007. Microbial extremophiles at the limits of life. Crit. Rev. Microbiol. 33 (3), 183−209.

Preiss, L., Hicks, D.B., Suzuki, S., Meier, T., Krulwich, T.A., 2015. Alkaliphilic bacteria with impact on industrial applications, concepts of early life forms, and bioenergetics of ATP synthesis. Front. Bioeng. Biotechnol. 3, 75.

Raddadi, N., Cherif, A., Daffonchio, D., Neifar, M., Fava, F., 2015. Biotechnological applications of extremophiles, extremozymes and extremolytes. Appl. Microbiol. Biotechnol. 99 (19), 7907–7913.

Rampelotto, P.H., 2010. Resistance of microorganisms to extreme environmental conditions and its contribution to astrobiology. Sustainability 2 (6), 1602–1623.

Ray, P.H., White, D.C., Brock, T.D., 1971. Effect of growth temperature on the lipid composition of *Thermus aquaticus*. J. Bacteriol. 108 (1), 227–235.

Reed, C.J., Lewis, H., Trejo, E., Winston, V., Evilia, C., 2013. Protein adaptations in archaeal extremophiles. Archaea 2013.

Rittershaus, E.S., Baek, S.H., Sassetti, C.M., 2013. The normalcy of dormancy: common themes in microbial quiescence. Cell Host Microbe 13, 643–651.

Rothschild, L.J., Mancinelli, R.L., 2001. Life in extreme environments. Nature 409 (6823), 1092–1101.

Sarethy, I.P., Saxena, Y., Kapoor, A., Sharma, M., Sharma, S.K., Gupta, V., et al., 2011. Alkaliphilic bacteria: applications in industrial biotechnology. J. Ind. Microbiol. Biotechnol. 38 (7), 769.

Sarmiento, F., Peralta, R., Blamey, J.M., 2015. Cold and hot extremozymes: industrial relevance and current trends. Front. Bioeng. Biotechnol. 3, 148.

Satyanarayana, T., Raghukumar, C., Shivaji, S., 2005. Extremophilic microbes: diversity and perspectives. Curr. Sci. 2005 (89), 78–90.

Setlow, P., 2014. The germination of spores of Bacillus species: what we know and don't know. J. Bacteriol. 196, 1297–1305.

Sharma, A., Kawarabayasi, Y., Satyanarayana, T., 2012. Acidophilic bacteria and archaea: acid stable biocatalysts and their potential applications. Extremophiles 16 (1), 1–19.

Siliakus, M.F., van der Oost, J., Kengen, S.W., 2017. Adaptations of archaeal and bacterial membranes to variations in temperature, pH and pressure. Extremophiles 21 (4), 651–670.

Sinensky, M., 1971. Temperature control of phospholipid biosynthesis in *Escherichia coli*. J. Bacteriol. 106 (2), 449–455.

Sinensky, M., 1974. Homeoviscous adaptation—a homeostatic process that regulates the viscosity of membrane lipids in *Escherichia coli*. Proc. Natl. Acad. Sci. 71 (2), 522–525.

Singh, O.V., Gabani, P., 2011. Extremophiles: radiation resistance microbial reserves and therapeutic implications. J. Appl. Microbiol. 110 (4), 851–861.

Van Den Burg, B., 2003. Extremophiles as a source for novel enzymes. Curr. Opin. Microbiol. 6 (3), 213–218.

Van de Mortel, M., Halverson, L.J., 2004. Cell envelope components contributing to biofilm growth and survival of *Pseudomonas putida* in low-water-content habitats. Mol. Microbiol 52, 735–750.

Yokoyama, A., Shizuri, Y., Hoshino, T., Sandmann, G., 1996. Thermocryptoxanthins: novel intermediates in the carotenoid biosynthetic pathway of *Thermus thermophilus*. Arch. Microbiol. 165 (5), 342–345.

CHAPTER 14

Adaptation strategies of piezophilic microbes

Somok Banerjee[1], Swatilekha Pati[1], Aveepsa Sengupta[1], Shakila Shaheen[1], Jayshree Sarma[1], Palla Mary Sulakshana[2], Shivendra Tenguria[3] and Ashutosh Kumar[1]

[1]Department of Microbiology, Tripura University (A Central University), Agartala, Tripura, India
[2]Department of Pharmacy, Raghu College of Pharmacy, Visakhapatnam, Andhra Pradesh, India
[3]Department of Pathology and Laboratory Medicine, Cedars-Sinai Medical Center (UCLA), Los Angeles, CA, United States

Introduction

The oceans occupy more than 70% of Earth's surface, with an average hydrostatic pressure of about 400 bar at the bottom of the ocean. Hydrostatic pressure has a significant effect on the structure and function of cells. In general, the maximum limit of pressure at which life can exist is found to be 100 MPa. The ability of an organism to adapt to pressure changes has highly influenced the distribution and evolution of living beings (Bartlett et al., 2007; Somero, 1990). Depending on the pressure on an organism's growth, they are classified into different groups (Fig. 14.1). The organisms that thrive in high pressure are known as piezophiles or barophiles. Barophile was originated from a Greek word in which "baro" means weight, was replaced with "piezo," which means "pressure" (Yayanos, 1995), for describing the microorganisms which can grow optimally at pressures higher than 0.1 MPa. Thus, the term "piezophile" is well suited to define the organisms that show optimum growth under high atmospheric pressure. Among them, peizophiles require 10–50 MPa, while hyperpiezophiles require >50 MPa for their optimal growth. Another group of organisms was designated as piezotolerant which are also capable of growing at or slightly higher than the atmospheric pressure and may grow irrespectively with pressure up to 10 MPa (Abe and Horikoshi, 2001; Fang et al., 2010). High hydrostatic pressure is unfavorable for the growth of most living beings and leads to a massive decrease in cellular activities that eventually causes cell death for several organisms which are classified as piezosensitive or nonpiezophilic organisms (Simonato et al., 2006; Lauro and Bartlett, 2008; Tamburini et al., 2013).

According to Simonato et al. (2006), the curiosity in piezophilic microbes started more than a century ago; however, due to various technological hurdles and lack of specialized equipment, very limited study of piezophiles are available. Nonetheless, the

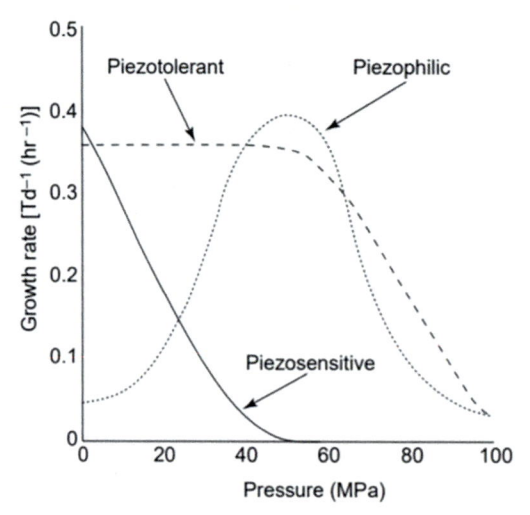

Figure 14.1 Microbial growth under different hydrostatic pressure. This diagram illustrates that the piezosensitive organisms cannot survive at high pressure, their growth decreases with increased pressure. Whereas the piezophilic organisms show optimum growth under high pressure and growth of the piezotolerant organisms is not affected much with increased pressure. *Adapted with permission from Abe, F., Horikoshi, K., 2001. The biotechnological potential of piezophiles. Trends Biotechnol. 19, 102–108.*

interest in the biotechnological aspects of piezophiles and their enzymes has been increasing during the last decades. This chapter mainly focuses on the effects of hydrostatic pressure on microbial cells and its adaptation mechanisms to sustain and also its application in the field of biotechnology.

Effects of pressure on microbial cells and macromolecules

Effect on the nucleic acids

The impact of hydrostatic pressure on the conformational stability of DNA and RNA has received less attention in comparison to other biomolecules. However, due to the increased concern in the hydration of biomolecules, the number of studies examining the influence of pressure on the structure of nucleic acids has increased (Macgregor, 1998). There is a lot of data about the effects of high hydrostatic pressure (HHP) on the nucleic acid. HHP may induce destablise the double stranded nucleic acid incase of melting temperature less than $50°C$ and may stablise the complex incase of melting temperature more than than $50°C$ into single stranded nucleic acid (Dubins et al 2001) Because of the complexity of the DNA double- to single-strand transition, these effects are complicated and are influenced by the base stacking interactions, negatively charged phosphate backbone, presence of compensating counter ions, as well as the degree of hydration. The state of hydration has a considerable impact on the structure of the double helix. HHP effect enhances the transformation efficiency of plasmid and DNA double-helical structure (Mentré and Hoa, 2001). Hydrostatic pressure has a great impact on protein—DNA complexes and has effects on their interactions, mainly ionic interactions that play an essential role in protein—DNA complexes. The hydrostatic pressure causes substantial charge neutralization that dissolves DNA—protein complexes (Macgregor, 2001).

Effect on proteins

To comprehend protein adaptation to pressure, one must first understand how pressure affects proteins. According to Bridgman, unfolding and compression are the main pressure impacts on the structure of the protein (Ichiye, 2018). Proteins get denatured when exposed to extreme pressure. The procedure that underpins such a process is yet unclear. Based on the molecular dynamics simulation studies considerable with globular proteins like apomyoglobin and lysozyme with evidence-based experimentations, the driving force for protein denaturation at high pressures is due to change in structured water leading to a decrease in hydrophobic effect (Grigera and McCarthy, 2010).

Effect on membrane lipids

The cell membrane is one of the most important biological structures. The lipid bilayer is a critical functional and structural component of a cell and shows the highest pressure sensitivity. The compressibility of lipids, particularly their hydrocarbon chains, is much higher (Brooks et al., 2011). The entire biomembrane is an extremely complex lamellar phospholipid bilayer matrix that is extremely complex which includes a wide range of lipids and proteins. Under high-pressure conditions, various characteristics of the lipid bilayer such as volume, fluidity, and phase transition were found to change. Furthermore, pressure variations can alter the folding—unfolding and the processes of membrane-embedded proteins aggregation (Ding et al., 2017). When lipids are compressed, they adapt to volume constraints by modifying their packing and conformation. As a result of increasing pressure, the fluidity of the lipid bilayer reduces, making it extremely water-impermeable, weakening lipid—protein interactions which are crucial for the basic function of the membrane proteins (Oger and Jebbar, 2010).

Effect on the cells

There are numerous effects of high pressure on living microorganisms that make it harder to specify its consequences on cell development and viability. When organisms are exposed to increasing hydrostatic pressure, various cellular processes are affected, such as synthesis of proteins and nucleic acids, the function of enzymes and metabolism, loss of membrane fluidity, loss of flagellar mobility, and variations in cellular architecture that ultimately cause the death of the piezosensitive bacteria like *Escherichia coli* (Oger and Jebbar, 2010).

Effect on microbial motility

In environments with sufficient nutrients, the motility of bacteria is ecologically important. The highly specialized tactic systems among several bacteria permit them to identify both physical and chemical gradients to migrate directly toward the source or far from the source (Meganathan and Marquis, 1973). Claude ZoBell first studied the impact of high pressure on bacterial motility. After various pressure exposures, he investigated the presence or lack of flagella on isolates of bacteria. Sophisticated networks of chemosensory and molecular motor control bacterial motility by which bacteria can pursue to physical as well as chemical gradients. The organisms of the family *Vibrionaceae*, including *Photobacterium profundum SS9*, propel themselves by rotating a helical filament, while a rotary motor implanted in the membrane of the cell generates the required torque by using the chemical gradients of Na^+ or H^+ (Meganathan and Marquis, 1973).

In *E. coli*, both rotation of flagella and the development of the flagellar filament are extremely pressure sensitive. Under high pressure in vitro, the irreversible depolymerization of filaments was observed. These in vitro findings contradict *E. coli* in vivo findings, demonstrating that even at pressures up to 60 MPa, does not trigger disaggregation of already constructed flagella (Bartlett et al., 2007). Hydrostatic pressure additionally acts as a physical barrier to bacterial motility; however, the elaborate process of this inhibition is still unknown. The pressure dependence of the revolution of a single flagellar motor and the mobility of swimming *E. coli* cells has been examined. As pressure increased, the frequency and speed of the swimming cells dropped. When the pressure approached 80 MPa, all cells ceased to swim and dispersed in the solution. Several cells quickly regained their original mobility just after the pressure was released (Nishiyama and Sowa, 2012). According to Tamura et al. (1997), only the depolymerization of the polar filament in *Salmonella typhimurium* has been demonstrated so far in deep-sea motility research (Tamura et al., 1997). Increased pressure was treated to a solution of *S. typhimurium* flagellar filaments. Polymerization of flagellin reached a peak at 98 MPa before transitioning to increased depolymerization at pressures as high as 245 MPa (Simonato et al., 2006).

Adaptation mechanisms in the piezophiles

Piezophilic organisms rely on hydrostatic pressure for their biological activity (Avagyan et al., 2019). Nonpiezophilic microorganisms have shown basic variations in regulatory mechanisms and cellular metabolic processes at the molecular level in contrast with increased pressures on piezophilic microorganisms. Analysis of nonpiezophilic organisms under high pressures reveals the pressure-sensitive cellular processes of which piezophilic bacteria must adapt to survive at high pressures (Bartlett et al., 2007).

It is important to understand the impacts of pressure on metabolic responses in nonpiezophilic organisms before discussing adaptive mechanisms in deep-sea microbes. High-pressure impacts on discrete functions in nonpiezophilic organisms, especially in the *E. coli* (a mesophile), have been studied multiple times. Such research has found significant variances in pressure sensitivity among distinct biological systems. Although deep-sea microorganisms may not be the only ones who can respond to pressure, they also acquire the ability to respond to pressure variations in an adaptable manner (Allen and Bartlett, 2002a,b). The effects of high pressure on *E. coli* cells are extremely pleiotropic. Rotation of flagellar filaments and polymerization occurs at pressures much below those that affect cell development (Meganathan and Marquis, 1973). Similarly, pressures of 20−50 MPa stop cells from dividing, leading single cells to form lengthy filaments (Zobell and Cobet, 1964). DNA synthesis, protein synthesis, and RNA synthesis have been observed to stop around 50, 58, and 70 MPa, respectively, based on the uptake of radiolabeled nucleotides and amino acids (Yayanos and Pollard, 1969). Furthermore, because the proton-translocating ATPase is pressure sensitive, cells under elevated pressure utilize much energy to pump protons across the membrane. Such processes are sensitive in *E. coli* but closely similar to piezophilic organisms that have developed molecular mechanisms to function under high pressures (Allen and Bartlett, 2002a).

The deep-sea microbes adapted very rapidly to uptake a wide range of nutritional concentrations (Wirsen and Molyneaux, 1999). The piezophilic microorganisms are exposed to high pressure and develop several mechanisms to cope with extreme pressure conditions. These adaptation mechanisms are as follows:

Genome

The cultivation techniques and ascertain the rate of growth under different pressures are performed for determining piezophile microbes. However, a study by Li et al. (1998) shows the presence of pressure-controlled operons ORFs 1−3 in *Shewanella* sp. Similarly, studies on the RecD gene (DNA-binding protein) of *P. profundum* strain *SS9* indicated its major role in piezophilic adaptation (Bidle and Bartlett, 1999; Abe et al., 1999) that shows the finding of a pressure-sensitive strain *SS9* as well as the ability of *E. coli* to grow under high pressure after transferring of RecD gene. Recently, Kurosaka and Abe (2018) showed that the gene Ypr153w of *S. cerevisiae* is responsible for the stability of its tryptophan permease Tat2 under hydrostatic pressure. This gene is also present in *Debaryomyces* sp. and *Candida* sp.

Protein

Protein expression can be influenced by increased hydrostatic pressure in both the atmospheric pressure-adapted and high pressure-adapted bacteria. Even organisms that have never lived in high pressure can also respond to high pressure by synthesizing "stress" proteins. According to a study by Welch et al. (1993), there was a decrease in growth rate and protein synthesis when anaerobically grown *E. coli* cells were isothermally switched from atmospheric pressure to 53 MPa. The high pressure causes the most heat shock proteins than most of the other situations outside of those that closely resemble a response of heat shock, as well as more cold shock proteins than the majority of other situations outside that closely resemble a response of cold shock. Temperature fluctuations and antibiotic treatments modify translational capacity, and heat shock and cold shock protein synthesis exhibit inverse reactions to these conditions (VanBogelen and Neidhardt, 1990).

An early step of translation is inhibited by high pressure and low temperature. It has been reported that the cold shock response is an adaptive response that promotes the gene expression involved in the initiation of translation (Jones et al., 1987). The membrane fluidity of the microbes has role in pressure adaptation (Casadei et al., 2002).

Apart from "extrinsic factors" like ions or specific ligands, proteins adapt to extreme conditions using the common repertory of the 20 natural amino acids. In terms of 3D structure and catalytic mechanism, homologous enzymes from extremophiles and mesophiles are very similar. Under their different physiological circumstances, their activity tends to be equivalent. Protein stability may be divided into two categories: the macrostability preserves the natural-folded shape, while the microstability regulates the dynamics of the protein that ensures optimal function (Jaenicke and Závodszky, 1990).

Piezophile proteins may have a greater overall volume of tiny internal cavities, making them more compressible and less prone to pressure distortion (Ichiye, 2018). Furthermore, the existence of more small cavities at HP allows water to penetrate and therefore enhance hydration, but the cavities are not large enough to promote denaturation of protein, as seen in MpDHFR, but they make the protein much more flexible. The presence of additional cavities which are small in size might reduce the number of molecules of water stored in every cavity (a single water molecule requires a volume of 15 Å^3 and each extra molecule requires an increase of roughly 45 Å^3) (Sonavane and Chakrabarti, 2008). In a low pressure environment, piezophiles are stressed and it is crucial to prevent hydration because the lower pressure causes the protein to take on an even more open shape and with a larger cavity volume, allowing water to go into the enzyme and affecting catalysis and flexibility (Nagae et al., 2012). It is also important to remember that cavities are more than just "packing defects," since they play a role in protein conformational changes, as well as regulate binding and catalysis (Salvador-Castell et al., 2020).

Volume reduction, fewer inner cavities, and few molecules of water inside the inner cavities seem to be the adaptive features of the piezophilic proteins. In addition, piezophiles have lesser proline and glycine residues that are thought to decrease the flexibility of protein by breaking and destabilizing helices and compressibility by reducing the space of conformation. These are considered to enhance the protein to remain stable in a high pressure (Brininger et al., 2018). A study by Chilukuri and Bartlett (1997) on different *Shewanella* strains reported a particular trend in the amino acid constitution of the single-stranded DNA binding (SSB) protein for piezo-adaptation. When the pressure was applied beyond their optimal range, there was a decline in proline and glycine proportion suggesting a reduction in the flexibility of the SSB protein from *Shewanella* sp. strain *PT99*. It has been shown that a proline to glycine replacement enhances protein stability and also reduces chain mobility (Royer et al., 1993).

Membrane modification

Piezophilic microbes maintain membrane fluidity and functionality under high hydrostatic pressure and decreased temperatures through three routes, that is, either by changing the acyl chain lengthwise, or by elevating the quantity of unsaturated fatty acids (FA) in their lipid membrane, or by causing a change in the phosphate group of the phospholipids (Oger and Cario, 2013; Siliakus et al., 2017). For instance, according to Allen et al. (1999) piezophilic bacteria, *P. profundum* strain *SS9* showed an increase mono and polyunsaturated FA when it was grown under low temperature or high pressure. It is sustained by another study which through genetic analysis showed that only mono-unsaturated FA are necessary for growth under elevated pressure and decreased temperature (Allen and Bartlett, 2002b). Moreover, two other piezophilic microbes namely *Shewanella* sp. *DB21MT-2* (having optimum growth under 70 MPa) and *Moritella* sp. *DB21MT-5* (optimum growth 80 MPa) contains a high quantity of mono-unsaturated FA octadecenoic acid (18:1) and tetradecenoic acid (14:1), correspondingly, in comparison to strains of *Shewanella benthica* and *Moritella marina* (Nogi and Kato, 1999; Kato et al., 1998). Increased length of the acyl chain enhances the rigidity of the membranes while larger polar head groups disrupt membrane packing which augments membrane fluidity. Few studies showed that the ability to modify membrane composition under high hydrostatic pressure is not restricted to the piezophilic microorganisms. Fernandes et al. (2004) showed that the expression of the ole1 gene (Stearoyl CoA desaturase) in *S. cerevisiae* is upregulated when grown under an elevated pressure of 200 MPa for 30 min maybe to increase the quantity of unsaturated FA.

Several researchers have pointed out the importance of the respiratory chain membrane proteins in piezophilic adaptation. For instance, the *E. coli cyd*D mutant exhibiting enhanced pressure sensitivity led to the investigation of the cytochrome constitution of various *Shewanella* sp. and it was shown that the *Shewanella* sp. strain DSS12 only expresses cytochrome *bd* under high hydrostatic pressures (Tamegai et al., 1998). Similarly, another study showed that the membrane-bound c-type cytochrome c-551 of *S. benthica* strain DB172F was expressed continuously from 0.1 to 60 MPa; however, cytochrome c-552 (cytoplasmic) was only expressed at 0.1 MPa pressure (Qureshi et al., 1998). Furthermore, the proportion of *cbb*-type quinol oxidase also increased with the rise in pressure which indicated that the respiratory chain of DB172F varies with high and low hydrostatic pressures.

Metabolic adaptation

High hydrostatic pressure also affects the metabolic processes. For instance, few studies reported the sensitivity of chloroplast and mitochondrial F_1F_{01} ATP-synthase to low pressures (50 MPa) which affect their structure and activity (Souza et al., 2004; Dreyfus et al., 1988). Such effects occurring at physiologically relevant hydrostatic pressures must be the driving force behind the piezo-adaptation of the F_1F_{01} ATP-synthase in the piezophiles. This conclusion is sustained by the presence of two complete sets of genes encoded by the *P. profundum SS9* genome for the F_1F_{01} ATP-synthase (Vezzi et al., 2005). A study by Fernandes et al. (2004) showed that the genes of *S. cerevisiae* are responsible for the upregulation of glycolysis, gluconeogenesis as well as glycogen metabolism when exposed to elevated pressure. Moreover, a group of genes was identified in the bacteria *P. profundum* strain *SS9* which are involved in Stickland reaction, found in obligate anaerobes including *Clostridiales* sp. and *Spirochaetales* sp. The expression of a few of these genes was increased when pressure elevated from 0.1 to 28 MPa (Graentzdoerffer et al., 2001).

Biotechnological applications

In recent times, extensive research on the piezophilic microbes and their enzymes has led to a number of biotechnological applications, for instance, the regulation of gene expression by the utilization of hydrostatic pressure. According to Simonato et al. (2006), this type of mechanism was first found in the piezophile *P. profundum*, where hydrostatic pressure was found to act on the ToxR-ToxS complex, which in turn regulates the expression of the outer membrane proteins, OmpH and OmpL. The membrane protein OmpH is dominant at 28 MPa which is the optimum pressure of the bacteria; however, when it is exposed to reducing the pressure the ratio of OmpH to OmpL decreases correspondingly.

Another study by Abe and Horikoshi (2001) showed the presence of a promoter in two species of *Shewanella* elevates the expression of the downstream genes under 50 MPa pressure. Now by placing this promoter upstream of specific genes, the pressure regulation expression of genes could be achieved.

There are a lot of proteins that are stable under elevated hydrostatic pressure and few studies have also shown that increased pressure helps in the prevention of protein denaturation even beyond the inactivation temperature. For instance, there was an increase in the half-life of the enzyme Taq polymerase from 5 min at 3 MPa pressure to 12 min at 45 MPa. Similar effects were also found in other DNA polymerases, hydrogenases as well as α-glucosidases (Abe and Horikoshi, 2001). According to Van den Berg (2003), by understanding this process, synthetic proteins could be designed which could enhance the efficiency of various processes in food processing and antibiotic production which requires high hydrostatic pressure.

A number of enzymes have been isolated from various microorganisms to date for several types of applications; however, these are not sufficient to cater to the new technological hurdles that arise every day (Dumorne et al., 2017). One of such problems is the stability of these enzymes under extreme industrial conditions which requires enzymes that are resistant to these conditions. This is where the piezophilic microbial enzymes come into play. According to Hikida et al. (2017), these types of enzymes are capable of conducting biocatalysis under high pressure, modifying specific enzymatic reactions, and are highly thermostable as well. An example of such an enzyme was commercialized DNA polymerase (Biolabs) isolated from extreme thermophilic piezophile, *Pyrococcus* sp. have shown 23 h of half-life at a temperature of 95°C. Furthermore, these enzymes may contain different properties in comparison to their surface homologs that may open multiple new prospects for the biotechnological industry (Schroeder et al., 2018).

In general, lipids of extremophilic microbes are unique and provide the cell membrane with high stability and impermeability. This property of high stability can be utilized by the biotechnological and pharmaceutical industries such as for the protection of therapeutic peptides from the extreme environment of the gastrointestinal tract (Benvegnu et al., 2009; Jacobsen et al., 2017). Moreover, the lipids of a number of piezophilic bacteria contain omega 3-PUFAs inside the membranes of their cell, which in general are precursors of many animal hormones. According to Schroeder et al. (2018), several clinical studies have been approved for treating hypertriglyceridemia by using the omega 3-PUFAs.

Genetic manipulation is one of the research areas where the genes of peizophiles can be introduced into host microbes like *S. cerevisiae* or antibiotic-producing actinomycetes to piezophilic mutant strains, may have an immense potential to produce unique bioactive metabolites with effective therapeutic potential, under high-pressure conditions.

Conclusion

Piezophiles contribute an important field of research along with having a great prospective for biotechnological application. The piezophilic microbes possess unique features to thrive in this harsh environment. These microorganisms represent a wide resource of molecules for thriving in such extreme conditions and thus can be exploited for numerous applications. The study about the molecular adaptations that these organisms undergo in order to thrive in such harsh environments has helped to extend our knowledge regarding the boundaries of physicochemical parameters and macromolecular stability of life. The genomic revolution has paved the way to know more about the molecular mechanisms of these properties in detail. The study of piezo-microbiology has contributed a lot to the developing science of astrobiology as well where the presence and survivability of life are being studied within the earth and beyond. However, more focused research in this field is required to unmask every possible mechanism of these microorganisms.

References

Abe, F., Kato, C., Horikoshi, K., 1999. Pressure-regulated metabolism in microorganisms. Trends Microbiol. 7, 44753.

Abe, F., Horikoshi, K., 2001. The biotechnological potential of piezophiles. Trends Biotechnol. 19, 102–108.

Allen, E.E., Bartlett, D.H., 2002a. Piezophiles: microbial adaptation to the deep-sea environment. Extremophiles. *Eolss Publishers Co. Ltd., Oxford.*

Allen, E.E., Bartlett, D.H., 2002b. Structure and regulation of the omega-3 polyunsaturated fatty acid synthase genes from the deep-sea bacterium *Photobacterium profundum* strain SS9. Microbiology 148, 1903–1913.

Allen, E.E., Facciotti, D., Bartlett, D.H., 1999. Monounsaturated but not polyunsaturated fatty acids are required for growth of the deep-sea bacterium *Photobacterium profundum* SS9 at high pressure and low temperature. Appl. Environ. Microbiol. 65, 1710–1720.

Avagyan, S., Vasilchuk, D., Makhatadze, G.I., 2019. Protein adaptation to high hydrostatic pressure: computational analysis of the structural proteome. Proteins 88, 584–592.

Bartlett, D.H., Lauro, F.M., Eloe, E.A., 2007. Microbial adaptation to high pressure. Physiol. Biochem. Extremophiles 333–348.

Benvegnu, T., Lemiègre, L., Cammas-marion, S., 2009. New generation of liposomes called archaeosomes based on natural or synthetic archaeal lipids as innovative formulations for drug delivery. Recent. Pat. Drug Deliv. Formul 33, 20620.

Bidle, K.A., Bartlett, D.H., 1999. RecD function is required for high-pressure growth of a deep-sea bacterium. J. Bacteriol. 181, 23307.

Brininger, C., Spradlin, S., Cobani, L., Evilia, C., 2018. The more adaptive to change, the more likely you are to survive: protein adaptation in extremophiles. Semin. Cell Dev. Biol. 84, 158–169.

Brooks, N.J., Ces, O., Templer, R.H., Seddon, J.M., 2011. Pressure effects on lipid membrane structure and dynamics. Chem. Phys. Lipids 164, 89–98.

Campanaro, S., Vezzi, A., Vitulo, N., Lauro, F.M., D'Angelo, M., Simonato, F., 2005. Laterally transferred elements and high pressure adaptation in *Photobacterium profundum* strains. BMC Genomics 6, 122–136.

Casadei, M.A., Manas, P., Niven, G., Needs, E., Mackey, B.M., 2002. Role of membrane fluidity in pressure resistance of Escherichia coli NCTC 8164. Appl. Environ. Microbiol. 68 (12), 5965–5972.

Chilukuri, L.N., Bartlett, D.H., 1997. Isolation and characterization of the gene encoding single-stranded-DNA-binding protein (SSB) from four marine *Shewanella* strains that differ in their temperature and pressure optima for growth. Microbiology. 143, 1163−1174.

Ding, W., Palaiokostas, M., Shahane, G., Wang, W., Orsi, M., 2017. Effects of high pressure on phospholipid bilayers. J. Phys. Chem. B 121 (41), 9597−9606.

Dreyfus, G., Guimaraes-Motta, H., Silva, J.L., 1988. Effect of hydrostatic pressure on the mitochondrial ATP synthase. Biochemistry 27, 6704−6710.

Dubins, D.N., Lee, A., Macgregor, R.B., Chalikian, T.V., 2001. On the stability of double stranded nucleic acids. J. Am. Chem. Soc. 123 (38), 9254−9259.

Dumorne, K., Córdova, D.C., Astorga-Elo, M., Renganathan, P., 2017. Extremozymes: a potential source for industrial applications. J. Microbiol. Biotechnol. 27, 64959.

Fang, J., Zhang, L., Bazylinski, D.A., 2010. Deep-sea piezosphere and piezophiles: geomicrobiology and biogeochemistry. Trends Microbiol. 18 (9), 413−422. Available from: https://doi.org/10.1016/j.tim.2010.06.006.

Fernandes, P.M., Domitrovic, T., Kao, C.M., Kurtenbach, E., 2004. Genomic expression pattern in *Saccharomyces cerevisiae* cells in response to high hydrostatic pressure. FEBS Lett. 556, 153−160.

Golub, M., Lehofer, B., Martinez, N., Ollivier, J., Kohlbrecher, J., Prassl, R., et al., 2017. High hydrostatic pressure specifically affects molecular dynamics and shape of low-density lipoprotein particles. Sci. Rep. 7, 46034.

Grigera, J.R., McCarthy, A.N., 2010. The behavior of the hydrophobic effect under pressure and protein denaturation. Biophys. J. 98, 1626−1631.

Graentzdoerffer, A., Pich, A., Andreesen, J.R., 2001. Molecular analysis of the grd operon coding for genes of the glycine reductase and of the thioredoxin system from Clostridium sticklandii. Arch. Microbiol. 175 (1), 8−18.

Hikida, Y., Kimoto, M., Hirao, I., Yokoyama, S., 2017. Crystal structure of deep vent DNA polymerase. Biochem. Biophys. Res. Commun. 483, 527.

Ichiye, T., 2018. Enzymes from piezophiles. Semin. Cell Dev. Biol. 84, 138−146.

Jacobsen, A.C., Jensen, S.M., Fricker, G., Brandl, M., Treusch, A.H., 2017. Archaeal lipids in oral delivery of therapeutic peptides. Eur. J. Pharm. Sci. 108, 10110.

Jaenicke, R., Závodszky, P., 1990. Proteins under extreme physical conditions. FEBS Lett. 268, 344−349.

Jones, P.G., VanBogelen, R.A., Neidhardt, F.C., 1987. Induction of proteins in response to low temperature in *Escherichia coli*. J. Bacteriol. 169, 2092−2095.

Kato, C., Li, L., Nogi, Y., Nakamura, Y., Tamaoka, J., Horikoshi, K., 1998. Extremely barophilic bacteria isolated from the Mariana Trench, Challenger Deep, at a depth of 11,000 m. Appl. Environ. Microbiol. 64, 1510−1513.

Kato, C., Nogi, Y., 2001. Correlation between phylogenetic structure and function: examples from deep-sea *Shewanella*. FEMS Microbiol. Ecol. 35, 223−230.

Kato, C., Qureshi, M.H., 1999. Pressure response in deep-sea piezophilic bacteria. J. Mol. Microbiol. Biotechnol. 1, 87−92.

Kurosaka, G., Abe, F., 2018. The YPR153W gene is essential for the pressure tolerance of tryptophan permease Tat2 in the yeast *Saccharomyces cerevisiae*. High. Press. Res. 38, 908.

Lauro, F.M., Bartlett, D.H., 2008. Prokaryotic lifestyles in deep sea habitats. Extremophiles 12, 15−25.

Li, L., Kato, C., Nogi, Y., Horikoshi, K., 1998. Distribution of the pressure-regulated operons in deep-sea bacteria. FEMS Microbiol. Lett. 159, 15966.

Macgregor Jr, R.B., 1998. Effect of hydrostatic pressure on nucleic acids. Biopolymers 48, 253−263.

Macgregor, R.B., 2002. The interactions of nucleic acids at elevated hydrostatic pressure. Biochim. Biophys. Acta 1595, 266−276.

Meganathan, R., Marquis, R.E., 1973. Loss of bacterial motility under pressure. Nature 246, 525−527.

Mentré, P., Hoa, G.H.B., 2001. Effects of high hydrostatic pressures on living cells: a consequence of the properties of macromolecules and macromolecule-associated water. Int. Rev. Cytol. 201, 1−84.

Nagae, T., Kato, C., Watanabe, N., 2012. Structural analysis of 3-isopropylmalate dehydrogenase from the obligate piezophile *Shewanella benthica* DB21MT-2 and the nonpiezophile *Shewanella oneidensis* MR-1. Acta Crystallogr. Sect. F. Struct. Biol. Cryst. Commun. 68, 265−268.

Nishiyama, M., Sowa, Y., 2012. Microscopic analysis of bacterial motility at high pressure. Biophys. J. 102, 1872–1880.

Nogi, Y., Kato, C., 1999. Taxonomic studies of extremely barophilic bacteria isolated from the Mariana Trench, and *Moritellayayanosii* sp. nov., a new barophilic bacterial species. Extremophiles 3, 71–77.

Oger, P.M., Jebbar, M., 2010. The many ways of coping with pressure. Res. Microbiol. 161, 799–809.

Oger, P.M., Cario, A., 2013. Adaptation of the membrane in Archaea. Biophys. Chem. 183, 4256.

Qureshi, M.H., Kato, C., Horikoshi, K., 1998. Purification of a ccb type quinol oxidase specifically induced in a deep-sea barophilic bacterium, *Shewanella* sp. strain DB-172F. Extremophiles 2, 93–99.

Royer, C.A., Hinck, A.P., Loh, S.N., Prehoda, K.E., Peng, X., Jonas, J., et al., 1993. Effects of amino acid substitutions on the pressure denaturation of staphylococcal nuclease as monitored by fluorescence and nuclear magnetic resonance spectroscopy. Biochemistry 32, 5222–5232.

Salvador-Castell, M., Oger, P., Peters, J., 2020. High-pressure adaptation of extremophiles and biotechnological applications. Physiological and Biotechnological Aspects of Extremophiles. *Academic Press*, pp. 105–122.

Schroeder, G., Bates, S.S., La Barre, S., 2018. Bioactive marine molecules and derivatives with biopharmaceutical potential. Blue Biotechnol. 61141.

Siliakus, M.F., van der Oost, J., Kengen, S.W.M., 2017. Adaptations of archaeal and bacterial membranes to variations in temperature, pH and pressure. Extremophiles 21, 65170.

Simonato, F., Campanaro, S., Lauro, F.M., Vezzi, A., D'Angelo, M., Vitulo, N., et al., 2006. Piezophilic adaptation: a genomic point of view. J. Biotechnol. 126, 11–25.

Somero, G.N., 1990. Life at low volume change: hydrostatic pressure as a selective factor in the aquatic environment. Am. Zool. 30, 123–135.

Sonavane, S., Chakrabarti, P., 2008. Cavities and atomic packing in protein structures and interfaces. PLoS Comput. Biol. 4, e1000188.

Souza, M.O., Creczynski-Pasa, T.B., Scofano, H.M., Graber, P., Mignaco, J.A., 2004. High hydrostatic pressure perturbs the interactions between CF0F1 subunits and induces a dual effect on activity. Int. J. Biochem. Cell. Biol. 36, 920–930.

Tamburini, C., Boutrif, M., Garel, M., Colwell, R.R., Deming, J.W., 2013. Prokaryotic responses to hydrostatic pressure in the ocean—a review. Environ. Microbiol. 15, 1262–1274.

Tamegai, H., Kato, C., Horikoshi, K., 1998. Pressure-regulated respiratory system in barotolerant bacterium *Shewanella* sp. strain DSS12. J. Biochem. Mol. Biol. Biophys. 1, 213–220.

Tamura, Y., Gekko, K., Yoshioka, K., Vonderviszt, F., Namba, K., 1997. Adiabatic compressibility of flagellin and flagellar filament of *Salmonella typhimurium*. Biochim. Biophys. Acta 1335, 120–126.

Van den Berg, B., 2003. Extremophiles as a source for novel enzymes. Curr. Opin. Microbiol. 6, 213–218.

VanBogelen, R.A., Neidhardt, F.C., 1990. Ribosomes as sensors of heat and cold shock in *Escherichia coli*. Proc. Natl. Acad. Sci. 87, 5589–5593.

Vezzi, A., Campanaro, S., D'Angelo, M., Simonato, F., Vitulo, N., Lauro, F.M., et al., 2005. Life at depth: *Photobacterium profundum* genome sequence and expression analysis. Science 307, 1459–1461.

Welch, T.J., Farewell, A., Neidhardt, F.C., Bartlett, D.H., 1993. Stress response of *Escherichia coli* to elevated hydrostatic pressure. J. Bacteriol. 175, 7170–7177.

Wirsen, C.O., Molyneaux, S.J., 1999. A study of deep-sea natural microbial populations and barophilic pure cultures using a high-pressure chemostat. Appl. Environ. Microbiol. 65, 5314–5321.

Yayanos, A.A., 1995. Microbiology to 10,500 meters in the deep sea. Annu. Rev. Microbiol. 49, 777–805.

Yayanos, A.A., Pollard, E.C., 1969. A study of the effects of hydrostatic pressure on macromolecular synthesis in *Escherichia coli*. Biophys. J. 9, 1464–1482.

Zobell, C.E., Cobet, A.B., 1964. Filament formation by *Escherichia coli* at increased hydrostatic pressures. J. Bacteriol. 87, 710–719.

Zobell, C.E., Johnson, F.H., 1949. The influence of hydrostatic pressure and growth and viability of terrestial and marine bacteria. J. Bacteriol. 57, 179–189.

CHAPTER 15

Survival and adaptation strategies of microorganisms in the extreme radiation

Soumyadip Ghosh[1],*, Shukla Banerjee[2],*, Aveepsa Sengupta[1], Vidyullatha Peddireddy[3], Anitha Mamillapalli[4], Aniruddha Banerjee[1], Bipin Kumar Sharma[1] and Ashutosh Kumar[1]

[1]Department of Microbiology, Tripura University (A Central University), Agartala, Tripura, India
[2]Dolphin PG Institute of Biomedical and Natural Sciences, Dehradun, Uttarakhand, India
[3]Department of Nutrition Biology, School of Interdisciplinary and Applied Sciences, Central University of Haryana, Mahendragarh, Haryana, India
[4]Department of Biotechnology, GITAM Institute of Science, GITAM (Deemed to be University), Visakhapatnam, Andhra Pradesh, India

Introduction

Typically living microorganisms are sensitive to drastic harsh changes in the environment. There are various types of extreme environmental conditions widely spread on the different regions of the earth at every altitude and latitude (Magazù et al., 2012). These extreme conditions can be temperature, pressure, salinity, pH, radiation, drought, etc. To survive in these environmental conditions, microbes must adapt their biomolecules and cellular integrity accordingly. Various extremophiles have a variety of mechanisms for their survival against such extreme environmental conditions (Rampelotto, 2010). Some microbes can tolerate high levels of ultraviolet radiation ($> 100 \, \text{J/m}^2$) and gamma radiation ($> 12 \, \text{kGy}$), X-rays, etc. The organisms like *Pyrococcus furiosus* and *Halobacterium* sp. can tolerate 3000 to 5000 Gy, and *Deinococcus geothermalis* can tolerate up to more than 12,000 Gy X-rays and Y-rays. In contrast, the radiation of 10 Gy is lethal to most vertebrates, including human beings, and typically microorganisms cannot survive at 200 Gy radiation (Daly, 2009). The reactive oxygen species (ROS) resulting from the desiccation and radiation damage the cellular proteins, lipids, nucleic acids, and carbohydrates (Slade and Radman, 2011). The primary and secondary metabolites that give the microorganisms a defensive mechanism against the radiation are some extremolytes and extremozymes (Gabani and Singh, 2013). To deal with various lethal radiations, such bacteria developed multiple

* These authors contributed equally.

Bacterial Survival in the Hostile Environment
DOI: https://doi.org/10.1016/B978-0-323-91806-0.00011-4

strategies. One such mechanism is the use of ions for radio-resistance. ROS generated by radiation are foraged by manganese-dependent antioxidant complexes. Many compatible solutes like di-myoinositol phosphate, mannosylglycerate, and trehalose accumulation helps the microorganisms to gain resistance against ionizing radiation (IR) (Webb and DiRuggiero, 2012). ROS generation induced by IR, causing a double-stranded DNA break. *Deinococcus* sp. contain some specific genes which respond to γ radiation and produce γ radiation-resistant genes. Yet another strategy of radio-resistance in bacteria is the extended synthesis-dependent strand annealing (ESDSA) pathway. ESDSA pathway helps in the reassembly of the broken DNA strands (Zahradka et al., 2006). These bacteria utilize a few more strategies for radio-resistance, like production of mycosporine-like amino acids (MAAs), scytonemin, and bacterioruberin for solid protection against UV radiation (Singh and Gabani, 2011).

Radiation and radio-resistance

Radiation is energy emission from a particular source and travels through space. This radiation has both electric and magnetic properties and is known as electromagnetic (EM) waves that can penetrate diverse materials. Radiation may be ionizing (X-ray, γ-rays) or nonionizing (microwave, infrared, ultraviolet); both radiations are the standard form of sunlight but differ based on wavelength and is present all around us. EM radiations are particles known as photons that move in a wave pattern. The EM spectrum is generally divided based on wavelength from shorter to higher; γ-ray, X-ray, ultraviolet ray, visible light, infrared radiation, microwaves, and radio waves. UV radiation is part of the EM spectrum between visible light and X-ray. UV radiation produces three types of rays—UV-A, UV-B, and UV-C. UV-A rays (315 to 400 nm) have the lowest energy with the longest wavelength, followed by UV-B (280−315 nm), and UV-C (100−280 nm) rays in the UV spectrum. UV-A has an immense effect on human skin and can cause biomolecules alteration, leading to mutations. This mutation may cause severe damage by transforming normal cells into cancerous cells. Ionizing particles like α particles, neutrons, etc., can damage the backbone of sugar and purine/pyrimidine base present in DNA and disrupts its structure (Jung et al., 2017).

Radio-resistants can survive and grow in high radiation levels (Fig. 15.1). Exposure to radiation leads the microorganisms to evolve and engage various defense mechanisms to reduce damaging effects (Singh and Gabani, 2011). Some microorganisms can possess unique DNA repair systems or bear polyploidy chromosomes. Cells are also protected from the oxidative stress generated through radiation-based water hydrolysis; in that situation, microbes can utilize their systems like enzymatic antioxidants with expressions of proteins related to ROS, nonenzymatic antioxidant systems with high concentrations of intracellular inorganic solutes, elevated the ratio of Mn/Fe and

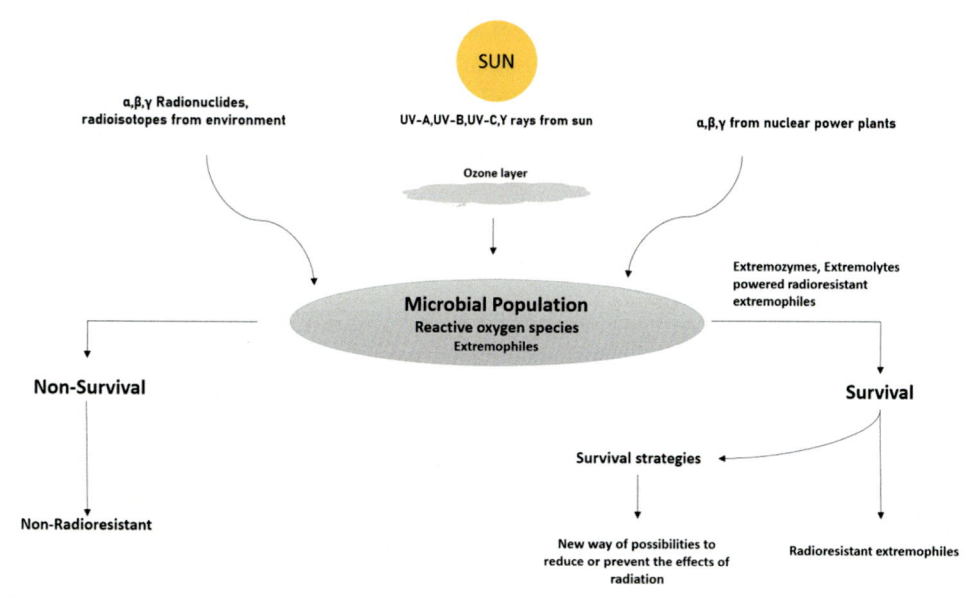

Figure 15.1 Radio waves (alpha, beta, and gamma), UV rays (A, B, and C), and nuclear power plant emissions affect microbial populations by generating reactive oxygen species and leading to mutations. Some of them can acquire adequate adaptive strategies for survival by forming extremozymes and extremolytes, termed radio-resistant extremophiles and those fail to develop extremophilic survival strategies that lead to death are termed nonradio-resistant.

various pigments (Gabani and Singh, 2013). Microorganisms that present and live in extremely harsh environmental conditions have evolved with multiple survival strategies by changing their physiological and biochemical characteristics depending on their need. Those diverse survival mechanisms of microbes to live in these radioactive environments are as follows:

Role of ions in the radiation resistance

The presence of Mn^{2+} in plants, animals, and microbes is significant. Mn^{2+} plays a vital role in oxygen production and provides protection against oxidative stress to microorganisms (Cellier, 2001). For aging the ROS generated by radiation is mainly formed by Mn–antioxidant complexes. In the presence of manganese, compatible solutes like trehalose, mannosylglycerate, and di-myoinositol phosphate get highly accumulated inside the cell, which is a supporting model of Mn^{2+}-dependent ROS foraging in the aerobic organism. And helps the microorganisms to get resistant against IR (Webb and DiRuggiero, 2012). Instead of all this, the Mn/Fe ratio present in microbial cells like *Deinococcus radiodurans* help to resist the high amount of radiation energy from the atmosphere (Paulino-Lima et al., 2016). These organisms accumulate a very high level of manganese and lower level of iron in their cells. In contrast, the radiation-sensitive organisms like *Shewanella* and *Pseudomonas putida* consist of low intracellular manganese concentrations, and iron concentrations are very high in their cell, which can conclude that Mn(II) accumulation is one of the significant reason for resistant against different radiation energy (Daly et al., 2004).

DNA repair for the radiation survival

Some microbes show a DNA repair mechanism that helps them to get resistant. After getting exposed to UV radiation, the DNA backbone and the sugar-phosphate bond get disrupted; by their particular activity, these organisms start to repair the DNA base pairs that helps in their survival (Slade and Radman, 2011). ROS response can be generated because of IR, and it can also create a double-stranded DNA break (DSBs). The radiation-resistant microbes have a tightly linked genome that forms a ring-like structure in the presence of IR. This structure keeps the DNA ends together, formed by the DSBs and helps in the radiation-based damage repair. DNA of these organisms can be successfully repaired up to 200 DSBs without losing their viability. On the other hand, radiation-sensitive organisms can lose their viability only in 12 DSBs (Jin et al., 2019). The recombinase A (RecA) gene present in the radiation-resistant organisms has a very high expression level that helps them to create SOS response and DNA repair. They also contains chaperones to assist protein folding and to repair radiation-affected misfolded proteins (Mrázek, 2002). The DNA repair system of these organisms is less complex than radiation-sensitive organisms; for example, the genes that code for the DNA repair proteins like photolyases, translesion polymerases, and DNA dioxygenase cannot be found in radiation-resistant organisms. Exposure of *Dra* to 6 kGy gamma irradiation dose, followed by the study of proteome changes by 2D gel electrophoresis and mass spectroscopy, showed upregulation of around 40 proteins involved in DNA repair, oxidative stress, and protein folding. The most important of them are single-stranded DNA-binding protein (SSBP), DNA damage response protein A (DdrA), DNA damage response protein B (DdrB), RecA, and pleiotropic protein promoting DNA repair (PprA). Protection of single-stranded DNA is important after radiation, and this is achieved by the SSBP, the first and the most abundantly upregulated protein. SSBP protects single-stranded DNA, thus protecting the information in the genome; as the first line of defense, the proteome of *Dra* undergoes both qualitative and quantitative changes after radiation exposure (Basu et al., 2012).

ESDSA is a pathway found in radiation-resistant microbes which helps them to reassemble the broken DNA strands. The action of the term elementary is quite similar to the synthesis–dependent strand annealing (SDSA). Elementary refers to the capability of DSBs to be repaired hundred times in the same cell (Krisko and Radman, 2013). The genomic DNA present in these organisms has to go through radiation-induced random DNA double-strand breakage and it produces different small DNA fragments. There is a two-step process in these organisms that helps them to reconstruct their DNA base pairs after breakage. The first one is an extended SDSA through which the genomic fragments in long liner intermediates can reassemble. And another one is after the assembly of long linear fragments; it can go through recombination to generate circular chromosomes. The RecF pathway present in these organisms is the main key component that helps in this assembly. Along with RecF, two more proteins called RecO and RecR are also essential for radio-resistance through ESDSA. Normal radio-sensitive organism is devoid of RecF, RecO, and RecR that cannot resist the high IR. UvrD helicase plays a crucial role in double-strand DNA break repair employing ESDSA. In wild-type cells, the RecBCD complex helps repair DSB. Still the RecBCD complex gets inactivated because of the mutations in sbcB (repressor of recBC), encoding the $3'-5'$ exonuclease-I and sbcC (or sbcD). In that case, the RecO, RecF, and RecR proteins act together and help to promote the repair of DSBs (Bentchikou et al., 2010).

Production of mycosporine-like amino acids

The ultraviolet-absorbing compounds are produced by different microorganisms like lichens, fungi, algae, and Cyanobacteria (Geraldes and Pinto, 2021). In the presence of MAAs, the damage formation due to UV rays gets retarded, and DNA receives protection from the construction of dimers. Due to the protective nature of these compounds, most cosmetic industries use MAAs to form sunscreen so that they can give UV protection to the skin (Gabani and Singh, 2013). The latest technology in microbial biotechnology, wherein a bottom-up approach is taken to maintain the function of the gene cluster with the removal of redundant elements. The gene cluster is called refactored gene clusters. Such an approach in *Escherichia coli* produces two intermediates 4-deoxygadusol and mycosporine-glycine. Three disubstituted analogs of MAA, shinorine, porphyra-334, and mycosporine-glycine-alanine, are also produced. MysH (2-oxoglutarate-dependent oxygenase) helps to convert MAAs into playthings (Chen et al., 2021). Palythine is a type of MAA frequently found in the algae, cnidarians, and planktonic organisms. These palythines help the organisms to get protection against UV rays. It has been determined that palythine has photodegradation rate constants under polychromatic irradiation in the presence of riboflavin in seawater medium (Conde et al., 2007).

Scytonemin biosynthesis and UV neutralization

In *Cyanobacteria* sp., the resistance toward UV radiations can be achieved by the cellular compartmentalization of scytonemin biosynthesis (Soule et al., 2009). In *Nostoc punctiforme*, a unique cluster of 18 genes can be observed, which is involved in scytonemin biosynthesis. The UV-A, a type of UV radiation can induce the conversion of chorismate to p-hydroxyphenyl pyruvate and tryptophan. Furthermore, *aroG* and *aroB* genes activate or amplify some of the core metabolic machinery. These downstream genes control the rate-limited enzymes. There are a few precursors which are also processed by NpR1259, Scy group of genes (A, B, and C) within the cytoplasmic compartment. Sometimes due to some unknown mechanism, these intermediates can get transferred to the space in the periplasmic region where they interact with the enzymes like TyrP, ScyD, ScyE, ScyF, and DsbA; resulting in the production of scytonemin in the reduced form. The secretion of scytonemin is in the auto-oxidized state which has a yellow—brown appearance. This results in the formation of sufficient quantities of a slime layer outside the cell which can block the incoming UV radiation (Soule et al., 2009).

Stress-activated protein kinases (SAPKs) are special protein kinases along with p38 or RK or CSBP kinase and c-Jun N-terminal kinase (JNK), which play a vital role against various environmental stresses like heat shock and UV radiation. Phosphorylation of SAPKs makes them active which in turn activate transcription factors ATF2, c-Jun, and Elk-1. The expression of these genes orchestrates responses toward UV radiation-induced stress. Scytonemin and its derivatives lead activation of SAPKs to help the cell to survive. In *Nostoc flagelliforme*, it was observed that complementary absorption of UV of MAAs and scytonemin can be accountable for sensitivity toward UV during photosynthesis (Ferroni et al., 2010).

Bacterioruberin

Some bacteria like *Rubrobacter radiotolerans* show extreme resistance to a lethal dose of IR. This resistance toward IR like UV radiation is also higher than the well-acclaimed radio-resistant bacterium *D. radiodurans* (Asgarnai et al., 2000). It was observed in another red-pigmented bacterium, *Halobacterium salinarum*, which contains bacterioruberin, is highly resistant against UV radiation and H_2O_2 (Asgarnai et al., 2000). This bacterium shows a high level of resistance against the lethal action of the DNA damaging agents like IR and UV light, which indicates a direct association between the bacterioruberin and DNA repair mechanisms (Shahmohammadi et al., 1998). In fact, bacterioruberin present in various radio-resistant microbes can possess a potential application in humans for repairing damaged DNA strands, caused by IR like UV radiation and can also prevent skin cancer (Singh and Gabani, 2011).

Radiation resistance in *D. radiodurans*

D. radiodurans is a unique species of bacteria having an extreme level of radiation resistance. It achieves radio-resistance by an efficient mechanism of proteomic protection rather than genomic damage protection. It is well known that a functional proteome can ensure easy cell recovery from substantial radiation damage caused to other cellular components by molecular repairing and various turnover processes which includes a systematic repair of disintegrated DNA. That is how cell death can be correlated with protein damage induced by radiation instead of DNA damage in both resistant and normal species.

It was observed that sequence wise the genome of *D. radiodurans* is a combination of *Thermus thermophiles* and *Bacillus subtilis* including few genes from other kingdom (Krisko and Radman, 2013). The origin of these interkingdom horizontal gene transfers could be from the high natural transformability of *D. radiodurans* facilitated by the DNA scavenging events. The amount of genetic information is directly dependent on the function of some dedicated proteins which carry out DNA repairing (Burrell et al., 1971; Bonura and Smith, 1976; Gérard et al., 2001). Therefore, it has been correlated that radiation resistance is one of the exceptional capabilities for repairing multiple times of DNA DSBs compared to normal species (Cox and Battista, 2005; Slade and Radman, 2011). From this observation, it can be said that improved "smart" repair proteins are much more efficient in *Dra* than in most other species. The exposure of DNA to radiation causes DNA double-stranded breakage, which leads to the fragmentation of DNA. These fragments get reassembled in a sequential manner in several steps. The major DNA repairing or the DSB repairing mechanism found in *Dra* is the ESDSA (Zahradka et al., 2006). In this mechanism, topoisomerases cut the DNA and this fragmented DNA is reassembled in *Dra*. The study on the pathways that are involved in this extraordinary capacity of recovering from the damage caused by radiation exposure showed the involvement of hundreds of proteins across diverse biological pathways. These proteins belong to homology-dependent DNA repair pathway, transcription, translation, metabolism, cell wall repair, chromosome architecture, and proteins involved in membrane transport (Ujaoney et al., 2021).

D. radiodurans is an example of a proteolytic bacterium with proper mechanisms for the degradation of proteins and catabolism of amino acids. It was observed that in *D. radiodurans* the IR can induce its proteolytic activities (Daly et al., 2010). Amino acids and peptides are produced by the degradation of proteins which are then reused in the energy demanding period of recovery which enables an efficient repairing of damage and survival (Omelchenko et al., 2005). It has also been suggested that during irradiation the amino acids and peptide pool have a major role in the scavenging of ROS (Daly et al., 2010). In *D. radiodurans*, this remarkable resistance toward radiation is also due to glucose metabolism. It was observed that in mutated *D. radiodurans* with deficiency of G6PDH are more sensitive toward induction of oxidative stress, mitomycin C, and H_2O_2 due to UV radiation (Zhang et al., 2005).

The protection of the genome of *D. radiodurans* after radiation exposure is also achieved by the expression of genes that belong to the radiation and desiccation resistance/response (RDR) regulon. A cis-regulatory sequence which is of size 17 bp called radiation desiccation response motif (RDRM) was identified. This sequence is found by the comparison of the genomes of *D. radiodurans* and *D. geothermalis* where the flanking genes of the RDRM motif showed upregulation. Inactivation of this motif leads to the radiation-sensitive nature of the organisms (Tanaka et al., 2004; Makarova et al., 2007). RDRM is essential for the upregulation of gamma radiation-induced expression of many genes. RDR regulon is involved in different stresses of gamma rays and UV rays, desiccation, and chemical mutagens like mitomycin C, methyl methanesulfonate, and ethidium bromide. Dose–dependent upregulation of the genes *ddrB*, *gyrB*, and Dr1143 is observed. This regulon is not activated by oxidative stresses of hydrogen peroxide and metalloids (zinc and tellurite) (Narasimha and Basu, 2021). These studies show coordination between regulatory sequences in DNA and its binding proteins.

Though many processes involving various proteins help in radiation protection, the role of manganese ion (Mn^{2+}) is crucial and has been proposed to give radiation protection to *D. radiodurans* (Daly et al., 2004). IR resistant *D. radiodurans* show high concentrations of Mn^{2+} complexes when compared to radiation-sensitive bacteria. The Mn^{2+} complexes are divided into two groups based on the EPR spectra. The H-Mn^{2+} give signal between 11,000 and 14,000 G, while L-Mn^{2+} give signals from 2000 G and beyond. The H-Mn^{2+} complexes protect proteins but fail to protect nucleotides. H-Mn^{2+} forms low molecular weight antioxidant metabolites. Studies show that the Mn-SOD or Cu/Zn-dependent SOD are not the major players in giving protection against radiation. In *D. radiodurans*, the Mn^{2+} ions are localized in DNA nucleoids which are involved in repair and replication. These localized granules may be a primitive mechanism involved in the protection of DNA when needed (Sharma et al., 2017).

Radiation resistance in eukaryotes

Many organisms do not endure IR. Humans cannot resist 3–10 Gy of radiation and that turns out to be the lethal dose. However, many species developed resistance to high doses of IR naturally overcoming the devastating effects and damage caused by IR. The tardigrades, also called water bears, are the well-explored eukaryotes with high radiation resistance. They can tolerate 2000–4000 Gy of radiation and hence can persist. They are also capable of surviving under severe desiccation. These spectacular tiny creatures employ a repertoire of superoxide dismutases and DNA repair proteins. Tardigrades possess a unique DNA safeguarding process and proteins for desiccation resistance which render them to inhibit and repair the massive harm triggered by extreme doses of IR. The development of IR perseverance in tardigrades could be due to their unique ability of resistance to desiccation (Chavez et al., 2019; Boothby et al., 2017).

Melanized fungi belonging to eukaryotes are an additional group gaining attention in the area of radiation ecology. These organisms can resist high levels of acute and chronic irradiation around 10,000 Gy with the assistance of melanin. These fungi are predominantly isolated from areas of high radiation ecosystems like Chernobyl and the Antarctic mountains due to their wide range of radiation resistance. International Space Station often find these organisms as contaminations in their areas. An experimental study on *Exophiala dermatitidis*, a melanized black fungus exhibited an upsurge in IR resistance. Genomic sequencing showed that in the evolved cells, nonhomologous end-joining was incapacitated, and the homologous recombinational DNA repair was augmented as a method to endure IR exposure. Nonhomologous end-joining is fundamentally mutagenic and may be inappropriate for the patch-up of the wide-ranging DSBs formed by extreme levels of IR exposure (Romsdahl et al., 2020).

Halophilic archaea with IR resistance have high desiccation and irradiation resistance and can withstand doses of ~3000 Gy. Haloarchaea exhibit polyploidy where some species retain several copies of chromosomes. Due to this polyploidy or genetic severance, these groups of archaea are able to restore the frequent DSBs produced by IR exposure via homologous recombination. Additionally, haloarchaea possess carotenoid-based pigmentation which might safeguard it against IR, through its antioxidant properties. Desiccation resistance in these species also aids in radiation resistance. Haloarchaea were isolated from fluid insertions within salt crystals after many years and possibly can survive more than that period (Stan-Lotter and Fendrihan, 2015; Fendrihan et al., 2006; McGenity et al., 2000).

Conclusion

Some microorganisms can survive under extreme and versatile conditions of the environment including radiation. Some proteins and DNA repair factors help them to survive in these harsh conditions of the environment. Elucidating all the radiation-responsive pigmented proteins, DNA-repair enzymes, and proteins involved in inducing topological changes that help for the resistance against extreme conditions, it will be possible to understand the basic biology of extremophiles. The knowledge helps in synthesizing them artificially for giving protection to the organisms which live in radioactive environments.

References

Asgarnai, E., Teroto, H., Asagoshi, K., Shahmohammadi, H.R., Ohyama, Y., Saito, T., et al., 2000. Purification and characterization of a novel DNA repair enzyme from the extremely radioresistant bacterium *Rubrobacter radiotolerans*. J. Radiat. Res. 41, 19−34.

Basu, B., Apte, S.K., 2012. Gamma radiation-induced proteome of *Deinococcus radiodurans* primarily targets DNA repair and oxidative stress alleviation. Mol. Cell. Proteomics, 11 (1).

Bentchikou, E., Servant, P., Coste, G., Sommer, S., 2010. A major role of the RecFOR pathway in DNA double-strand-break repair through ESDSA in *Deinococcus radiodurans*. PLoS Genet. 6 (1), e1000774. Available from: https://doi.org/10.1371/journal.pgen.1000774.

Bonura, T., Smith, K.C., 1976. The involvement of indirect effects in cell-killing and DNA double-strand breakage in γ-irradiated *Escherichia coli K-12*. Int. J. Rad. Biol. 29, 293−296.

Boothby, T.C., et al., 2017. Tardigrades use intrinsically disordered proteins to survive desiccation. Mol. Cell. 65, 975−984.

Burrell, A.D., Feldschreiber, P., Dean, C.J., 1971. DNA−membrane association and the repair of double breaks in X-irradiated *Micrococcus radiodurans*. Biochim. Biophys. Acta 247, 38−53.

Cellier, M., 2001. Bacterial genes controlling manganese accumulation. Microbial Transport Systems 325−345.

Chavez, C., Cruz-Becerra, G., Fei, J., Kassavetis, G.A., Kadonaga, J.T., 2019. The tardigrade damage suppress or protein binds to nucleosomes and protects DNA from hydroxylradicals. Elife 8, e47682.

Chen, M., Rubin, G.M., Jiang, G., Raad, Z., Ding, Y., 2021. Biosynthesis and heterologous production of mycosporine-like amino acid palythines. J. Org. Chem. Available from: https://doi.org/10.1021/acs.joc.1c00368. May.

Conde, F.R., Churio, M.S., Previtali, C.M., 2007. Experimental study of the excited-state properties and photostability of the mycosporine-like amino acid palythine in aqueous solution. Photochem. Photobiol. Sci. 6 (6), 669−674. Available from: https://doi.org/10.1039/B618314J.

Cox, M.M., Battista, J.R., 2005. *Deinococcus radiodurans*—the consummate survivor. Nat. Rev. Microbiol. 3 (11), 882−892.

Daly, M.J., Gaidamakova, E.K., Matrosova, V.Y., Kiang, J.G., Fukumoto, R., Lee, D.-Y., et al., 2010. Small-molecule antioxidant proteomeshields in *Deinococcus radiodurans*. PLoS One 5, e12570.

Daly, M.J., Gaidamakova, E.K., Matrosova, V.Y., Vasilenko, A., Zhai, M., Venkateswaran, A., et al., 2004. Accumulation of Mn(II) in *Deinococcus radiodurans* facilitates gamma-radiation resistance. Science 306 (5698), 1025−1028. Available from: https://doi.org/10.1126/science.1103185.

Daly, M.J., 2009. A new perspective on radiation resistance based on *Deinococcus radiodurans*. Nat. Rev. Microbiol. 7 (3), 237−245. Available from: https://doi.org/10.1038/nrmicro2073.

Fendrihan, S., Legat, A., Pfaffenhuemer, M., Gruber, C., Weidler, G., Gerbl, F., et al., 2006. Extremely halophilic archaea and the issue of long-term microbial survival. In: *Life in Extreme Environments*, Springer, Dordrecht, pp. 125−140.

Ferroni, L., Klisch, M., Pancaldi, S., Hader, D., 2010. Complementary UV-absorption of mycosporine-like amino acids and scytonemin is responsible for the UV-insensitivity of photosynthesis in *Nostoc flagelliforme*. Mar. Drugs 8, 106−121.

Gabani, P., Singh, O.V., 2013. Radiation-resistant extremophiles and their potential in biotechnology and therapeutics. Appl. Microbiol. Biotechnol. 97 (3), 512−555. Available from: https://doi.org/10.1007/s00253-012-4642-7.

Gérard, E., Jolivet, E., Prieur, D., Forterre, P., 2001. DNA protection mechanisms are not involved in the radioresistance of the hyperthermophilic archaea *Pyrococcus abyssi* and *P. furiosus*. Mol. Genet. Genom. 266, 72−78.

Geraldes, V., Pinto, E., 2021. Mycosporine-like amino acids (MAAs): biology, chemistry and identification features. Pharmaceuticals 14 (1), 63. Available from: https://doi.org/10.3390/ph14010063.

Jin, M., Xiao, A., Zhu, L., Zhang, Z., Huang, H., Jiang, L., 2019. The diversity and commonalities of the radiation-resistance mechanisms of *Deinococcus* and its up-to-date applications. AMB. Express 9 (1), 138. Available from: https://doi.org/10.1186/s13568-019-0862-x.

Jung, K.-W., Lim, S., Bahn, Y.-S., 2017. Microbial radiation-resistance mechanisms. J. Microbiol. 55 (7), 499−507. Available from: https://doi.org/10.1007/s12275-017-7242-5.

Krisko, A., Radman, M., 2013. Biology of extreme radiation resistance: the way of *Deinococcus radiodurans*. Cold Spring Harb. Perspect. Biol. 5 (July). Available from: https://doi.org/10.1101/cshperspect.a012765.

Magazù, S., Migliardo, F., Gonzalez, M.A., Mondelli, C., Parker, S.F., Vertessy, B.G., 2012. Molecular mechanisms of survival strategies in extreme conditions. Life: Open Access J. 2 (4), 364−376. Available from: https://doi.org/10.3390/life2040364.

Makarova, K.S., Omelchenko, M.V., Gaidamakova, E.K., Matrosova, V.Y., Vasilenko, A., Zhai, M., et al., 2007. *Deinococcus geothermalis*: the pool of extreme radiation resistance genes shrinks. PLoS One, 2 (9), e955.

McGenity, T.J., et al., 2000. Origins of halophilic microorganisms in ancient salt deposits. Environ. Microbiol. 2, 243−250.

Mrázek, J., 2002. New technology may reveal mechanisms of radiation resistance in *Deinococcus radiodurans*. Proc. Natl. Acad. Sci. USA 99 (17), 10943−10944. Available from: https://doi.org/10.1073/pnas.182429699.

Narasimha, A., Basu, B., 2021. New insights into the activation of radiation desiccation response regulon in *Deinococcus radiodurans*. J. Biosci. 46, 10.

Omelchenko, M.V., Wolf, Y.I., Gaidamakova, E.K., Matrosova, V.Y., Vasilenko, A., Zhai, M., et al., 2005. Comparative genomics of *Thermus thermophiles* and *Deinococcus radiodurans*: divergent routes of adaptation to thermophily and radiation resistance. BMC Evol. Biol. 5, 57.

Paulino-Lima, I.G., Fujishima, K., Navarrete, J.U., Galante, D., Rodrigues, F., Azua-Bustos, A., et al., 2016. Extremely high UV-C radiation resistant microorganisms from desert environments with different manganese concentrations. J. Photochem. Photobiol. B: Biol. 163 (October), 327−336. Available from: https://doi.org/10.1016/j.jphotobiol.2016.08.017.

Rampelotto, P.H., 2010. Resistance of microorganisms to extreme environmental conditions and its contribution to astrobiology. Sustainability 2 (6), 1602−1623. Available from: https://doi.org/10.3390/su2061602.

Romsdahl, J., et al., 2020. Adaptive evolution of a melanized fungus reveals robust augmentation of radiation resistance by abrogating non-homologous end-joining. Environ. Microbiol. 23. Available from: https://doi.org/10.1111/1462-2920.15285. Published online October 19, 2020.

Shahmohammadi, H.R., Asgarani, E., Terato, H., Saito, T., Ohyama, Y., Gekko, K., et al., 1998. Protective roles of bacteriotuberin and intracellular KCl in the resistance of *Halobacterium salinarium* against DNA damaging agents. J. Radiat. Res. 39, 251−262.

Sharma, A., et al., 2017. Across the tree of life, radiation resistance is governed by antioxidant Mn^{2+}, gauged by paramagnetic resonance. Proc. Natl. Acad. Sci. USA 114 (44), E9253−E9260. Available from: https://doi.org/10.1073/pnas.1713608114.

Singh, O., Gabani, P., 2011. Extremophiles: radiation resistance microbial reserves and therapeutic implications. J. Appl. Microbiol. 110, 851−861. Available from: https://doi.org/10.1111/j.1365-2672.2011.04971.x.

Slade, D., Radman, M., 2011. Oxidative stress resistance in *Deinococcus radiodurans*. Microbiol. Mol. Biol. Rev.: MMBR 75 (1), 133−191. Available from: https://doi.org/10.1128/MMBR.00015-10.

Soule, T., Palmer, K., Gao, Q., Potrafka, R.M., Stout, V., Garcia-Pichel, F., 2009. A comparative genomics approach to understanding the biosynthesis of sunscreen scytonemin in *Cyanobacteria*. BMC Genom. 10, 336−346.

Stan-Lotter, H., Fendrihan, S., 2015. Halophilic archaea: life with desiccation, radiation and oligotrophy over geological times. Life (Basel) 5, 1487−149650.

Tanaka, M., Earl, A.M., Howell, H.A., Park, M.J., Eisen, J.A., Peterson, S.N., et al., 2004. Analysis of Deinococcus radiodurans's transcriptional response to ionizing radiation and desiccation reveals novel proteins that contribute to extreme radioresistance. Genetics, 168 (1), 21−33.

Ujaoney, A.K., Padwal, M.K., Basu, B., 2021. An *in vivo* interaction network of DNA-repair proteins: a snapshot at double strand break repair in *Deinococcus radiodurans*. J. Proteome Res. 20 (6), 3242−3255.

Webb, K.M., DiRuggiero, J., 2012. Role of Mn2 + and compatible solutes in the radiation resistance of thermophilic bacteria and archaea. Archaea 2012, 845756. Available from: https://doi.org/10.1155/2012/845756.

Zahradka, K., Slade, D., Bailone, A., Sommer, S., Averbeck, D., Petranovic, M., et al., 2006. Reassembly of shattered chromosomes in *Deinococcus radiodurans*. Nature 443, 569−573.

Zhang, C., Wei, J., Zheng, Z., Ying, N., Sheng, D., Hua, Y., 2005. Proteomic analysis of *Deinococcus radiodurans* recovering from g-irradiation. Proteomics 5, 138−143. Extreme Radiation Resistance of *D. radiodurans*.

CHAPTER 16

Adaptation strategies of thermophilic microbes

Swatilekha Pati[1], Somok Banerjee[1], Aveepsa Sengupta[1], Jayshree Sarma[1], Shakila Shaheen[1], Shivendra Tenguria[2] and Ashutosh Kumar[1]
[1]Department of Microbiology, Tripura University (A Central University), Agartala, Tripura, India
[2]Department of Pathology and Laboratory Medicine, Cedars-Sinai Medical Center (UCLA), Los Angeles, CA, United States

Introduction

Over the past several years, scientists have been fascinated by the microorganisms that thrive in extreme habitats, which is lethal for other organisms to sustain life (Rampelotto., 2013). Such microorganisms are known as extremophiles, further classified into the psychrophiles, low-temperature tolerant microorganisms, thermophiles, high-temperature tolerant microbes, halophiles, the microbes requiring high concentration of salt for growth, alkaliphiles, the extreme pH tolerant ones, acidophiles, and lastly barophiles, the organisms that can thrive under pressure.

This chapter mainly focuses on the group of microorganisms that can survive under high-temperature conditions or thermophiles (Wang et al., 2015). Temperature is one of the most significant factors that control living beings' activities and evolution. The word "thermophile" has been derived from two Greek words, "thermotita" (means: heat) and "philia" (means: love). Thus thermophiles can be defined as heat-loving organisms requiring high temperatures for survival and growth (Mehta and Satyanarayana, 2013).

Based on temperature as a parameter, microbes are classified into psychrophiles, mesophiles (grow optimally at $\approx 37°C$), and thermophiles (grow optimally at $>50°C$). The thermophiles can be further divided into moderate thermophiles, extreme thermophiles, and hyperthermophiles. Among these, thermophiles organisms are the most studied of all extremophiles (Gupta et al., 2014).

About 4 billion years ago, anaerobic chemoautotrophic thermophiles were the first to thrive on earth. These organisms were found underwater near hydrothermal vents that optimally grow at a temperature of around $50-125°C$ approximately (Lusk, 2019). In general, hot springs are also suitable for colonizing these heat-loving organisms. The study of thermophiles was mainly started by Thomas Brock in the 1960s, after the ground-breaking discovery of microbes, their diversity, and characterization in the hot springs of Yellowstone National Park (Brock, 1967; Marsh and Larsen, 1953).

Bacterial Survival in the Hostile Environment
DOI: https://doi.org/10.1016/B978-0-323-91806-0.00012-6

The diversified nature was because of certain main factors such as pH, temperature, and dissolved amount of hydrogen sulfide (Purcell et al., 2007).

Besides their evolution and diversity, thermophilic organisms can produce thermostable enzymes, which have various applications in the biotechnological fields (Gomes et al., 2016). For example, Taq DNA polymerase produced by *Thermus aquatius* is applicable in the field of biotechnology (Eichler, 2001). Archaea are the most prominent hyperthermophilic organisms; an example is *Pyrolobus fumarii* who can grow up to a temperature of 113°C. Enzymes of hyperthermophiles have higher activity approaching up to 142°C in the case of amylopullulanase. Thermophiles are also seen among the phototrophic bacteria (Green bacteria, cyanobacteria) and eubacteria (Thiobacillus, Bacillus, Lactic acid bacteria, Actinomycetes) (Rothschild and Mancinelli, 2001). The most cultivable, chemolithotrophic thermophilic or hyperthermophilic microorganism is crenarchaeote (nitrate-reducing), *Pyrolobus fumarii*. They have a growth limit of 113°C and divide at 105°C (Charlier and Droogmans, 2005).

Thermal environments on this globe, ranging from the terrestrial volcanic sites which have temperatures slightly higher than the optimum to submarine hydrothermal vents, volcanoes, or fumaroles where the temperature generally exceeds 300°C to subterranean areas such as sun-baked soil surfaces and oil reserves with temperatures of about 65°C (Wiegel and Canganella, 2001). Other than natural thermal environments, there are artificial thermal environments, for example, compost piles having temperatures around 60−70°C that can reach up to a maximum level of 100°C, in industries (Oshima and Moriya, 2008).

Taxonomical diversity

Thermophilic organisms are widespread and present among all three domains of life, that is, Bacteria, Archaea, and Eukarya.

Bacteria

Bacteria are found in a vast range of maximum thermal environments. Though they cannot tolerate as high temperatures as archaea, they can survive at temperatures above 80°C which mainly include autotrophs, phototrophs, heterotrophs, and chemolithotrophs. Some examples include Cyanobacteria carrying out oxygenic photosynthesis can be found in hot springs of temperature ranging up to 74°C, where they are present as the dominating oxygenic phototrophic group because algae cannot tolerate such temperatures. Other organisms such as *Synechococcus lividus* thrive at high temperatures of 60−74°C, are often found to coexist with some anoxygenic phototrophs such as *Chloroflexus*. Some anoxygenic photoautotrophs are also found in hot springs, for example purple sulfur and green sulfur bacteria. In thermal environments containing sulfide, these microorganisms convert sulfide into sulfur by oxidization and accumulating either inside or outside the cell, eventually getting oxidized to sulfate in the case of Proteobacteria (purple sulfur bacteria) and Chlorobi (green sulfur bacteria), respectively. For instance, a thermophilic proteobacterium found at 58°C is *Thermochromatium tepidum*. *Aquifex aeolicus* and *Thermotoga maritima* can grow at 90°C or above (Sokolova et al., 2007).

Archaea

The major kingdoms of the archaeal domain are Euryarchaeota, Crenarchaeota, Korearchaeota, Nanoarchaeota and Thaumarchaeota (Pan, 2012). The kingdom Crenarchaeota contains many extreme thermophilic organisms. The kingdom Euryarchaeota mainly consists of physiologically distinct groups, among which the methanogens are the most studied ones. *Nanoarchaeum equitans* of kingdom Nanoarchaeota was first found within the deep-sea vents of North Iceland and it is an obligate symbiont of Ignicoccus (Huber et al., 2002). In thermophilic Archaea, the membrane phospholipids are comprised of isoprenoid chains that are saturated and are connected to the glycerol backbone by the chemically resistant ether linkage, rather than ester linkage that links the fatty acids and glycerol moieties in most eukaryal and bacterial membranes (Pikuta et al., 2007).

Eukarya

Eukaryotic thermophiles are not as extreme as Archaea or Bacteria; only some fungal and algal species can tolerate moderately high temperatures of about 60°C. Over 100 years ago, the first fungus to be slightly thermophilic, *Mucorpusillus*, was isolated from bread. Also, *Mycelia sterilia* and the fungi imperfecti can grow at around 60°C. These fungi require warm, aerobic, humid environments for their growth. They are commonly found in compost piles or in accumulated organic matter where these fungi play a vital role in the mineralization of the organic carbon. Some fungi living in hot environments are Dactylaria, associated with an acidophilic alga named *Cyanidium*. *Galdiera* sp., *Cyanidium caldarium*, and *Cyanidochyzon* sp. are some examples of thermophilic Rhodophytes (Ferrera and Reysenbach, 2007).

Effect of temperature on microbial cells

Temperature is termed one of the most important environmental factors which affect life. Almost every microbial activity gets regulated by temperature (Wiegel, 1990). Higher temperatures reaching up to 100°C can denature nucleic acids and proteins, increasing membrane fluidity to a lethal level (Rothschild and Mancinelli, 2001).

The effect of temperature can be observed clearly from the transforming microbial velocity and their compositions (Noll et al., 2010). For instance, the increase in temperature causes changes in the extracellular layers of bacteria. The outer membrane of gram-negative bacteria is sensitive to mild heat. Few studies have reported changes in morphology and structure along with reduction of lipopolysaccharide in the outer membrane, which lead to change in the membrane permeability, loss of periplasmic proteins, and increased sensitivity to hydrophobic antibiotics (Mackey, 1983; Tsuchido et al., 1989). Moreover, the effect of mild temperature on the outer membrane further increases in the presence of Tris buffer.

Gram-positive bacteria such as *Staphylococcus aureus* have a more rigid cell wall than gram-negative bacteria. Allwood and Russell (1969) reported shrinkage of the cell and leakage of intracellular components in *S. aureus*. Several studies have been conducted on the effect of high temperature on the stability of ribosomes and rRNA. According to a report, moderate heating degrades rRNA along with leakage of materials from the metabolic pool (Allwood and Russell, 1968). Microorganisms can maintain a steep salt gradient or pH on their membrane. Although they are incapable of insulating themselves from the aqueous and hot surroundings, hyperthermophiles need to adapt to every level of cellular development; molecular machinery needs to be active and stable at elevated temperatures. The smaller molecules, substrates, and potentially unfavorable reaction intermediates need to be guarded against the hot vigorous cytoplasmic fluids (Charlier and Droogmans, 2005).

Denaturation of protein and coagulation occurs when cells are subjected to high temperatures. For instance, Rosenberg et al. (1971) showed a link between the thermodynamic parameters of denaturation of proteins and the death rates of numerous organisms, indicating cell death in the mesophilic bacteria.

Adaptation mechanism of thermophiles
Modification of cell membrane

The cell membrane is the outer layer of the cell composed of lipids and proteins that protects the cytoplasm from the hostile environment. The phospholipids are hydrophobic and therefore they regulate the fluidity or viscosity and permeability of the cell membrane. Elevated temperature affects the cell membrane when exposed to an extreme environment. In general, at high temperatures, both the fluidity and permeability of the cell membrane of bacteria increase rapidly, resulting in the disarranged packing of lipids necessary for maintaining the liquid crystalline form. As the temperature gradually increases above the optimum temperature, the membrane phospholipids shift from crystalline liquid to fluid and, subsequently, to a nonlamellar form (Escribá, 2006). The thermophiles adapt to extreme temperatures partly by modifying the composition of the lipids in their membrane. The membrane of a thermophile must be in a liquid-crystal state to be functional. It should not be permeable, even to ions and smaller molecules, to enable the proton motive force generation and pH homeostasis (Pikuta et al., 2007). Permeability is one of the main factors contributing to the heat tolerance of thermophiles. The cell membrane acts as a permeability barrier to regulate the outflow and inflow of lower-molecular-weight compounds. Fatty acyl ester lipid membrane and their permeability are highly dependent on temperature, and their phase-transition temperature is dependent on the composition of fatty acids. Therefore when the growth temperature changes, the fatty acid composition of membrane lipids is rapidly regulated (Mehta et al., 2016).

The adaptation mechanism employed by thermophilic bacteria differs from that of archaea. The thermophilic bacteria maintain the membrane fluidity by increasing the quantity of different fatty acids, including saturated, long-chain, branched, and polar carotenoid content (Oshima and Miyagawa, 1974; Yokoyama et al., 1996). Among the hyperthermophilic bacteria which fall under the highest temperature bracket, few of the bacteria amend a large amount of branched-chain fatty acids, mainly composed of iso- branched-chain fatty acids. Thermophilic bacteria capable of growing under high temperatures (generally above 70°C) employ phospholipids that are more or less similar to the fatty acids present in archaea such as the tetraethers, tetraesters, and diethers. For instance, several studies have shown the presence of membrane-spanning tetraether fatty acids derived from diabolic acid in the *Thermotogales* sp. The tetraesters were detected in *Tridenchthonius africanus* (Damsté et al., 2007; Huber et al., 1989). Few studies reported the presence of diethers in *Thermodesulfobacterium commune* and *Aquifex pyrophilus* (Langworthy et al., 1983; Jahnke et al., 2001). Compared to the archaeal tetraethers, the tetraether and tetraesters found in thermophilic bacteria are formed by tail—tail condensation of two branched-chain iso lipids extending from both the leaflets of the cell membrane.

In the thermophilic archaeal membrane, the tetraethers are present in high amounts and are mainly the only core lipid. This type of lipid forms highly stable monolayers due to the restricted movement of the hydrocarbon chains. According to a few studies, the highest temperature at which a bacteria can grow is 100°C (*Geovibrio ferrireducens*), while that of thermophilic archaea is 122°C (*Methanopyrus kandleri*) (Kashefi et al., 2002; Takai et al., 2008). Cario et al. (2015) showed that with an increase in temperature, there is an increase in the tetraether/diether lipid ratio in *Thermococci*. Therefore the presence of the tetraether lipids is advantageous for growth at high temperatures, as can also be seen in the case of thermophilic bacteria using tetraethers. Gliozzi et al. (1983) reported that the stability of the tetraether-containing membranes could be further enhanced by the inclusion of pentacyclic, which leads to the rise in transition temperature.

In general, bacteria maintain the viscosity and permeability of the cell membrane only at temperatures just beyond the phase transition temperature. However, the viscosity and permeability of the archaeal membrane are maintained throughout the temperature range of 0—100°C, and the membrane remains in the liquid crystalline form.

Protein modification

Proteins in thermophiles appear to be more basic and smaller, contributing to their stability (Rampelotto, 2010). Maintaining protein stability is one of the challenges faced by thermophiles at higher temperatures. The thermophilic microbes maintaining the stability of their proteins has remained a topic of interest for a long period; however, it is limited only to the status of individual proteins. In recent times, advancements in proteomics have revealed more concrete evidence regarding the thermostability of proteins.

According to a study by Gu and Hilser (2009), protein adaptation at elevated temperature is mainly concentrated in the proteins synthesizing enzymes and the regulator proteins, including protease and triosephosphate isomerase. Moreover, the proteomic analysis showed that the number of proteins detected at elevated temperatures is a lot lesser than proteins present at optimum growth temperature, which indicates that only the thermostable proteins having necessary functions required for survival at high temperatures are present (Li et al., 2010; Wang et al., 2007).

Few studies reported that some of the thermophiles possess a higher number of disulfide bridges than the mesophiles, which may indicate that the thermophiles use disulfide bonding to stabilize the proteins (Mallick et al., 2002; Beeby et al., 2005). For instance, the proteomic analysis showed that more than 47% of the cysteine residues in *Pyrobaculum aerophilum* were developing the disulfide bridges. In comparison, in *Escherichia coli*, only 8% of the cysteine residues were present. Furthermore, 2-D gel electrophoresis revealed that the number of protein complexes formed in *P. aerophilum* through disulfide bonds is notably higher than in *E. coli*. The protein structures found in thermophilic archaea aids the widespread disulfide bonds present in thermophiles. Moreover, according to Boutz et al. (2007) the protein complexes in thermophiles tend to form compact structures and the disulfide bonds support this process.

The utilization of disulfide bonding for the stabilization of proteins is astonishing since the intracellular environment remains at a reducing state, preventing disulfide bonding. Comparative proteomic analysis revealed two things, one is the presence of a disulfide oxidoreductase protein in thermophiles, and the other is that the thermophiles possess intracellular disulfide bonding in massive amounts, which is the key mechanism involved in the formation of intracellular disulfide bonds (Beeby et al., 2005; Pedone et al., 2004).

Electrostatic interactions are a crucial factor contributing towards thermostability of proteins. This is supported by observations like higher number of salt bridges seen in structures of many hyperthermophilic proteins (Karshikoff and Ladenstein, 2001). Cysteins and disulfide bridges are known to be susceptible towards temperature higher than 100°C, according to studies in mesophilic enzymes. However, recent studies have proved that some proteins containing disulfide bridges remain stable at temperature higher than 100°C, which indicates that disulfide bridge might be a strategy for thermal stabilisation, and solvent accesibility and environmental conditions contribute towards protecting disulfide bridges from destruction (Vieille and Zeikus, 2001). ZAvodszky et al. (1998) reported that the motility of the hyperthermostable proteins is lower at room temperatures than that of the thermophiles or mesophiles since the conformational motility of the atoms increases with the temperature rise. An additional method employed to enhance protein stability is by chaperones' action or working, which contribute to refolding the denatured proteins (Rampelotto, 2010).

Many extremophiles possess heat shock proteins (HSPs) which are considered one of the primary responses to environmental stresses such as high or low temperatures, nutrient stress, and dehydration (Laksanalamai and Robb, 2004). They are classified based on molecular weight, that is, high-molecular-weight HSPs (60−100 kDa) and low−molecular-weight HSPs (HSP with a molecular weight of 15−42 kDa). Most of the HSPs also perform the role of chaperones which prevent protein aggregation and participates in the folding of newly synthesized and denatured proteins. The homologs of these proteins are present in genomes of all classes of extremophiles. It is known that there is an interaction between proteins to form a complex network that allows proper communication among the molecules. Almost 20% of the proteins found in the thermophiles are engaged in these complexes, including super, homomeric, and heteromeric complexes. An analysis on *Thermoanaerobacter tengcongensis* revealed that by blue native electrophoresis, protein complexes could be separated, and some complex components were found. These components varied as temperature changes of the culture, specifying that interactions of proteins in the thermophiles are dependent on temperature changes. For instance, HSPs increased in the complexes with the increase in temperature, showing that HSPs played a role in protecting these interacting proteins at elevated temperatures. Analysis of thermal stability of GLK (Glucokinase) in *T. tengcongensis* confirmed this hypothesis (Wang et al., 2015).

Genomic modification

DNA is unstable at elevated temperatures, and to adapt to high temperature, thermophiles have evolved a stringent DNA-repair system to balance their genomic stability (Wang et al., 2015). Changes in the environment, including temperature change, induce gene evolution in bacteria which prepares them to survive at high temperatures. This type of adaptation can be achieved through horizontal gene transfer (HGT), deletion, or mutation (Averhoff and Müller, 2010).

Several studies show the horizontal transfer of DNA among different species is one of the key factors responsible for bacterial adaptation at high temperatures (Boto, 2014; Feng et al., 2014; Li et al., 2013). Through HGT, the homologous gene sequence is either replaced by the new sequence, or the new sequence is integrated by transduction, conjugation, and transformation. Zhaxybayeva et al. reported that through HGT approximately 24% of genes in *T. maritima* were introduced from archaea (Zhaxybayeva et al., 2009). Most of the genes acquired through HGT showed thermophilic properties required for survivability under elevated temperatures. For example, reverse gyrase is a thermophilic adaptation trait considered to be transferred from thermophilic archaea to bacteria. It is a type of DNA topoisomerase responsible for inducing positive supercoiling of dsDNA. In an experiment on *Thermococcus koda-karensis*, when this gene was removed from the DNA, the bacteria started to grow slowly under increased temperatures, indicating its importance in inducing thermo-phily (Atomi et al., 2004; Perugino et al., 2009). According to van Wolferen et al. (2013), the thermophilic microbes may not sustain without gene transfer among species.

A study reported that after comparing the size of DNA between *Thermus thermophi-lus* (thermophile) and *Deinococcus radiodurans* (nonthermophile), it was revealed that in *T. thermophilus* systematic loss of genes takes place including urease complex, rhamnose metabolic pathway, fructose transport, and glycerol metabolic pathway (Omelchenko et al., 2005). This removal of genes is thought to be a cost reduction mechanism for the thermophilic microbes to sustain high temperatures (Burra et al., 2010, Das et al., 2006; Dutta and Chaudhuri, 2010). Recent studies have revealed that some gene expressions play a crucial role in the adaptation mechanism of thermophiles. There is upregulation of these genes either at proteomic or transcriptomic levels. These genes possess specific functions at increasing temperatures. HSPs are crucial proteins that are upregulated as a response to heat stress and protect the cell against damage. *Pyrococcus furiosus*, shows a trigger in upregulation of Hsp20 and Hsp60 while the temperature shifts to 105°C from 90°C, and HSPs were upregulated in *Sulfolobus solfataricus* in response to a change in temperature to 90°C from 80°C in 5 min. Also, upregulation of chaperone proteins was observed at the proteomic level in response to an increase in temperature in many thermophilic species which included *Thermoto gamaritima* (Wang et al., 2015).

Drake (2009) reported that by analyzing the genomic mutations in a few thermo-philes it was revealed that the thermophiles displayed low-base mutations in compar-ison to the nonthermophiles. Few of the mutations may be advantageous for their growth. For instance, genomic analysis of a thermophilic descendent of *E. coli* strain MG1655 showed that the mutations that took place in the genes including *glpF*, *fabA* provided thermotolerance ability to the thermophilic bacteria (Blaby et al., 2012). Transformation of the produced strain (thermo-tolerant), that is, EVG1064, with wild-type allele of glpF, minimized fitness at risen temperatures. On the other hand, mutation in the gene, fabA rises the amount of saturation in the mem-brane lipids. Proteins related to the glycolysis pathway are considered to be a key factor contributing to thermophilic characteristics, by providing instant energy to thermophiles to overcome heat stress. Hence, the expression level of these proteins in thermophilic microorganisms shows their response physiologically. Genes involved in the glycolysis pathway, that is, GAPDH, PGK, and PGL are known to be upregulated at the proteomic or transcriptomic levels in response to higher temperatures including *T. tengcogensis* and *T. gamaritima*. In addition, there is an upregulation of glycolysis and the TCA cycle is known to be downregulated, as a response to a change from a high to low redox state in *G. thermoglucosidasius*. This might be concluded accordingly as the redox state could be reduced as a response toward heat stress (Wang et al., 2015).

Modification of DNA and RNA

In general, circular DNA is more resistant to high temperatures compared to linear DNA. DNA supercoiling in thermophiles further raises the melting temperature (López-García, 1999). In vitro studies have shown that thermophilic microbes often contain a high concentration of salts which are known to stabilize dsDNA and proteins (Hensel and König, 1988). Few studies showed that polyamines are also responsible for DNA modification in thermophiles. According to Terui et al. (2005), thermophiles contain polyamines that are more effective on dsDNA stabilization while branched polyamines are more efficient on ssDNA and tRNA. Balancing the primary structure of DNA and RNA at higher temperatures is considered to be problematic. The thermophilic chromosome is exposed to higher amounts of damage such as depurination, damage caused by oxidative stress, deamination, breaks in double and single strands. All of these damages are seen to be increased with temperature. Since mutation rates in hyperthermophiles do not vary much compared to mesophiles, they must acquire a proper repair mechanism. Homologs of eukaryal/bacterial DNA repair mechanisms and enzymes: Uracil N-glycosylase, RecA-RAD51, photoreactivation, O^6-alkylguanine-DNA transferase are found in thermophiles. Many enzymes of DNA metabolism of thermophiles show distinct characteristics which might be important for DNA repair at elevated temperatures (Charlier and Droogmans, 2005).

All the thermophiles including archaea, bacteria, and eukarya possess reverse gyrase which shows helicase and topoisomerase I activity and hence induces positive super turns of DNA through hydrolysis of ATP (Bouthier De La Tour et al., 1990; Forterre et al., 1996). Divalent and monovalent salts improve nucleic acid stability as these salts screen out the negative charges of KCl, $MgCl_2$, and phosphate groups, protecting the DNA from hydrolysis and depurination. In addition, DNA can be stabilized by the use of DNA-binding proteins and genome compaction into chromatin (Rampelotto, 2010). Aminoacyl tRNAs are one of the key elements in the synthesis of proteins.

However, they are sensitive to high temperatures. Therefore, according to Ibba and Söll (2001) to overcome the translational hurdle, certain thermophiles contain noncanonical enzymes or pathways. Enhanced synthesis of the polypeptide on ribosomes can also overcome this difficulty. However, studies are scarce on such a subject. Stepanov and Nyborg (2002) reported that a fast turnover together with high aminoacyl tRNA synthesis can be a possible way of compensating aminoacyl tRNA decay under high temperatures.

An increase in the GC content might contribute to thermostabilization, but this is not seen in the case of thermophiles. No correlation is known between the growing temperature of the organism and the G:C ratio of a microbial genome. DNA topology was seen to be different in hyperthermophiles during heat stress such as an increase in linking number might aid in transcription under conditions of cold shock. Mesophiles tend to be requiring energy of negative supercoiling, which is produced by the working of histone wrapping or gyrase for the opening of strands and enabling the process of transcription and DNA replication. Salt concentration plays a crucial role in stabilizing rRNA and tRNA. In contrast to genomic DNA, there is a correlation between the growth of the organism and the GC content of rRNA. This correlation is confined to double-stranded RNA. In addition, the higher affinity of ribosomal proteins toward rRNA is another contributing factor toward thermal stability. In the rRNA of *S. solfataricus* (hyperthermophile), posttranscriptional modifications that are usual in RNA are seen in abundance, especially for ribose methylation at 2′-hydroxyl position (Charlier and Droogmans, 2005).

Application of thermophiles and their enzymes

Mankind had been using microorganisms for producing food for thousands of years. With the emergence of industrialization, microorganisms were also utilized for the production of biofuel (ethanol) and raw materials for chemicals during the First World War (Songstad et al., 2009; Jones and Woods, 1986). In general, mesophilic microbes are used for several biotechnological and commercial purposes. However, the thermophilic microbes and their thermostable enzymes have industrial importance concerning their biotechnological applications (Mehta et al., 2016). Below are a few of the applications of thermophilic microbes and their enzymes.

Role in bioremediation

The thermophiles and their enzymes have been used for the bioremediation of several environmental pollutants including petroleum, textile dyes, hydrocarbons, and other pollutants present in wastewater. According to several studies, thermophiles are more advantageous compared to mesophiles due to high reaction kinetics and the ability to survive in high-temperature oil reservoirs. Wang et al. (2006) reported the capacity of *G. thermodenitrificans* NG80−2 (isolated from an oil reservoir) in utilizing crude oil as a carbon source and it can potentially be used in the oil remediation wastes at high temperatures. Another study showed that *Geobacillus pallidus* XS2 and XS3 were efficient in degrading 85% of crude oil and 70% of Polyaromatic hydrocarbons (Zheng et al., 2011).

In the context of bioremediation of dyes, Ertuğrul et al. (2008) reported the capability of thermophilic cyanobacteria *Phormidium* sp. in removing 50%−80% of dye under high temperatures. Another study reported the use of thermostable enzyme laccase isolated from *G. thermocatenulatus Ms5* in removing synthetic dyes such as Bromophenol Blue, Congo Red, and Remazole Brilliant Blue R (Verma and Shirkot, 2014). Other than biodegradation of crude oil and textile dyes, thermophiles and their enzymes are also applied in the removal of other significant pollutants present in wastewater. For instance, thermophilic cyanobacterium *Fischerella* sp. is reported to be efficient in the removal of nitrates and phosphates from nuclear plant wastewater (Radway et al., 1992, 1994). Moreover, according to Patel et al. (2017), some of these thermophiles are capable of degrading the phenolic compounds present in wastewater. Recently, Karatay et al. (2017) reported that *Phormidium* sp. which is also a thermophilic cyanobacterium is capable of uptake 165.1 mg g^{-1} of phenol from wastewater having pH 7.

Role in biotransformation

Many microbes have the capability of lignocellulose degradation or hydrogen production. However, there has been no conclusive research showing all these capabilities in a single microbe (Mehta et al., 2016). The only exception is the thermophiles. They are capable of producing cellulolytic biohydrogen from lignocellulosic biomass since the rate of cellulose degradation presumably increases at high temperatures (Wiegel et al., 1985; Blumer-Schuette et al., 2008). A number of studies reported the use of a coculture of cellulolytic *Clostridium thermocellum* with noncellulolytic bacteria and cellulolytic *C. saccharolyticus* (hyperthermophile) for producing CBP-based hydrogen (Liu et al., 2008; Ivanova et al., 2009). In contrast, several other species of the genus *Thermoanaerobacterium* are reported to have the capability of utilizing a number of macromolecules followed by H$_2$ formation (Ganghofner et al., 1998; Ren et al., 2008; Sompong et al., 2008).

Some studies show the capability of thermophilic bacteria including *Thermoanaerobacter ethanolicus*, *C. thermohydrosulfuricum*, and *Thermoanaerobium brockii* isolated from brewery effluents, geysers, etc., produce ethanol from D-Xylose (Mehta et al., 2016). In this regard, the thermophiles are advantageous since they have a wide range of substrates and are capable of degrading both hexoses and pentoses at the same time; few thermophilic microbes can naturally degrade complex carbohydrates (Okano et al., 2010).

Role in bioproduction

The thermophilic microbes produce a wide range of bioactive compounds among which a few inhibits the growth of other microbes (Biondi et al., 2008). For instance, the methanolic extract of seven species of thermophilic cyanobacteria isolated from a hot spring in Iran displayed antibacterial activity against several species of Bacillus (Heidari et al., 2012). Fish and Codd (1994) reported the production of extracellular compounds in *Phormidium* sp. which showed antimicrobial activity against both bacteria and fungi. Another thermophilic algae namely *Cosmarium* sp. exhibited cytotoxic and antioxidant effects against both gram-positive and gram-negative bacteria with the biomass and extracellular polysaccharide, having a minimum inhibitory concentration of $28-85\ \mu g\ mL^{-1}$ and $50-150\ \mu g\ mL^{-1}$, respectively (Challouf et al., 2012).

Cyanobacteria possess c-phycocyanins(C-PCs, primary photosynthetic pigment) which are part of the light-harvesting phycobiliproteins (Eriksen, 2008). The C-PCs are thermostable, fluorescent proteins having a wide range of applications in medicine by fluorescent tagging, in the pharmaceutical field, and even in foods for antiinflammatory and antioxidant characteristics (Carfagna et al., 2018). Several thermophilic cyanobacteria from order *Cyanidales* and *Thermosynechococcaceae* are currently utilized for thermostable C-PCs (Liang et al., 2018; Cennamo et al., 2012).

Several thermophilic microbes are reported to synthesize biofuels. Currently, biofuels are considered as one of the primary substitutes for nonrenewable fossil fuels (Patel et al., 2016). Few studies reported, thermophilic algae *Micractinium* sp. produces lipids under N_2-starvation and showed an efficient FAME profile for the production of biodiesel under mixotrophic conditions (Abu-Ghosh et al., 2018; Engin et al., 2018). Similarly, the study of Karatay and Dönmez (2011) shows that the thermophilic strains of *Synechococcus* sp., *Cyanobacterium* sp., and *Phormidium* sp., were capable of producing a significant amount of palmitic and stearic acid at 90.6% and 38.2%, 42.8%, and 46.9%, 45.0%, and 67.7%, respectively.

Role in the medical field

A study by Liao et al. (2012) showed two novel compounds namely Asperjinone and Terrein, isolated from *Aspergillus terreus* are capable of reimposing drug sensitivity and possess the key to enhance breast cancer treatment. The compound Terrein showed enhanced cytotoxicity against breast cancer cell line MCF-7 and was also capable of suppressing the growth of ABCG2 expressing cancer cells by causing apoptosis through the activation of the caspase-7 pathway and the inhibition of the Akt pathway (Table 16.1).

Table 16.1 Some extremozymes isolated from thermophiles and their uses.

Extremozyme	Source/Organism	Use	Reference
Xylanase	*Bacillus halodurans*	Bleaching paper	Rothschild and Mancinelli (2001) and Kohli et al. (2020)
Lipase	*Stearothermophilus 5, Geobacillus* sp. *SBS-4S, Thermosyntrophalipolytica*	Dairy industry, Detergent industry, Pharmaceutical industry	Hohli et al. (2020)
Protease	*Bacillus brevis*	Bakery	Banerjee et al. (1999)
Amylase	*Bacillus stearothermophilus*	Brewing, Baking industry	Sarmiento et al. (2015)
Nucleoside phosphorylase	*Aeropyrumpernix K1*	Antiviral therapies	Raddadi et al. (2015)
Glucoamylase	*Geobacillus thermoleovorans NP33, G. thermoleovorans NP54, Thermomucorindicaeseudaticae*	Starch saccharification	Satyanarayana et al. (2004)
Esterase	*Alicyclobacillus acidocaldarius*	Detergent industry	Van Den Burg (2003) and Kohli et al. (2020)
amylopullulanases	*Geobacillus thermoleovorans*	Producing maltose (Starch processing enzymes)	Sarmiento et al. (2015)
Chitinase	*Bacillus licheniformis*	Health and food products	Van Den Burg (2003) Kohli et al. (2020)
Pullulan hydrolase type III	*Thermococcus aggregans*	Starch hydrolyzing	Bertoldo and Antranikian (2002)
Alcohol dehydrogenases	*S. solfataricus*	Chemical industry	Raddadi et al. (2015)
Cellulase	*Clostridium thermocellum, Thermobifida fusca Cel9A*	Animal feed industry, Wine industry	Kohli et al. (2020)
Taq polymerase	*Thermus aquaticus*	Polymerase chain reaction	Charlesworth and Burns (2016)
Pf amylase (mutated)	*Pyrococcus furiosus*	Maltoheptaose production, useful for Cosmetic industry, food industry	Reed et al. (2013)
Urease	*Campylobacter laridis, Bacillus* sp. *strain TB-90*	Automobile industry, Agriculture industry	Kohli et al. (2020)
Alkaline protease	*Bacillus brevis*	Detergent industry	Gupta et al. (2014)
Thermolysin (a protease)	*Bacillus thermoproteolyticus*	Synthesis of dipeptides	Raddadi et al. (2015)
α-Amylases	*Bacillus licheniformis*	Used in starch conversion	Bruins et al. (2001)
β-Glucosidase	*Thermoanaerobacter brockii*	Chemical industry	Kohli et al. (2020)
Aminoacylase	*T. litoralis*	Biotransformation processes	Egorova and Antranikian (2005)

Conclusion

Extremophiles set an example as living beings that can push their limits for surviving in extreme environments. The wide range of metabolic and phylogenetic diversity of thermophilic adaptations starting from their genomic to cellular and molecular levels itself describes the remarkable flexibility of life on this planet which is already providing us with new insights for various applications in biotechnology and industry.

Some of the examples of thermostable enzymes as stated in this chapter, having increased robustness in comparison to the enzymes previously used in industry, confirm the great potential of these microorganisms and their specific environmental adaptations. However, a more focused study on thermophiles and their enzymes need to be done because it holds the immense potential of revolutionizing various biotechnological utilizations. Thus it can be used to solve the bottlenecks and in turn, influence the relationship with the environment positively for more greener technologies in the future.

References

Abu-Ghosh, S., Dubinsky, Z., Banet, G., Iluz, D., 2018. Optimizing photon dose and frequency to enhance lipid productivity of thermophilic algae for biofuel production. Bioresour. Technol. 260, 374–379.

Allwood, M.C., Russell, A.D., 1969. Thermally induced changes in the physical properties of *Staphylococcus aureus*. J. Appl. Bacteriol. 32 (1), 68–78.

Allwood, M.C., Russell, A.D., 1968. Thermally-induced ribonucleic acid degradation and leakage of substances from the metabolic pool in *Staphylococcus aureus*. J. Bacteriol. 95, 345–349.

Atomi, H., Matsumi, R., Imanaka, T., 2004. Reverse gyrase is not a prerequisite for hyperthermophilic life. J. Bacteriol. 186 (14), 4829–4833.

Averhoff, B., Müller, V., 2010. Exploring research frontiers in microbiology: recent advances in halophilic and thermophilic extremophiles. Res. Microbiol 161 (6), 506–514.

Banerjee, U.C., Sani, R.K., Azmi, W., Soni, R., 1999. Thermostable alkaline protease from *Bacillus brevis* and its characterization as a laundry detergent additive. Process. Biochem. 35 (1–2), 213–219.

Beeby, M., O'Connor, B.D., Ryttersgaard, C., Boutz, D.R., Perry, L.J., Yeates, T.O., 2005. The genomics of disulfide bonding and protein stabilization in thermophiles. PLoS Biol. 3 (9), e309.

Bertoldo, C., Antranikian, G., 2002. Starch-hydrolyzing enzymes from thermophilic archaea and bacteria. Curr. Opin. Chem. Biol. 6 (2), 151–160.

Biondi, N., Tredici, M.R., Taton, A., Wilmotte, A., Hodgson, D.A., Losi, D., et al., 2008. Cyanobacteria from benthic mats of Antarctic lakes as a source of new bioactivities. J. Appl. Microbiol 105 (1), 105–115.

Blaby, I.K., Lyons, B.J., Wroclawska-Hughes, E., Phillips, G.C., Pyle, T.P., Chamberlin, S.G., et al., 2012. Experimental evolution of a facultative thermophile from a mesophilic ancestor. Appl. Environ. Microbiol 78 (1), 144–155.

Blumer-Schuette, S.E., Kataeva, I., Westpheling, J., Adams, M.W., Kelly, R.M., 2008. Extremely thermophilic microorganisms for biomass conversion: status and prospects. Curr. Opin. Biotechnol. 19 (3), 210–217.

Boto, L., 2014. Horizontal gene transfer in the acquisition of novel traits by metazoans. Proc. R. Soc. B: Biol. Sci. 281 (1777), 20132450.

Bouthier De La Tour, C., Portemer, C., Nadal, M., Stetter, K.O., Forterre, P., Duguet, M., 1990. Reverse gyrase, a hallmark of the hyperthermophilic archaebacteria. J. Bacteriol. 172 (12), 6803–6808.

Boutz, D.R., Cascio, D., Whitelegge, J., Perry, L.J., Yeates, T.O., 2007. Discovery of a thermophilic protein complex stabilized by topologically interlinked chains. J. Mol. Biol. 368 (5), 1332–1344.

Brock, T.D., 1967. Life at high temperatures: evolutionary, ecological, and biochemical significance of organisms living in hot springs is discussed. Science 158 (3804), 1012–1019.

Bruins, M.E., Janssen, A.E., Boom, R.M., 2001. Thermozymes and their applications. Appl. Biochem. Biotechnol. 90 (2), 155–186.

Burra, P.V., Kalmar, L., Tompa, P., 2010. Reduction in structural disorder and functional complexity in the thermal adaptation of prokaryotes. PLoS One 5 (8), e12069.

Carfagna, S., Landi, V., Coraggio, F., Salbitani, G., Vona, V., Pinto, G., et al., 2018. Different characteristics of C-phycocyanin (C-PC) in two strains of the extremophilic Galdieria phlegrea. Algal Res. 31, 406–412.

Cario, A., Grossi, V., Schaeffer, P., Oger, P.M., 2015. Membrane homeoviscous adaptation in the piezo-hyperthermophilic archaeon *Thermococcus barophilus*. Front. Microbiol 6, 1152.

Cennamo, P., Marzano, C., Ciniglia, C., Pinto, G., Cappelletti, P., Caputo, P., et al., 2012. A survey of the algal flora of anthropogenic caves of CampiFlegrei (Naples, Italy) archeological district. J. Cave Karst Stud. 74 (3).

Challouf, R., Dhieb, R.B., Omrane, H., Ghozzi, K., Ouda, H.B., 2012. Antibacterial, antioxidant and cytotoxic activities of extracts from the thermophilic green alga, Cosmarium sp. Afr. J. Biotechnol 11 (82), 14844–14849.

Charlesworth, J., Burns, B.P., 2016. Extremophilic adaptations and biotechnological applications in diverse environments. AIMS Microbiol. 2 (3), 251–261.

Charlier, D., Droogmans, L., 2005. Microbial life at high temperature, the challenges, the strategies. Cell. Mol. Life Sci. 62 (24).

Damsté, J.S.S., Rijpstra, W.I.C., Hopmans, E.C., Schouten, S., Balk, M., Stams, A.J., 2007. Structural characterization of diabolic acid-based tetraester, tetraether and mixed ether/ester, membrane-spanning lipids of bacteria from the order Thermotogales. Arch. Microbiol 188 (6), 629–641.

Das, S., Paul, S., Bag, S.K., Dutta, C., 2006. Analysis of *Nanoarchaeum equitans* genome and proteome composition: indications for hyperthermophilic and parasitic adaptation. BMC Genomics 7 (1), 1–16.

Drake, J.W., 2009. Avoiding dangerous missense: thermophiles display especially low mutation rates. PLoS Genet. 5 (6), e1000520.

Dutta, A., Chaudhuri, K., 2010. Analysis of tRNA composition and folding in psychrophilic, mesophilic and thermophilic genomes: indications for thermal adaptation. FEMS Microbiol Lett. 305 (2), 100–108.

Egorova, K., Antranikian, G., 2005. Industrial relevance of thermophilic Archaea. Curr. Opin. Microbiol 8 (6), 649–655.

Eichler, J., 2001. Biotechnological uses of archaeal extremozymes. Biotechnol. Adv. 19 (4), 261–278.

Engin, I.K., Cekmecelioglu, D., Yücel, A.M., Oktem, H.A., 2018. Evaluation of heterotrophic and mix-otrophic cultivation of novel *Micractinium* sp. ME05 on vinasse and its scale up for biodiesel production. Bioresour. Technol. 251, 128–134.

Eriksen, N.T., 2008. Production of phycocyanin—a pigment with applications in biology, biotechnology, foods and medicine. Appl. Microbiol Biotechnol. 80 (1), 1–14.

Ertuğrul, S., Bakır, M., Dönmez, G., 2008. Treatment of dye-rich wastewater by an immobilized thermophilic cyanobacterial strain: *Phormidium* sp. Ecol. Eng. 32 (3), 244–248.

Escribá, P.V., 2006. Membrane-lipid therapy: a new approach in molecular medicine. Trends Mol. Med. 12 (1), 34–43.

Feng, S., Powell, S.M., Wilson, R., Bowman, J.P., 2014. Extensive gene acquisition in the extremely psychrophilic bacterial species *Psychroflexus torquis* and the link to sea-ice ecosystem specialism. Genome BiolEvol 6, 133–148.

Ferrera, I., Reysenbach, A.-L., 2007. Thermophiles. Encyclopedia of Life Sciences. John Wiley & Sons, Hoboken, NJ. Available from: https://doi.org/10.1002/9780470015902.a0000406.

Fish, S.A., Codd, G.A., 1994. Bioactive compound production by thermophilic and thermotolerant cyanobacteria (blue-green algae). World J. Microbiol Biotechnol. 10 (3), 338–341.

Forterre, P., Bergerat, A., Lopex-Garcia, P., 1996. The unique DNA topology and DNA topoisomerases of hyperthermophilic archaea. FEMS Microbiol. Rev. 18 (2–3), 237–248.

Ganghofner, D., Kellermann, J., Staudenbauer, W.L., Bronnenmeier, K., 1998. Purification and properties of an amylopullulanase, a glucoamylase, and an α-glucosidase in the amylolytic enzyme system of *Thermoanaerobacteriumthermosaccharolyticum*. Bioscience, Biotechnol, Biochem. 62 (2), 302–308.

Gliozzi, A., Paoli, G., De Rosa, M., Gambacorta, A., 1983. Effect of isoprenoid cyclization on the transition temperature of lipids in thermophilic archaebacteria. Biochim. Biophys. Acta (BBA)-Biomembranes 735 (2), 234–242.

Gomes, E., Souza, A.R.D., Orjuela, G.L., Silva, R.D., Oliveira, T.B.D., Rodrigues, A., 2016. Applications and benefits of thermophilic microorganisms and their enzymes for industrial biotechnology. Gene expression systems in fungi: advancements and applications. Springer, Cham, pp. 459–492.

Gu, J., Hilser, V.J., 2009. Sequence-based analysis of protein energy landscapes reveals nonuniform thermal adaptation within the proteome. Mol. Biol. Evolution 26 (10), 2217–2227.

Gupta, G.N., Srivastava, S., Khare, S.K., Prakash, V., 2014. Extremophiles: an overview of microorganism from extreme environment. Int. J. Agriculture, Environ. Biotechnol. 7 (2), 371–380.

Heidari, F., Riahi, H., Yousefzadi, M., Asadi, M., 2012. Antimicrobial activity of cyanobacteria isolated from hot spring of geno. Middle-East J. Sci. Res. 12 (3), 336–339.

Hensel, R., König, H., 1988. Thermoadaptation of methanogenic bacteria by intracellular ion concentration. FEMS Microbiol. Lett. 49 (1), 75–79.

Huber, H., Hohn, M.J., Rachel, R., Fuchs, T., Wimmer, V.C., Stetter, K.O., 2002. A new phylum of Archaea represented by a nanosized hyperthermophilic symbiont. Nature 417 (6884), 63–67.

Huber, R., Woese, C.R., Langworthy, T.A., Fricke, H., Stetter, K.O., 1989. Thermosipho africanus gen. nov., represents a new genus of thermophilic eubacteria within the "Thermotogales.". Syst. Appl. Microbiol. 12 (1), 32–37.

Ibba, M., Söll, D., 2001. The renaissance of aminoacyl-tRNA synthesis. EMBO Rep. 2 (5), 382–387.

Ivanova, G., Rákhely, G., Kovács, K.L., 2009. Thermophilic biohydrogen production from energy plants by *Caldicellulosiruptor saccharolyticus* and comparison with related studies. Int. J. Hydrog. Energy 34 (9), 3659–3670.

Jahnke, L.L., Eder, W., Huber, R., Hope, J.M., Hinrichs, K.U., Hayes, J.M., et al., 2001. Signature lipids and stable carbon isotope analyses of octopus spring hyperthermophilic communities compared with those of Aquificales representatives. Appl. Environ. Microbiol. 67 (11), 5179–5189.

Jones, D.T., Woods, D., 1986. Acetone-butanol fermentation revisited. Microbiol. Rev. 50 (4), 484–524.

Karatay, S.E., Dönmez, G., 2011. Microbial oil production from thermophile cyanobacteria for biodiesel production. Appl. Energy 88 (11), 3632–3635.

Karatay, S.E., Dönmez, G., Aksu, Z., 2017. Effective biosorption of phenol by the thermophilic cyanobacterium Phormidium sp. Water Sci. Technol. 76 (12), 3190–3194.

Karshikoff, A., Ladenstein, R., 2001. Ion pairs and the thermotolerance of proteins from hyperthermophiles: a 'traffic rule'for hot roads. Trends Biochem. Sci. 26 (9), 550–557.

Kashefi, K., Holmes, D.E., Reysenbach, A.L., Lovley, D.R., 2002. Use of Fe (III) as an electron acceptor to recover previously uncultured hyperthermophiles: isolation and characterization of *Geothermobacterium ferrireducens* gen. nov., sp. nov. Appl. Environ. Microbiol. 68 (4), 1735–1742.

Kohli, I., Joshi, N.C., Mohapatra, S., Varma, A., 2020. Extremophile—an adaptive strategy for extreme conditions and applications. Curr. Genomics 21 (2), 96–110.

Laksanalamai, P., Robb, F.T., 2004. Small heat shock proteins from extremophiles: a review. Extremophiles 8 (1), 1–11.

Langworthy, T.A., Holzer, G., Zeikus, J.G., Tornabene, T.G., 1983. Iso-and anteiso-branched glycerol diethers of the thermophilic anaerobe *Thermodesulfotobacterium* commune. Syst. Appl. Microbiol. 4 (1), 1–17.

Li, H., Ji, X., Zhou, Z., Wang, Y., Zhang, X., 2010. Thermus thermophilus proteins that are differentially expressed in response to growth temperature and their implication in thermoadaptation. J. Proteome Res. 9 (2), 855–864.

Li, X., Xing, J., Li, B., Yu, F., Lan, X., Liu, J., 2013. Phylogenetic analysis reveals the coexistence of interfamily and interspecies horizontal gene transfer in Streptococcus thermophilus strains isolated from the same yoghurt. Mol. Phylogenetics Evolution 69 (1), 286–292.

Liang, Y., Kaczmarek, M.B., Kasprzak, A.K., Tang, J., Shah, M.M.R., Jin, P., et al., 2018. *Thermosynechococcaceae* as a source of thermostable C-phycocyanins: properties and molecular insights. Algal Res. 35, 223–235.

Liao, W.Y., Shen, C.N., Lin, L.H., Yang, Y.L., Han, H.Y., Chen, J., et al., 2012. Asperjinone, a nor-neolignan, and terrein, a suppressor of ABCG2-expressing breast cancer cells, from thermophilic *Aspergillus terreus*. J. Nat. Products 75 (4), 630–635.

Liu, Y., Yu, P., Song, X., Qu, Y., 2008. Hydrogen production from cellulose by co-culture of *Clostridium thermocellum* JN4 and *Thermoanaerobacteriumthermosaccharolyticum* GD17. Int. J. Hydrog. Energy 33 (12), 2927–2933.

López-García, P., 1999. DNA supercoiling and temperature adaptation: a clue to early diversification of life? J. Mol. Evol. 49 (4), 439–452.

Lusk, B.G., 2019. Thermophiles; or, the modern Prometheus: the importance of extreme microorganisms for understanding and applying extracellular electron transfer. Front. Microbiol. 10, 818.

Mackey, B.M., 1983. Changes in antibiotic sensitivity and cell surface hydrophobicity in *Escherichia coli* injured by heating, freezing, drying or gamma radiation. FEMS Microbiol. Lett. 20 (3), 395–399.

Mallick, P., Boutz, D.R., Eisenberg, D., Yeates, T.O., 2002. Genomic evidence that the intracellular proteins of archaeal microbes contain disulfide bonds. Proc. Natl. Acad. Sci. 99 (15), 9679–9684.

Marsh, C.L., Larsen, D.H., 1953. Characterization of some thermophilic bacteria from the hot springs of Yellowstone National Park. J. Bacteriol. 65 (2), 193–197.

Mehta, D., Satyanarayana, T., 2013. Diversity of hot environments and thermophilic microbes. Thermophilic Microbes in Environmental and Industrial Biotechnology. Springer, Dordrecht, pp. B–60.

Mehta, R., Singhal, P., Singh, H., Damle, D., Sharma, A.K., 2016. Insight into thermophiles and their wide-spectrum applications. 3 Biotech. 6 (1), 81.

Mehta, R., Singhal, P., Singh, H., Damle, D., Sharma, A.K., 2016. Insight into thermophiles and their wide-spectrum applications. 3 Biotech. 6 (1), 1–9.

Noll, M., Klose, M., Conrad, R., 2010. Effect of temperature change on the composition of the bacterial and archaeal community potentially involved in the turnover of acetate and propionate in methanogenic rice field soil. FEMS Microbiol. Ecol. 73 (2), 215–225.

Okano, K., Tanaka, T., Ogino, C., Fukuda, H., Kondo, A., 2010. Biotechnological production of enantiomeric pure lactic acid from renewable resources: recent achievements, perspectives, and limits. Appl. Microbiol. Biotechnol. 85 (3), 413–423.

Omelchenko, M.V., Wolf, Y.I., Gaidamakova, E.K., Matrosova, V.Y., Vasilenko, A., Zhai, M., et al., 2005. Comparative genomics of thermus thermophilus and *Deinococcus radiodurans*: divergent routes of adaptation to thermophily and radiation resistance. BMC Evolut. Biol. 5 (1), 1–22.

Oshima, M., Miyagawa, A., 1974. Comparative studies on the fatty acid composition of moderately and extremely thermophilic bacteria. Lipids 9 (7), 476–480.

Oshima, T., Moriya, T., 2008. A preliminary analysis of microbial and biochemical properties of high-temperature compost. Ann. N. Y. Acad. Sci. 1125 (1), 338–344.

Pan, M., 2012. Thermococcus kodakarensis DNA replication machinery. University of Maryland, College Park.

Patel, A., Matsakas, L., Rova, U., Christakopoulos, P., 2016. A perspective on biotechnological applications of thermophilic microalgae and cyanobacteria. Bioresour. Technol. 278, 424–434.

Patel, A., Sartaj, K., Arora, N., Pruthi, V., Pruthi, P.A.P.A., 2017. Biodegradation of phenol via meta cleavage pathway triggers de novo TAG biosynthesis pathway in oleaginous yeast. J. Hazard. Mater. 340, 47–56.

Pedone, E., Ren, B., Ladenstein, R., Rossi, M., Bartolucci, S., 2004. Functional properties of the protein disulfide oxidoreductase from the archaeon *Pyrococcus furiosus*: a member of a novel protein family related to protein disulfide-isomerase. Eur. J. Biochem. 271 (16), 3437–3448.

Perugino, G., Valenti, A., D'amaro, A., Rossi, M., Ciaramella, M., 2009. Reverse gyrase and genome stability in hyperthermophilic organisms. Biochem. Soc. Trans. 37 (1), 69–73.

Pikuta, E.V., Hoover, R.B., Tang, J., 2007. Microbial extremophiles at the limits of life. Crit. Rev. Microbiol 33 (3), 183–209.

Purcell, D., Sompong, U., Yim, L.C., Barraclough, T.G., Peerapornpisal, Y., Pointing, S.B., 2007. The effects of temperature, pH and sulphide on the community structure of hyperthermophilic streamers in hot springs of northern Thailand. FEMS Microbiol. Ecol. 60 (3), 456–466.

Raddadi, N., Cherif, A., Daffonchio, D., Neifar, M., Fava, F., 2015. Biotechnological applications of extremophiles, extremozymes and extremolytes. Appl. Microbiol. Biotechnol. 99 (19), 7907–7913.

Radway, J.C., Weissman, J.C., Wilde, E.W., Benemann, J.R., 1994. Nutrient removal by thermophilic Fischerella (Mastigocladuslaminosus) in a simulated algaculture process. Bioresour. Technol. 50 (3), 227–233.

Radway, J.C., Weissman, J.C., Wilde, E.W., Benemann, J.R., 1992. Exposure of Fischerella [Mastigocladus] to high and low temperature extremes: strain evaluation for a thermal mitigation process. J. Appl. Phycol. 4 (1), 67−77.

Rampelotto, P.H., 2013. Extremophiles and extreme environments. Life 3 (3), 482−485.

Rampelotto, P.H., 2010. Resistance of microorganisms to extreme environmental conditions and its contribution to astrobiology. Sustainability 2 (6), 1602−1623.

Reed, C.J., Lewis, H., Trejo, E., Winston, V., Evilia, C., 2013. Protein adaptations in archaeal extremophiles. Archaea 2013.

Ren, N., Cao, G., Wang, A., Lee, D.J., Guo, W., Zhu, Y., 2008. Dark fermentation of xylose and glucose mix using isolated Thermoanaerobacteriumthermosaccharolyticum W16. Int. J. Hydrog. Energy 33 (21), 6124−6132.

Rosenberg, B., Kemeny, G., Switzer, R.C., Hamilton, T.C., 1971. Quantitative evidence for protein denaturation as the cause of thermal death. Nature 232 (5311), 471−473.

Rothschild, L.J., Mancinelli, R.L., 2001. Life in extreme environments. Nature 409 (6823), 1092−1101.

Sarmiento, F., Peralta, R., Blamey, J.M., 2015. Cold and hot extremozymes: industrial relevance and current trends. Front. Bioeng. Biotechnol. 3, 148.

Satyanarayana, T., Noorwez, S.M., Kumar, S., Rao, J.L.U.M., Ezhilvannan, M., Kaur, P., 2004. Development of an ideal starch saccharification process using amylolytic enzymes from thermophiles. Biochem. Soc. Trans. 32 (Pt 2), 276−278.

Sokolova, T., Hanel, J., Onyenwoke, R.U., Reysenbach, A.L., Banta, A., Geyer, R.J.M.G., et al., 2007. Novel chemolithotrophic, thermophilic, anaerobic bacteria Thermolithobacterferrireducens gen. nov., sp. nov. and Thermolithobactercarboxydivorans sp. nov. Extremophiles 11 (1), 145−157.

Sompong, O., Prasertsan, P., Karakashev, D., Angelidaki, I., 2008. Thermophilic fermentative hydrogen production by the newly isolated Thermoanaerobacteriumthermosaccharolyticum PSU-2. Int. J. Hydrog. Energy 33 (4), 1204−1214.

Songstad, D.D., Lakshmanan, P., Chen, J., Gibbons, W., Hughes, S., Nelson, R., 2009. Historical perspective of biofuels: learning from the past to rediscover the future. Vitro Cell. Developmental Biology-Plant 45 (3), 189−192.

Stepanov, V.G., Nyborg, J., 2002. Thermal stability of aminoacyl-tRNAs in aqueous solutions. Extremophiles 6 (6), 485−490.

Takai, K., Nakamura, K., Toki, T., Tsunogai, U., Miyazaki, M., Miyazaki, J., et al., 2008. Cell proliferation at 122 C and isotopically heavy CH4 production by a hyperthermophilic methanogen under high-pressure cultivation. Proc. Natl. Acad. Sci. 105 (31), 10949−10954.

Terui, Y., Ohnuma, M., Hiraga, K., Kawashima, E., Oshima, T., 2005. Stabilization of nucleic acids by unusual polyamines produced by an extreme thermophile, Thermus thermophilus. Biochem. J. 388 (2), 427−433.

Tsuchido, T., Aoki, I., Takano, M., 1989. Interaction of the fluorescent dye lN-phenylnaphthylamine with Escherichia coli cells during heat stress and recovery from heat stress. Microbiology 135 (7), 1941−1947.

Van Den Burg, B., 2003. Extremophiles as a source for novel enzymes. Curr. Opin. Microbiol. 6 (3), 213−218.

van Wolferen, M., Ajon, M., Driessen, A.J., Albers, S.V., 2013. How hyperthermophiles adapt to change their lives: DNA exchange in extreme conditions. Extremophiles 17 (4), 545−563.

Verma, A., Shirkot, P., 2014. Purification and characterization of thermostable laccase from thermophilic Geobacillusthermocatenulatus MS5 and its applications in removal of textile dyes. Scholars Acad. J. Biosci. 2 (8), 479−485.

Vieille, C., Zeikus, G.J., 2001. Hyperthermophilic enzymes: sources, uses, and molecular mechanisms for thermostability. Microbiol. Mol. Biol. Rev. 65 (1), 1−43.

Wang, J., Zhao, C., Meng, B., Xie, J., Zhou, C., Chen, X., et al., 2007. The proteomic alterations of *Thermoanaerobacter tengcongensis* cultured at different temperatures. Proteomics 7 (9), 1409−1419.

Wang, L., Tang, Y., Wang, S., Liu, R.L., Liu, M.Z., Zhang, Y., et al., 2006. Isolation and characterization of a novel thermophilic Bacillus strain degrading long-chain n-alkanes. Extremophiles 10 (4), 347−356.

Wang, Q., Cen, Z., Zhao, J., 2015. The survival mechanisms of thermophiles at high temperatures: an angle of omics. Physiology 30 (2), 97−106.

Wiegel, J., 1990. Temperature spans for growth: hypothesis and discussion. FEMS Microbiol. Rev. 6 (2−3), 155−169.

Wiegel, J., Canganella, F., 2001. Extreme thermophiles. eLS. Wiley, Chichester, 392. Available from: https://doi.org/10.1038/npg.els.0000392; http://www.els.net/WileyCDA/ElsArticle/refId-a0000392. html.

Wiegel, J., Ljungdahl, L.G., Demain, A.L., 1985. The importance of thermophilic bacteria in biotechnology. Crit. Rev. Biotechnol. 3 (1), 39−108.

Yokoyama, A., Shizuri, Y., Hoshino, T., Sandmann, G., 1996. Thermocryptoxanthins: novel intermediates in the carotenoid biosynthetic pathway of Thermus thermophilus. Arch. Microbiol. 165 (5), 342−345.

ZAvodszky, P., Kardos, J., Svingor, Á., Petsko, G.A., 1998. Adjustment of conformational flexibility is a key event in the thermal adaptation of proteins. Proc. Natl. Acad. Sci. 95 (13), 7406−7411.

Zhaxybayeva, O., Swithers, K.S., Lapierre, P., Fournier, G.P., Bickhart, D.M., DeBoy, R.T., et al., 2009. On the chimeric nature, thermophilic origin, and phylogenetic placement of the Thermotogales. Proc. Natl. Acad. Sci. U. S. A. 106 (14), 5865−5870.

Zheng, C., He, J., Wang, Y., Wang, M., Huang, Z., 2011. Hydrocarbon degradation and bioemulsifier production by thermophilic Geobacillus pallidus strains. Bioresour. Technol. 102 (19), 9155−9161.

Index

Printed in the United States
by Baker & Taylor Publisher Services